Random Differential Equations
in Science and Engineering

This is Volume 103 in
MATHEMATICS IN SCIENCE AND ENGINEERING
A series of monographs and textbooks
Edited by RICHARD BELLMAN, *University of Southern California*

The complete listing of books in this series is available from the Publisher
upon request.

Random Differential Equations in Science and Engineering

T. T. SOONG

Departments of Engineering Science and Civil Engineering
State University of New York
Buffalo, New York

ACADEMIC PRESS New York and London 1973

A Subsidiary of Harcourt Brace Jovanovich, Publishers

ACADEMIC PRESS, INC.
111 Fifth Avenue, New York, New York 10003

United Kingdom Edition published by
ACADEMIC PRESS, INC. (LONDON) LTD.
24/28 Oval Road, London NW1

LIBRARY OF CONGRESS CATALOG CARD NUMBER: 72-84372

AMS (MOS) 1970 Subject Classifications: 60G05, 93E15

PRINTED IN THE UNITED STATES OF AMERICA

TO
my wife Dottie

Contents

Preface

This work is intended to serve as a textbook for a one-year course in stochastic calculus, random differential equations, and their applications. I have endeavored to present a body of knowledge which follows that in an introductory course in stochastic processes and, in my mind, serves a twofold purpose.

The primary objective is to give the reader a working knowledge of random differential equations. In the last twenty to thirty years, the theory of stochastic processes has exerted a profound impact on the modeling and analysis of physical problems. Just as classical differential equations are at the heart of characterizing deterministic physical processes, random differential equations are essential to modern analysis. The study of random differential equations has led to startling advances on practically every front in engineering and science. This impact will certainly continue to be felt.

It is also hoped that the contents of this book will bridge the gap between introductory materials on stochastic processes and advanced topics in science and technology involving probabilistic methods. In ordinary analysis it would be difficult indeed for one to study differential equations without a solid foundation in calculus, or to pursue advanced mathematical topics, such as calculus of variations, without a basic understanding of the theory of differential equations. Yet, this situation seems to prevail in many applied areas where the theory of stochastic processes is involved. It appears that there are two main reasons for this.

Probabilistic methodology in science and engineering, although very much evident today, has a rather recent origin, and it is fast expanding. As a result of this rapid development, workers in applied areas began to attack specialized problems using the theory of stochastic processes often without adequate understanding of the underlying mathematical workings. This phenomenon, I feel, is largely responsible for the existence of the gap between introductory materials and advanced topics in stochastic processes.

The second reason is that, for topics in stochastic processes, the transition between theory and applications has not been a smooth one. This is perhaps a universal problem in applied mathematics as a whole; but it seems to stand out more visibly in the theory of stochastic processes. Probabilistic concepts can involve extraneous and abstract mathematics. Applied researchers, on the other hand, are interested in obtaining results for complex but practical problems. In the process, it is often necessary to follow a simplified or even intuitive approach at the sacrifice of mathematical rigor.

With these considerations in mind, this book is slanted toward methodology and problem solving, but every effort has been made to achieve a balance between mathematical theory and applications. It is written for the reader who has had a good introductory course in probability theory, random variables, and stochastic processes. He is also assumed to be familiar with basic concepts in ordinary deterministic calculus and differential equations. The aim of this work is not only to give the reader a working knowledge in random differential equations, but also to provide him with a degree of understanding and maturity so that he may pursue further studies in this fast expanding field.

Acknowledgments

My thanks are due to a number of my teachers, colleagues, and students. Sincere thanks are due to Professor John L. Bogdanoff of Purdue University and to Professor Frank Kozin of Brooklyn Polytechnic Institute. Their lectures and lecture notes on "Engineering Statistical Analysis" at Purdue guided my thinking and organization of Chapters 3 and 4. I am also indebted to Professor Richard Bellman, who provided encouragement. His stimulating comments led to a number of improvements. I wish to express my appreciation to Dr. Peter A. Ruymgaart of the Technological University of Delft, The Netherlands, who read parts of the manuscript and offered valuable comments. Most of the material contained in Appendix A is due to him. My thanks also go to my colleagues and students at the State University of New York at Buffalo who went through several stages of revisions of this material with me and made many contributions. Special thanks go to Dr. D. A. Prelewicz of Washington University, to Dr. J. W. Dowdee of TRW Systems, and to Dr. S. N. Chuang of Southern Research Institute.

Finally, I owe a very special debt of gratitude to my wife, Dottie, who edited and typed many versions of this book. Above all, without her constant encouragement, patience, and sense of humor, this book would still have been in the form of scattered notes in my desk drawer.

Chapter 1

Introduction

A more representative title of this book might be "A Mean Square Treatment of Initial Value Problems in Random Ordinary Differential Equations." This title would give a better description of its scope and limitations.

Random differential equations are defined here as differential equations involving random elements. From the standpoint of applications, the mean square theory of stochastic calculus and random differential equations is attractive because of its simplicity and because its applications to practical problems follow in broad outline corresponding ordinary deterministic procedures. Although not always explicitly stated, the formulation and analysis of stochastic problems in science and engineering have heavily relied upon analysis in the mean square sense. It is thus the objective of this book to give a careful presentation of the mean square calculus, to develop a mean square theory of random ordinary differential equations, and to explore their applications to problems in a number of scientific areas.

Central to our discussion in this book is the basic system of equations

$$dx_i(t)/dt = f_i(x_1(t), \ldots, x_n(t); y_1(t), \ldots, y_m(t); t)$$
$$i = 1, 2, \ldots, n. \tag{1.1}$$

1

with initial conditions

$$x_i(t_0) = x_{i0} \tag{1.2}$$

where $y_j(t)$, $j = 1, 2, \ldots, m$, and x_{i0}, $i = 1, 2, \ldots, n$, may be specified only in a stochastic sense. Problems described by equations of this type now arise in an amazingly wide variety of applied areas. This has been largely due to a general shift from deterministic to stochastic point of view in our modeling and analysis of physical processes.

A classical example of this shift in interest and in the method of approach is found in physics with the development of quantum theory in the nineteenth century. In recent years, stochastic approach has achieved a high level of maturity and its circle of influence contains physics, chemistry, biology, economics, management, sociology, psychology, medicine, and all engineering disciplines. Because of our increasingly stringent demands on quality and accuracy in dealing with increasingly complex systems, applied workers have become aware of the importance of stochastic function theory as a weapon of rational approach toward a wide variety of problems.

To show more specifically how random differential equations arise in physical situations, let us first classify Eqs. (1.1) into three categories according to their physical origins and mathematical flavoring.

The first and the simplest class is one where only the initial conditions x_{i0} are random. This class has its origin in statistical physics and kinetic theory. Recently, these equations also play a prominent role in engineering mechanics, chemical kinetics, drug administration, and other applied areas. Take, for example, a problem in space trajectory analysis. Eqs. (1.1) describe ballistic trajectories of space vehicles whose initial conditions represent injection conditions at the end of the powered-flight phase. Due to random error sources in the injection guidance system and due to disturbances during the powered-flight phase, the initial conditions can best be described in terms of random variables with a prescribed joint probability distribution. The resulting trajectory analysis is thus a stochastic one.

The functions $y_j(t)$ may enter Eqs. (1.1) either through their coefficients or through the nonhomogeneous terms. The second class of random differential equations is characterized by the presence of random nonhomogeneous or input terms. Langevin's investigation in 1908 of the motion of a particle executing Brownian motion perhaps marked the beginning of the study of differential equations of this type. Since that time, this development has been progressing at an amazing pace. Fruitful applications are evident in practically all applied areas, such as filtering and predictions, statistical communication theory, random vibration, operational research, and systems and structural analysis. In all these studies, inability to determine the future values of an input from a knowledge of the past has forced the introduction of probabilistic methods. Indeed, differential equations of this type represent systems of all descriptions subject to random disturbances.

Differential equations with random coefficients constitute the third type of random differential equations. The study of equations of this type has a rather recent origin. Today, it has become a subject of intensive research. The reason for this vigorous research activity is clear if we pause and consider the problem of mathematical modeling of practically all physical phenomena in nature. The translation from a physical problem to a set of mathematical equations is never perfect. This is due to a combination of uncertainties, complexities, and ignorance on our part which inevitably cloud our mathematical modeling process. In order to take them into account, mathematical equations with random coefficients offer a natural and rational approach. Applied problems that have been considered using this approach include wave propagations in inhomogeneous media, systems and structures with parametric excitations, and dynamics of imperfectly known systems in physics, engineering, biology, medicine, and economics.

A simple example in this regard is found in physics. For thermodynamic reasons, a great deal of work has been devoted to the study of crystal lattice dynamics in which there is a small percentage of impurity atoms. Randomness arises from the fact that atom masses may take two or more values based upon a probabilistic law. As another example, we mention structural analyses where imperfections in geometry and material are so complicated that they can best be represented by stochastic processes. It is clear that the random quantities described above enter the governing differential equations through their coefficients.

In this book, random differential equations of these types and their applications are considered following a review of the probability theory, random variables, and stochastic processes; and following a careful presentation of a stochastic calculus. The organization of this book is described below in more detail.

Organization

Chapters 2 and 3 are concerned with the fundamental principles in probability, random variables, and stochastic processes. Since the reader is assumed to be familiar with the basic concepts in these subjects, these two chapters are written primarily as a brief introduction and a summary of results useful in our latter development. Chapter 2 summarizes some main results in the set-theoretic construction of probability theory and introduces basic concepts in the theory of random variables on a rather elementary level. Chapter 3 is concerned with the definitions and analysis of stochastic processes, and with their classification. The classification is, of course, not unique. The one we choose to discuss is centered around two of the most important classes of stochastic processes, namely, the Markov processes and the stationary processes.

The reader is encouraged to go through Chapters 2 and 3 both as a review and as a reference point for the subsequent development in this book. For readers who are not thoroughly familiar with this background material, these two chapters can also be used as a guide to further reading. References which deal with the contents of these two chapters in more detail are cited therein.

Stochastic convergence of a sequence of random variables and mean square calculus are discussed in detail in Chapter 4. Second-order random variables and second-order stochastic processes form the basis of our development. Ideas and concepts advanced in this chapter are central to our treatment of random differential equations, to which the rest of this book is devoted.

In the study of random differential equations, as in the case of their deterministic counterparts, our primary concerns are the existence and uniqueness of their solutions and the properties of these solutions. These questions are considered in general terms in Chapter 5. Motivated by their usefulness in applications, differential equations of the Ito type are given a careful examination. We also mention here that, in seeking pertinent properties of the solution process, one is often interested in the properties of certain random variables associated with it. They characterize, for example, extreme values, number of threshold crossings, first passage time. In view of their practical significance, the problem of determining the properties of some of these random variables is considered in Appendix B.

From the mathematical as well as the physical point of view, the characterization of random differential equations is strongly dependent upon the manner in which the randomness enters the equations. Accordingly, as we indicated earlier, it is fruitful to present our treatment in three stages. Chapter 6 considers a class of differential equations where only the initial conditions are random. Differential equations of this type are treated first because it is simple in principle and because their basic features revealed here have important implications in more complicated situations.

Chapter 7 deals with differential equations when random inputs are added. While this is basically a well-developed area, we hope to bring out a number of interesting and difficult aspects, particularly in nonlinear cases.

Differential equations having random coefficients constitute the main topic in Chapter 8. The treatment of differential equations of this type is far from being complete and is fast expanding. A serious attempt is made in this chapter to cover this difficult area with balance and unification.

Many other interesting questions arise in connection with differential equations with random coefficients. The theory of stability, for example, now involves stochastic elements and stochastic stability is important in many applications, particularly in the area of control. This is considered in Chapter 9.

The main reason for including Appendix A in this book is to emphasize that the path we have chosen to follow in our study of random differential equations is by no means the only one. This appendix gives a brief introduction of a sample approach to the analysis of random differential equations, and it compares the sample theory with the mean square theory. To be sure, our goal in writing this appendix is not to give an exhaustive account of the sample theory. It will serve its purpose well if it can help the reader to develop a deeper appreciation of the many fascinating and difficult aspects of the theory of random differential equations.

In closing, let us refer back to the opening remark of this Introduction and observe that the vast domain of random differential equations is only partially covered in this book. No systematic account is given to boundary-value problems; neither have we considered the eigenvalue problems associated with random differential equations. Furthermore, although comments and examples are offered in regard to random partial differential equations and integral equations, a unified treatment of these topics is beyond the scope of this book.

Chapter 2

Probability and Random Variables: A Review

The mathematical theory of probability and the basic concepts of random variables form the basis of our development in this book. As indicated in the Introduction, the reader is assumed to be familiar with these concepts, so that in this chapter we only summarize some basic definitions and results in probability theory and random variables which we need in what follows. For more detailed discussions the reader is referred to the literature [1–6].

2.1. Elements of Set Theory

Events and combinations of events occupy a central place in probability theory. The mathematics of events is closely tied to the theory of sets, and this section constitutes a summary of some elements of this theory.

A *set* is a collection of arbitrary objects. These objects are called *elements* of the set and they can be of any kind with any specified properties. We may consider, for example, a set of numbers, a set of points, a set of functions, a set of persons, or a set of mixture of things. The capital letters A, B, C, Φ, Ω, ... shall be used to denote sets and lower case letters a, b, c, φ, ω, ... to denote their elements. A set of sets is called a *class* and classes will be denoted by capital script letters $\mathscr{A}$, $\mathscr{B}$, $\mathscr{C}$, ...

6

A set containing no elements is called an *empty set* and is denoted by 0. We distinguish between sets containing a finite number of elements and those having an infinite number. They are called, respectively, *finite sets* and *infinite sets*. An infinite set is called *enumerable* or *countable* if all its elements can be arranged in such a way that there is a one-to-one correspondence between them and all positive integers; thus, a set containing all positive integers 1, 2, ... is a simple example of an enumerable set. A *nonenumerable* or *noncountable* set is one where the above mentioned one-to-one correspondence cannot be established. A simple example of a non-enumerable set is the set of all points on a straight line segment.

If every element of a set A is also an element of a set B, the set A is called a *subset* of B and this is represented symbolically by

$$A \subset B \quad \text{or} \quad B \supset A \tag{2.1}$$

It is clear that an empty set is a subset of any set. In the case when both $A \subset B$ and $B \subset A$ hold, the set A is then *equal* to B, and we write

$$A = B \tag{2.2}$$

We now give meaning to a particular set we shall call a *space*. In our development, we shall consider only sets which are subsets of a fixed (nonempty) set. This "largest" set containing all elements of all the sets under consideration is called the *space*, and it is denoted by the symbol Ω. The class of all the sets in Ω is called the *space of sets* in Ω. It will be denoted by $\mathscr{S}(\Omega)$.

Consider a subset A in Ω. The set of all elements in Ω which are not elements of A is called the *complement* of A, and we denote it by A'. The following relations clearly hold:

$$\Omega' = 0, \qquad 0' = \Omega, \qquad (A')' = A \tag{2.3}$$

2.1.1. Set Operations

Let us now consider operations of sets $A, B, C, \ldots$ which are subsets of space Ω. We are primarily concerned with addition, subtraction, and multiplication of these sets.

The *union* or *sum* of A and B, denoted by $A \cup B$, is the set of all elements belonging to A or B or both.

The *intersection* or *product* of A and B, written as $A \cap B$ or simply AB, is the set of all elements which are common to A and B.

If $AB = 0$, the sets A and B contain no common elements, and we call A and B *mutually exclusive* or *disjoint*. The symbol "$+$" shall be reserved to denote the union of two disjoint sets.

The definitions of the union and the intersection can be directly generalized to those involving an arbitrary number (finite or countably infinite) of sets. Thus, the set

$$A_1 \cup A_2 \cup \cdots \cup A_n = \bigcup_{j=1}^{n} A_j \tag{2.4}$$

stands for the set of all elements belonging to one or more of the sets A_j, $j = 1, 2, \ldots, n$. The intersection

$$A_1 A_2 \cdots A_n = \bigcap_{j=1}^{n} A_j \tag{2.5}$$

is the set of all elements common to *all* A_j, $j = 1, 2, \ldots, n$. The sets A_j, $j = 1, 2, \ldots, n$, are called mutually exclusive if

$$A_i A_j = 0 \qquad \text{for every} \quad i, j \ (i \neq j) \tag{2.6}$$

It is easy to verify that the union and the intersection operations of sets are associative, commutative, and distributive, that is,

$$(A \cup B) \cup C = A \cup (B \cup C) = A \cup B \cup C, \qquad A \cup B = B \cup A$$
$$(AB)C = A(BC) = ABC, \qquad AB = BA, \qquad A(B \cup C) = AB \cup AC \tag{2.7}$$

Clearly, we also have

$$A \cup A = AA = A, \qquad A \cup 0 = A, \qquad A0 = 0$$
$$A \cup \Omega = \Omega, \qquad A\Omega = A, \qquad A \cup A' = \Omega, \qquad AA' = 0 \tag{2.8}$$

Moreover, the following useful relations hold:

$$A \cup BC = (A \cup B)(A \cup C), \qquad (AB)' = A' \cup B', \qquad A \cup B = A \cup A'B$$
$$(A \cup B)' = A'B', \qquad \left(\bigcup_1^n A_j \right)' = \bigcap_1^n A_j', \qquad \left(\bigcap_1^n A_j \right)' = \bigcup_1^n A_j' \tag{2.9}$$

The last two relations are referred to as *De Morgan's law*.

Finally, we define the *difference* between two sets. The difference $A - B$ is the set of all elements belonging to A but not to B. From the definition, we have the simple relations

$$A - 0 = A, \qquad \Omega - A = A', \qquad A - B = AB' \tag{2.10}$$

2.1.2. Borel Field

Given a space Ω, a *Borel field*, or a *σ-field* $\mathscr{B}$ of subsets of Ω is a class of, in general noncountable, subsets A_j, $j = 1, 2, \ldots$, having the following properties:

1. $\Omega \in \mathscr{B}$.
2. If $A_1 \in \mathscr{B}$, then $A_1' \in \mathscr{B}$.
3. If $A_j \in \mathscr{B}$, $j = 1, 2, \ldots$, then $\bigcup_1^\infty A_j \in \mathscr{B}$.

The first two properties imply that

$$0 = \Omega' \in \mathscr{B} \tag{2.11}$$

and, with the aid of De Morgan's law, the second and the third properties lead to

$$\bigcap_1^\infty A_j = \left(\bigcup_1^\infty A_j' \right)' \in \mathscr{B} \tag{2.12}$$

A Borel field is thus a class of sets, including the empty set 0 and the space Ω, which is closed under all countable unions and intersections of its sets. It is clear, of course, that a class of all subsets of Ω is a Borel field. However, in the development of the basic concepts of probability, this particular Borel field is too large and impractical. We in general consider the *smallest* class of subsets of Ω which is a Borel field and it contains all sets and elements under consideration.

2.2. Axioms of Probability and Probability Spaces

The basic concepts of probability theory are centered around the idea of a *random experiment E*, whose outcomes are *events*. Consider an elementary example where the throw of a die constitutes a random experiment. Each throw gives as an outcome one of the integers $1, 2, \ldots, 6$. The event that, say, the integer one is realized is called a *simple event* since it can only happen when the outcome is one. The event that an odd number occurs is regarded as an *observable event* or a *compound event* because it represents a collection of *simple events*, namely, the integers 1, 3, and 5. The choice of the word "observable" can be justified by means of another example.

Consider, for example, a random experiment whose outcomes are all real numbers on a line segment. Every real number on this line segment is thus a simple event. However, because of physical limitations in observing the outcomes, only intervals and not individual real numbers can be observed. These intervals are collection of simple events and they are called, justifiably, *observable events*.

Given a random experiment E, the collection of all possible events is called a *sample space*, whose elements are the *simple events*. *Observable events* thus enter the sample space as its subsets. The definitions of events and sample space provide a framework within which the analysis of events can be performed, and all definitions and relations between events in probability theory can be described by sets and set operations in the theory of sets. Consider the space Ω of elements ω and with subsets $A, B, \ldots$ Some of these corresponding set and probability terminologies are given in Table 2.1.

TABLE 2.1

Set theory	Probability theory
Space Ω	Sample space, sure event
Empty set 0	Impossible event
Element ω	Simple event
Set A	Observable or compound event
A	A occurs
A'	A does not occur
$A \cup B$	At least one of A and B occurs
AB	Both A and B occur
$A - B$	Only A of A and B occurs

In our subsequent discussion of probability theory, we shall assume that both Ω and 0 are observable events. We also require that the collection of all observable events associated with a random experiment E constitutes a Borel field $\mathscr{B}$, which implies that all events formed through countable unions and intersections of observable event are observable, and they are contained in $\mathscr{B}$.

We now introduce the notion of a *probability function*. Given a random experiment E, a finite number $P(A)$ is assigned to every event A in the Borel field $\mathscr{B}$ of all observable events. The number $P(A)$ is a function of set A and is assumed to be defined for all sets in $\mathscr{B}$. The set function $P(A)$ is called the *probability measure* of A or simply the *probability* of A. It is assumed to have the following properties (axioms of probability):

1. $P(A) \geq 0$.
2. $P(\Omega) = 1$.
3. For a countable collection of mutually disjoint $A_1, A_2, \ldots$ in $\mathscr{B}$,

$$P\left\{\sum_j A_j\right\} = \sum_j P(A_j)$$

Hence, Axioms 1–3 define a countably additive and nonnegative set function $P(A)$, $A \in \mathscr{B}$. The following properties associated with the probability function can be easily deduced from the axioms stated above.

$$P(0) = 0 \qquad (2.13)$$

If $A \subset B$,

$$P(A) \leq P(B) \qquad (2.14)$$

We also see that, in general,

$$P(A \cup B) = P(A) + P(B) - P(AB) \qquad (2.15)$$

Other identities can be obtained based on set relations (2.7)–(2.10).

A mathematical description of a random experiment is now complete. As we have seen above, it consists of three fundamental constituents: A sample space Ω, a Borel field $\mathscr{B}$ of observable events, and the probability function P. These three quantities constitute a probability space associated with a random experiment, and it is denoted by $(\Omega, \mathscr{B}, P)$.

2.3. Random Variables

Consider a random experiment E whose outcomes ω are elements of Ω in the underlying probability space $(\Omega, \mathscr{B}, P)$. These outcomes are in general represented by real numbers, functions, or objects of various kinds. In order to construct a model for a random variable, we assume that for all experiments it is possible to assign a real number $X(\omega)$ for each ω following a certain set of rules. The "number" $X(\omega)$ is really a real-valued *point function* defined over the domain of the basic probability space.

The point function $X(\omega)$ is called a *random variable* (r.v.) if (1) it is a finite real-valued function defined on a sample space Ω of a random experiment for which a probability function is defined on the Borel field $\mathscr{B}$ of events and (2) for every real number x, the set $\{\omega: X(\omega) \leq x\}$ is an event in $\mathscr{B}$. The relation $X = X(\omega)$ takes every element ω in Ω of the probability space unto a point X on the real line $R^1 = (-\infty, \infty)$.

Notationally, the dependence of a r.v. X on ω will be omitted for convenience. In the sequel, all random variables will be written in capital letters $X, Y, Z, \ldots$. The values that a r.v. X can take on will be denoted by the corresponding lower case x or $x_1, x_2, \ldots$.

We will have many occasions to consider a sequence of r.v.'s X_j, $j = 1, 2, \ldots, n$. In these cases we shall assume that they are defined on the same probability space. The r.v.'s $X_1, X_2, \ldots, X_n$ will then map every element ω of Ω in the probability space unto a point of the n-dimensional Euclidean space R^n. We note here that an analysis involving n r.v.'s is equivalent to considering a *random vector* having the n r.v.'s as components. The notion of a random vector will be used frequently in what follows, and we will denote them by boldface capital letters $\mathbf{X}, \mathbf{Y}, \mathbf{Z}, \ldots$.

2.3.1. Distribution Functions and Density Functions

Given a random experiment with its associated r.v. X and given a real number x, let us consider the probability of the event $\{\omega: X(\omega) \leq x\}$, or simply $P(X \leq x)$. This probability is a number, clearly dependent upon x. The function

$$F_X(x) = P(X \leq x) \tag{2.16}$$

is defined as the *distribution function* of X. In the above, the subscript X identifies the r.v. This subscript will usually be omitted when no confusion seems possible.

The distribution function always exists. By definition, it is a nonnegative, continuous to the right, and nondecreasing function of the real variable x. Moreover, we have

$$F_X(-\infty) = 0 \quad \text{and} \quad F_X(+\infty) = 1 \tag{2.17}$$

A r.v. X is called a *continuous* r.v. if its associated distribution function is continuous and differentiable almost everywhere. It is a *discrete* r.v. when the distribution function assumes the form of a staircase with a finite or countably infinite jumps. For a continuous r.v. X, the derivative

$$f_X(x) = dF_X(x)/dx \tag{2.18}$$

in this case exists and is called the *density function* of the r.v. X. It has the properties that

$$f_X(x) \geq 0, \quad \int_a^b f_X(x)\,dx = F_X(b) - F_X(a), \quad \int_{-\infty}^{\infty} f_X(x)\,dx = 1 \tag{2.19}$$

On the other hand, the density function of a discrete r.v. or of a r.v. of the mixed type does not exist in the ordinary sense. However, it can be constructed with the aid of the Dirac delta function. Consider the case where a r.v. X takes on only discrete values $x_1, x_2, \ldots, x_n$. A consistent definition of its density function is

$$f_X(x) = \sum_{j=1}^{n} p_j \, \delta(x - x_j) \tag{2.20}$$

where $\delta(x)$ is the Dirac delta function and

$$p_j = P(X = x_j), \qquad j = 1, 2, \ldots, n \tag{2.21}$$

This definition of $f_X(x)$ is consistent in the sense that it has all the properties indicated by Eqs. (2.19), and the distribution function $F_X(x)$ is recoverable from Eq. (2.20) by integration.

Figure 2.1 illustrates graphically the distribution functions and density functions of a typical set of random variables. Figure 2.1a gives the typical shapes of these functions of a continuous r.v. The distribution and density functions of a discrete r.v. are shown in Fig. 2.1b, where the density function in this case takes the form of a sequence of impulses whose strengths are determined by the probabilities $p_j = P(X = x_j)$. Figure 2.1c shows these functions for a r.v. of the mixed type.

The definitions of the distribution function and the density function can be readily extended to the case of two or more random variables. Consider first the case of two random variables. The r.v.'s X and Y induce a probability distribution in the Euclidean plane R^2 and this distribution is described by the *joint distribution function*

$$F_{XY}(x, y) = P\{(X \le x) \cap (Y \le y)\} \tag{2.22}$$

The function $F_{XY}(x, y)$ is nonnegative, nondecreasing, and continuous to the right with respect to each of the real variables x and y. We also note that

$$F_{XY}(-\infty, -\infty) = F_{XY}(-\infty, y) = F_{XY}(x, -\infty) = 0$$
$$F_{XY}(+\infty, +\infty) = 1, \quad F_{XY}(x, +\infty) = F_X(x), \quad F_{XY}(+\infty, y) = F_Y(y) \tag{2.23}$$

In the above, the function $F_X(x)$, for example, is the distribution function of X alone. In the context of several random variables, the distribution function of each random variable is commonly called the *marginal distribution function*.

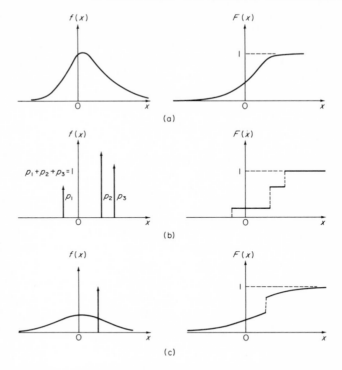

Fig. 2.1. Examples of distribution functions and density functions. (a) A continuous random variable. (b) A discrete random variable. (c) A random variable of the mixed type.

If the second-order partial derivative

$$\partial^2 F_{XY}(x, y)/\partial x\, \partial y$$

exists, we denote it by

$$f_{XY}(x, y) = \partial^2 F_{XY}(x, y)/\partial x\, \partial y \qquad (2.24)$$

and call it the *joint density function* of X and Y. We observe that $f_{XY}(x, y)$ is always nonnegative and it has the properties that

$$\int_{-\infty}^{\infty} \int_{-\infty}^{\infty} f_{XY}(x, y)\, dx\, dy = 1, \qquad \int_{-\infty}^{x} \int_{-\infty}^{y} f_{XY}(u, v)\, du\, dv = F_{XY}(x, y)$$

$$\qquad (2.25)$$

$$\int_{-\infty}^{\infty} f_{XY}(x, y)\, dy = f_X(x), \qquad \int_{-\infty}^{\infty} f_{XY}(x, y)\, dx = f_Y(y)$$

The density functions $f_X(x)$ and $f_Y(y)$ are now called the *marginal density functions* of X and Y, respectively.

The discussion above can be immediately generalized to cover the case of many random variables. Thus, the joint distribution function of a sequence of n r.v.'s, $\{X_n\}$, is defined by

$$F_{X_1 X_2 \cdots X_n}(x_1, x_2, \ldots, x_n) = P\{(X_1 \leq x_1) \cap (X_2 \leq x_2) \cdots \cap (X_n \leq x_n)\}$$

$$(2.26)$$

The corresponding joint density function is

$$f_{X_1 X_2 \cdots X_n}(x_1, x_2, \ldots, x_n) = \partial^n F_{X_1 X_2 \cdots X_n}(x_1, x_2, \ldots, x_n)/\partial x_1 \, \partial x_2 \cdots \partial x_n$$

$$(2.27)$$

if the indicated partial derivative exists. Various properties possessed by these functions can be readily inferred from those indicated for the two-random-variable case. As we have mentioned previously, the finite sequence $\{X_n\}$ may be regarded as the components of an n-dimensional random vector **X**. The distribution function and the density function of **X** are identical with (2.26) and (2.27), but they may be written in the more compact forms $F_\mathbf{X}(\mathbf{x})$ and $f_\mathbf{X}(\mathbf{x})$, where **x** is a vector with components $x_1, x_2, \ldots, x_n$.

Let us mention again that we will omit the subscripts in writing the distribution functions and density functions whenever convenient. In the sequel, the existence of the density functions shall always be assumed.

2.3.2. Conditional Distributions and Statistical Independence

Let A and B be two arbitrary events and let $P(B) > 0$. The number

$$P(A \mid B) = P(AB)/P(B) \qquad (2.28)$$

is called the *conditional probability* of A given B or the probability of A *conditional upon* B. In the case of two r.v.'s X and Y, we see that

$$P\{x_1 < X \leq x_2 \mid y_1 < Y \leq y_2\}$$

$$= \int_{y_1}^{y_2} \int_{x_1}^{x_2} f(x, y) \, dx \, dy \Big/ \int_{y_1}^{y_2} \int_{-\infty}^{\infty} f(x, y) \, dx \, dy$$

$$= \int_{y_1}^{y_2} \int_{x_1}^{x_2} f(x, y) \, dx \, dy \Big/ \int_{y_1}^{y_2} f_Y(y) \, dy \qquad (2.29)$$

We define the *conditional distribution function* of X given Y, denoted by $F_{XY}(x \mid y)$, as the conditional probability

$$F_{XY}(x \mid y) = P\{X \leq x \mid Y = y\} \qquad (2.30)$$

Its relationship with the joint distribution of X and Y can be obtained by means of Eq. (2.29) by setting $x_1 = -\infty$, $x_2 = x$, $y_1 = y$, $y_2 = y + \Delta y$ and taking the limit as $\Delta y \to 0$. The result is

$$F_{XY}(x \mid y) = \int_{-\infty}^{x} f(u, y) \, du / f_Y(y) \tag{2.31}$$

provided that $f_Y(y) > 0$.

A consistent definition of the *conditional density function* of X given Y, $f_{XY}(x \mid y)$, is then the derivative of its corresponding conditional distribution function. Thus,

$$f_{XY}(x \mid y) = dF_{XY}(x \mid y)/dx = f(x, y)/f_Y(y) \tag{2.32}$$

which follows the general form expressed in Eq. (2.28).

Rearranging Eq. (2.32) gives

$$f(x, y) = f_{XY}(x \mid y) f_Y(y) \tag{2.33}$$

or, by interchanging the variables,

$$f(x, y) = f_{YX}(y \mid x) f_X(x) \tag{2.34}$$

Equations (2.33) and (2.34) give the relationships between the conditional density functions and the joint density function of two random variables.

In the particular case when

$$f_{XY}(x \mid y) = f_X(x) \tag{2.35}$$

the random variables X and Y are called *statistically independent* or simply *independent*. Equation (2.35) indicates that the probabilistic behavior of the r.v. X is independent of that of Y. For this case, Eq. (2.33) becomes

$$f_{XY}(x, y) = f_X(x) f_Y(y) \tag{2.36}$$

which can also be used as a definition of statistical independence.

The extension of the definition above to the case of more than two random variables is straightforward. Starting from the basic definition

$$P(A \mid BC) = P(AB \mid C)/P(B \mid C)$$

$$= P(ABC)/P(B \mid C)P(C), \qquad P(B \mid C), P(C) > 0 \tag{2.37}$$

for the three events A, B, and C, we have

$$P(ABC) = P(A \mid BC)P(B \mid C)P(C) \tag{2.38}$$

or, for three r.v.'s X, Y, and Z,

$$f_{XYZ}(x, y, z) = f_{XYZ}(x \mid y, z) f_{YZ}(y \mid z) f_Z(z) \tag{2.39}$$

Hence, for the general case of a sequence $\{X_n\}$ of n r.v.'s, we can write

$$f(x_1, x_2, \ldots, x_n)$$
$$= f(x_1 \mid x_2, \ldots, x_n) f(x_2 \mid x_3, \ldots, x_n) \cdots f(x_{n-1} \mid x_n) f(x_n) \tag{2.40}$$

In the event that the r.v.'s X_1, X_2, $\ldots$, X_n are *mutually independent*, Eq. (2.40) becomes

$$f(x_1, x_2, \ldots, x_n) = f_{X_1}(x_1) f_{X_2}(x_2) \cdots f_{X_n}(x_n) \tag{2.41}$$

2.3.3. Moments

Some of the most useful information concerning a random variable is revealed by its moments, particularly those of the first and second order. The nth *moment* of a r.v. X, α_n, is defined by

$$\alpha_n = E\{X^n\} = \int_{-\infty}^{\infty} x^n f(x) \, dx \tag{2.42}$$

if $\int_{-\infty}^{\infty} \mid x \mid^n f(x) \, dx$ is finite. The symbol E is used above to indicate an expected-value operation.

The first moment α_1 gives the statistical average of the r.v. X. It is synonymously called *mean, mathematical expectation,* or *statistical average.* We will usually denote it by m_X or simply m.

The *central moments* of a r.v. X are the moments of X with respect to its mean. Hence, the nth *central moment* of X, μ_n, is defined as

$$\mu_n = E\{(X - m)^n\} = \int_{-\infty}^{\infty} (x - m)^n f(x) \, dx \tag{2.43}$$

The second central moment measures the spread or dispersion of the r.v. X about its mean. It is also called the *variance* and is commonly denoted by σ_X^2 or simply σ^2. In terms of the moments α_n, σ^2 is expressed by

$$\sigma^2 = \alpha_2 - \alpha_1^2 = \alpha_2 - m^2 \tag{2.44}$$

Another measure of the spread is the *standard deviation*. It is denoted by σ_X or σ and it is equal to the square root of the variance of X.

The moments of two or more random variables are defined in a similar fashion. The joint moments of the r.v.'s X and Y, α_{nm}, are defined by

$$\alpha_{nm} = E\{X^n Y^m\} = \int_{-\infty}^{\infty} \int_{-\infty}^{\infty} x^n y^m f(x, y) \, dx \, dy \qquad (2.45)$$

when they exist. It is seen that α_{10} and α_{01} are simply the respective means of X and Y. If the r.v.'s X and Y are independent, the moments become

$$\alpha_{nm} = \int_{-\infty}^{\infty} \int_{-\infty}^{\infty} x^n y^m f_X(x) f_Y(y) \, dx \, dy = \alpha_{n0} \alpha_{0m} \qquad (2.46)$$

We note, however, that Eq. (2.46) does not necessarily imply independence of X and Y.

The *joint central moments* of the r.v.'s X and Y, μ_{nm}, are given by .

$$
\begin{aligned}
\mu_{nm} &= E\{(X - m_X)^n (Y - m_Y)^m\} \\
&= \int_{-\infty}^{\infty} \int_{-\infty}^{\infty} (x - m_X)^n (y - m_Y)^m f(x, y) \, dx \, dy \qquad (2.47)
\end{aligned}
$$

Clearly, μ_{20} and μ_{02} are the respective variances. The central moment μ_{11} is called the *covariance* of X and Y. It is easy to show that

$$\mu_{20} = \alpha_{20} - \alpha_{10}^2, \qquad \mu_{02} = \alpha_{02} - \alpha_{01}^2, \qquad \mu_{11} = \alpha_{11} - \alpha_{10}\alpha_{01} \qquad (2.48)$$

In dealing with the covariance, it is sometimes convenient to consider a normalized quantity. The normalized covariance is given by

$$\varrho = \mu_{11}/(\mu_{20}\mu_{02})^{1/2}, \qquad |\varrho| \leq 1 \qquad (2.49)$$

and is called the *correlation coefficient* of X and Y.

The covariance or the correlation coefficient is of great importance in the analysis of two random variables. It is a measure of their linear interdependence in the sense that its value is a measure of accuracy with which one random variable can be approximated by a linear function of the other. In order to see this, let us consider the problem of approximating a r.v. X by a linear function of a second r.v. Y, $aY + b$, where a and b are chosen so that the mean square error e, defined by

$$e = E\{[X - (aY + b)]^2\} \qquad (2.50)$$

is minimized. Straightforward calculations show that this minimum is attained by setting

$$a = \sigma_X \varrho / \sigma_Y, \qquad b = m_X - a m_Y \tag{2.51}$$

and the minimum of the mean square error is $\sigma_X^2 (1 - \varrho^2)$. We thus see that an exact fit in the mean square sense is achieved when $|\varrho| = 1$ and the linear approximation is the worst when $\varrho = 0$.

The r.v.'s X and Y are said to be *uncorrelated* if

$$\varrho = 0 \tag{2.52}$$

which implies

$$\mu_{11} = 0 \qquad \text{or} \qquad E\{XY\} = E\{X\}\, E\{Y\} \tag{2.53}$$

In the particular case when

$$E\{XY\} = 0 \tag{2.54}$$

the r.v.'s are called *orthogonal*.

It is clearly seen from Eq. (2.53) that two independent r.v.'s with finite second moments are uncorrelated. We point out again that the converse of this statement is not necessarily true.

In closing this subsection, we state without proof two inequalities involving the moments of random variables which will be used extensively in the latter chapters.

Tchebycheff Inequality:

$$P\{|X - m| \geq k\sigma\} \leq 1/k^2 \tag{2.55}$$

for any $k > 0$.

Schwarz Inequality:

$$E\{|XY|\} \leq [E\{X^2\}\, E\{Y^2\}]^{1/2} \tag{2.56}$$

A generalization of the Schwarz inequality is

$$E\{|XY|\} \leq [E\{|X|^n\}]^{1/n} [E\{|Y|^m\}]^{1/m} \tag{2.57}$$

where

$$n, m > 1, \qquad 1/n + 1/m = 1 \tag{2.58}$$

Inequality (2.57) is known as the *Hölder inequality*.

2.3.4. Characteristic Functions

Let X be a r.v. with density function $f(x)$. The *characteristic function* of X, $\phi_X(u)$ or $\phi(u)$, is defined by

$$\phi(u) = E\{e^{iuX}\} = \int_{-\infty}^{\infty} e^{iux} f(x)\, dx \tag{2.59}$$

where u is arbitrary and real-valued. It is seen that the characteristic function $\phi(u)$ and the density function $f(x)$ form a Fourier transform pair. Since a density function is absolute integrable, its associated characteristic function always exists. It is continuous as a function of u and $|\phi(u)| \leq 1$. Furthermore, it follows from the theory of Fourier transforms that the density function is uniquely determined in terms of the characteristic function by

$$f(x) = (1/2\pi) \int_{-\infty}^{\infty} e^{-iux} \phi(u)\, du \tag{2.60}$$

Equation (2.60) points out one of the many uses of a characteristic function. In many physical problems, it is often more convenient to determine the density function of a random variable by first determining its characteristic function and then performing the Fourier transform as indicated by Eq. (2.60).

Another important property enjoyed by the characteristic function is that it is simply related to the moments. The MacLaurin series of $\phi(u)$ gives

$$\phi(u) = \phi(0) + \phi'(0)u + \phi''(0)\frac{u^2}{2!} + \cdots \tag{2.61}$$

Now, we see that

$$\phi(0) = \int_{-\infty}^{\infty} f(x)\, dx = 1$$

$$\phi'(0) = i \int_{-\infty}^{\infty} x f(x)\, dx = i\alpha_1 \tag{2.62}$$

$$\vdots$$

$$\phi^{(n)}(0) = i^n \int_{-\infty}^{\infty} x^n f(x)\, dx = i^n \alpha_n$$

If the expectations $E\{|X|^n\}$ exist for some $n \geq 1$, Eqs. (2.61) and (2.62) lead to

$$\phi(u) = \sum_{j=0}^{n} \frac{\alpha_j}{j!}(iu)^j + o(u^n) \tag{2.63}$$

where $o(u^n)$ stands for the terms such that

$$\frac{o(u^n)}{u^n} \to 0 \quad \text{as} \quad u \to 0 \tag{2.64}$$

Equations (2.62) or Eq. (2.63) shows that the moments of a r.v. are easily derivable from the knowledge of its characteristic function.

The *joint characteristic function* of two random variables X and Y, $\phi_{XY}(u, v)$ or $\phi(u, v)$, is defined by

$$\phi(u, v) = E\{e^{i(uX+vY)}\} = \int_{-\infty}^{\infty} \int_{-\infty}^{\infty} e^{i(ux+vy)} f(x, y) \, dx \, dy \tag{2.65}$$

where u and v are two arbitrary real variables. Analogous to the one-random-variable case, the joint characteristic function $\phi(u, v)$ is often called upon for the determination of the joint density function $f(x, y)$ of X and Y and their joint moments. The density function $f(x, y)$ is uniquely determined in terms of $\phi(u, v)$ by the two-dimensional Fourier transform

$$f(x, y) = (1/4\pi^2) \int_{-\infty}^{\infty} \int_{-\infty}^{\infty} e^{-i(ux+vy)} \phi(u, v) \, du \, dv \tag{2.66}$$

and the moments $E\{X^n Y^m\} = \alpha_{nm}$, if they exist, are related to $\phi(u, v)$ by

$$\frac{\partial^{n+m}}{\partial u^n \, \partial v^m} \phi(u, v) \,|_{u,v=0} = i^{n+m} \int_{-\infty}^{\infty} \int_{-\infty}^{\infty} x^n y^m f(x, y) \, dx \, dy$$
$$= i^{n+m} \alpha_{nm} \tag{2.67}$$

The MacLaurin series expansion of $\phi(u, v)$ thus takes the form

$$\phi(u, v) = \sum_{j=0}^{n} \sum_{k=0}^{m} \frac{\alpha_{jk}}{j!k!} (iu)^n (iv)^m + o(u^n v^m) \tag{2.68}$$

The above development can be generalized to the case of more than two random variables in an obvious manner.

2.4. Functions of Random Variables

The transformation of a sequence of random variables into another sequence will play a central role in our discussions throughout this book. In this section we consider a class of algebraic transformations which will serve as basic tools for analyzing more complicated transformations involving, for example, differentiations and integrations.

2.4.1. Sum of Random Variables

Let $\{X_n\}$ be a sequence of n r.v.'s having the joint density function $f(x_1, x_2, \ldots, x_n)$. The sum

$$Z = \sum_{j=1}^{n} X_j \qquad (2.69)$$

is a r.v. and we are interested in finding its probability distribution.

It suffices to determine the distribution of the sum of two random variables. The result for this case can then be applied successively to give the distribution of the sum of any number of random variables.

Consider

$$Z = X + Y \qquad (2.70)$$

The distribution function $F_Z(z)$ is determined by the probability of the event $\{Z \leq z\}$. It is thus equal to the probability that the event $\{X + Y \leq z\}$ occurs in the domain of the r.v.'s X and Y. The region in the xy-plane satisfying

$$x + y \leq z$$

is shown in Fig. 2.2 as the shaded area indicated by R. Hence, in terms of $f_{XY}(x, y)$,

$$F_Z(z) = P\{Z \leq z\} = P\{X + Y \leq z\}$$

$$= \int_R \int f_{XY}(x, y) \, dx \, dy = \int_{-\infty}^{\infty} \int_{-\infty}^{z-y} f_{XY}(x, y) \, dx \, dy \qquad (2.71)$$

Differentiating Eq. (2.71) with respect to z we obtain the desired result

$$f_Z(z) = \int_{-\infty}^{\infty} f_{XY}(z - y, y) \, dy \qquad (2.72)$$

An important special case of the above is one where X and Y are independent. We then have

$$f_{XY}(x, y) = f_X(x) f_Y(y) \qquad (2.73)$$

The substitution of Eq. (2.73) into Eq. (2.72) hence gives

$$f_Z(z) = \int_{-\infty}^{\infty} f_X(z - y) f_Y(y) \, dy \qquad (2.74)$$

This result shows that the density function of the sum of two independent random variables is the convolution of their respective density functions.

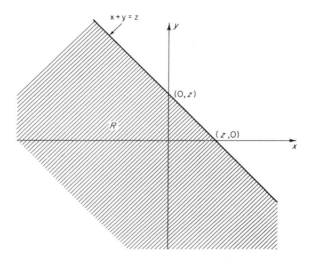

Fig. 2.2. The region R: $\{x + y \leq z\}$.

While the density function of the sum of n independent random variables can be obtained by repeated application of Eq. (2.74), a more straightforward and labor-saving procedure is to make use of the characteristic functions. Let the r.v.'s $X_1, X_2, \ldots, X_n$ be mutually independent with marginal density functions $f_{X_1}(x_1), f_{X_2}(x_2), \ldots, f_{X_n}(x_n)$ and marginal characteristic functions $\phi_{X_1}(u_1), \phi_{X_2}(u_2), \ldots, \phi_{X_n}(u_n)$. By definition, if

$$Z = \sum_{j=1}^{n} X_j \tag{2.75}$$

the characteristic function of Z, $\phi_Z(v)$, is

$$\phi_Z(v) = E\{e^{ivZ}\} = E\left\{\exp\left[iv\left(\sum_{j=1}^{n} X_j\right)\right]\right\}$$

$$= \prod_{j=1}^{n} E\{\exp(ivX_j)\} = \prod_{j=1}^{n} \int_{-\infty}^{\infty} \exp(ivx_j) f_{X_j}(x_j) \, dx_j$$

$$= \prod_{j=1}^{n} \phi_{X_j}(v) \tag{2.76}$$

Consequently, the function $\phi_Z(v)$ is directly obtainable as the product of the characteristic functions of X_j. The density function of Z can then be found by means of Eq. (2.60).

Other useful results concerning the sum of random variables are the moment relations. If $X_1, X_2, \ldots, X_n$ are r.v.'s with respective means

$m_1, m_2, \ldots, m_n$ and respective variances $\sigma_1{}^2, \sigma_2{}^2, \ldots, \sigma_n{}^2$, the mean m of the sum

$$\sum_{j=1}^{n} X_j$$

is given by

$$m = \sum_{j=1}^{n} m_j \tag{2.77}$$

Furthermore, if the r.v.'s X_j are mutually independent, we have the useful formula

$$\sigma^2 = \sum_{j=1}^{n} \sigma_j{}^2 \tag{2.78}$$

where σ^2 is the variance of the sum.

This additive property also applies to the third central moment of the sum of independent random variables, but not to the central moment of the fourth order or higher.

2.4.2. n Functions of n Random Variables

We consider next the relationship between the joint density functions of two sets of n random variables related by a one-to-one mapping.

Let

$$\mathbf{X} = \begin{bmatrix} X_1 \\ X_2 \\ \cdot \\ \cdot \\ \cdot \\ X_n \end{bmatrix} \quad \text{and} \quad \mathbf{Y} = \begin{bmatrix} Y_1 \\ Y_2 \\ \cdot \\ \cdot \\ \cdot \\ Y_n \end{bmatrix} \tag{2.79}$$

be two n-dimensional random vectors and let $\mathbf{h}$ be a deterministic transformation of $\mathbf{Y}$ into $\mathbf{X}$, that is,

$$\mathbf{X} = \mathbf{h}(\mathbf{Y}) \tag{2.80}$$

We are interested in the determination of the joint density function $f_{\mathbf{X}}(\mathbf{x})$ of $\mathbf{X}$ in terms of the joint density function $f_{\mathbf{Y}}(\mathbf{y})$ of $\mathbf{Y}$. It is assumed that the transformation $\mathbf{h}$ is continuous with respect to each of its arguments, has continuous partial derivatives, and defines a one-to-one mapping. Thus, the inverse transform $\mathbf{h}^{-1}$ defined by

$$\mathbf{Y} = \mathbf{h}^{-1}(\mathbf{X}) \tag{2.81}$$

exists. It also has continuous partial derivatives.

In order to determine $f_{\mathbf{X}}(\mathbf{x})$, we observe that, if a closed region R_y in the range of sample values of $\mathbf{Y}$ is mapped into a closed region R_x in the range of sample values of $\mathbf{X}$ under the transformation $\mathbf{h}$, the conservation of probability gives

$$\int_{R_x} f_{\mathbf{X}}(\mathbf{x})\,d\mathbf{x} = \int_{R_y} f_{\mathbf{Y}}(\mathbf{y})\,d\mathbf{y} \tag{2.82}$$

where the integrals represent n-fold integrals with respect to the components of $\mathbf{x}$ and $\mathbf{y}$, respectively. Following the usual rule of change of variables, we have

$$\int_{R_y} f_{\mathbf{Y}}(\mathbf{y})\,d\mathbf{y} = \int_{R_x} f_{\mathbf{Y}}[\mathbf{y} = \mathbf{h}^{-1}(\mathbf{x})]\,|\,J\,|\,d\mathbf{x} \tag{2.83}$$

where J is the Jacobian of the transformation. It is given by

$$J = \left| \frac{\partial \mathbf{y}^T}{\partial \mathbf{x}} \right| = \begin{vmatrix} \dfrac{\partial y_1}{\partial x_1} & \dfrac{\partial y_2}{\partial x_1} & \cdots & \dfrac{\partial y_n}{\partial x_1} \\ \vdots & & & \vdots \\ \dfrac{\partial y_1}{\partial x_n} & \dfrac{\partial y_2}{\partial x_n} & \cdots & \dfrac{\partial y_n}{\partial x_n} \end{vmatrix} \tag{2.84}$$

Equations (2.82) and (2.83) then lead to the desired formula

$$f_{\mathbf{X}}(\mathbf{x}) = f_{\mathbf{Y}}[\mathbf{y} = \mathbf{h}^{-1}(\mathbf{x})]\,|\,J\,| \tag{2.85}$$

The result given above can also be applied to the case when the dimension of $\mathbf{X}$ is smaller than that of $\mathbf{Y}$. Consider, for example, the transformation

$$X_j = g_j(Y_1, Y_2, \ldots, Y_n), \qquad j = 1, 2, \ldots, m, \quad m < n \tag{2.86}$$

In order to use the result given by Eq. (2.85), we first augment the m-dimensional vector $\mathbf{X}$ by another $(n - m)$-dimensional vector $\mathbf{Z}$. The vector $\mathbf{Z}$ can be constructed, for example, by defining

$$Z_k = g_{m+k}(Y_1, Y_2, \ldots, Y_n) = Y_{m+k}, \qquad k = 1, 2, \ldots, n - m \tag{2.87}$$

On combining Eqs. (2.86) and (2.87), we obtain the joint density function of $\mathbf{X}$ and $\mathbf{Z}$ by means of Eq. (2.85). The joint density function of $\mathbf{X}$ alone is then obtained by integration with respect to the components of $\mathbf{Z}$.

In the above, we have assumed implicitly that the augmented transformation is continuous with continuous partial derivatives, and it defines a one-to-one mapping.

2.5. The Gaussian Random Variable and the Central Limit Theorem

The most important distribution of random variables which we encounter in physical problems is the *Gaussian* or *normal* distribution. A r.v. X is *Gaussian* or *normal* if its density function $f(x)$ has the form

$$f(x) = \frac{1}{\sigma(2\pi)^{1/2}} \exp\left[-\frac{(x-m)^2}{2\sigma^2}\right] \tag{2.88}$$

Its distribution function is the integral

$$F(x) = \frac{1}{\sigma(2\pi)^{1/2}} \int_{-\infty}^{x} \exp\left[-\frac{(u-m)^2}{2\sigma^2}\right] du \tag{2.89}$$

It is seen that a Gaussian r.v. is completely characterized by the two parameters m and σ; they are, as can be easily shown, the mean and the standard deviation of X, respectively.

Alternatively, a Gaussian r.v. can be defined in terms of its characteristic function. It is of the form

$$\phi(u) = \exp\{ium - \tfrac{1}{2}\sigma^2 u^2\} \tag{2.90}$$

As indicated by Eqs. (2.62), the moments of X can be easily derived from Eq. (2.90) by successively differentiating $\phi(u)$ with respect to u and setting $u = 0$. The results, expressed in terms of its central moments, are

$$\mu_n = \begin{cases} 0, & n \text{ odd} \\ 1 \cdot 3 \cdots (n-1)\sigma^n, & n \text{ even} \end{cases} \tag{2.91}$$

In addition to having these simple properties, another reason for the popularity of a Gaussian r.v. in physical applications stems from the central limit theorem stated below. Rather than giving the theorem in general terms, it serves our purpose quite well by stating a more restricted version due to Lindeberg [7].

Central Limit Theorem. Let $\{X_n\}$ be a sequence of mutually independent and identically distributed r.v.'s with means m and variances σ^2. Let

$$X = \sum_{j=1}^{n} X_j \tag{2.92}$$

and let the normalized r.v. Y be defined as

$$Y = (X - nm)/\sigma n^{1/2} \tag{2.93}$$

Then the distribution function of Y, $F_Y(y)$, converges to the zero mean, unit variance Gaussian distribution as $n \to \infty$ for every fixed y.

Proof. We first remark that the r.v. Y is simply the "normed" r.v. X with mean zero and variance one. In terms of the characteristic functions $\phi_X(u)$ of the r.v.'s X_j, the characteristic function $\phi_Y(u)$ of Y has the form

$$\phi_Y(u) = E\{e^{iuY}\}$$
$$= [\exp(-ium/\sigma n^{1/2})\phi_X(u/\sigma n^{1/2})]^n \qquad (2.94)$$

In view of the expansion (2.63), we can write

$$\phi_Y(u) = \left\{\exp(-ium/\sigma n^{1/2})\left[1 + ium/\sigma n^{1/2} + \frac{(\sigma^2 + m^2)}{2!}\left(\frac{iu}{\sigma n^{1/2}}\right)^2 + \cdots\right]\right\}^n$$
$$= \left[1 - u^2/2n + o\left(\frac{u^2}{n}\right)\right]^n \to \exp[-u^2/2] \qquad (2.95)$$

as $n \to \infty$. In the last step we have used the elementary identity in calculus,

$$\lim_{n \to \infty} (1 + c/n)^n = e^c \qquad (2.96)$$

for any real c.

Equation (2.95) shows that $\phi_Y(u)$ approaches the characteristic function of a zero mean and unit variance Gaussian distribution in the limit. The proof is thus complete.

The central limit theorem thus indicates that often a physical phenomenon has a Gaussian distribution when it represents the sum of a large number of small independent random effects. This indeed is a model or at least an idealization of many physical phenomena we encounter.

Consider a sequence $\{X_n\}$ of n r.v.'s. They are said to be *jointly Gaussian* if its associated joint density function has the form

$$f_{\mathbf{X}}(\mathbf{x}) = f_{X_1 X_2 \cdots X_n}(x_1, x_2, \ldots, x_n)$$
$$= (2\pi)^{-n/2} \, | \Lambda |^{-1/2} \exp[-\tfrac{1}{2}(\mathbf{x} - \mathbf{m})^T \Lambda^{-1}(\mathbf{x} - \mathbf{m})] \qquad (2.97)$$

where

$$\mathbf{m}^T = [m_1 \ m_2 \ \cdots \ m_n] = [E\{X_1\} \ E\{X_2\} \ \cdots \ E\{X_n\}]$$

and $\Lambda = [\mu_{ij}]$ is the $n \times n$ covariance matrix of $\mathbf{X}$ with

$$\mu_{ij} = E\{[X_i - m_i][X_j - m_j]\} \qquad (2.98)$$

The superscripts T and -1 denote, respectively, the matrix transpose and the matrix inverse. Again, we see that a joint Gaussian distribution is completely characterized by the first- and second-order joint moments.

Parallel to our discussion of a single Gaussian r.v., an alternative definition of a sequence of jointly Gaussian distributed random variables is that it has the joint characteristic function

$$\phi_{\mathbf{X}}(\mathbf{u}) = \exp[i\mathbf{m}^{\mathrm{T}}\mathbf{u} - \tfrac{1}{2}\mathbf{u}^{\mathrm{T}}\Lambda\mathbf{u}] \tag{2.99}$$

where $\mathbf{u}^{\mathrm{T}} = [u_1 \, u_2 \, \dots \, u_n]$. This definition is sometimes preferable since it avoids the difficulties when the covariance matrix Λ becomes singular.

The joint moments of $\mathbf{X}$ can be obtained by differentiating the joint characteristic function $\phi_{\mathbf{X}}(\mathbf{u})$ and setting $\mathbf{u} = \mathbf{0}$. The expectation

$$E\{X_1^{m_1} X_2^{m_2} \cdots X_n^{m_n}\},$$

for example, is given by

$$E\{X_1^{m_1} X_2^{m_2} \cdots X_n^{m_n}\} = i^{-(m_1 + m_2 + \cdots + m_n)} \left[\frac{\partial^{m_1 + m_2 + \cdots + m_n}}{\partial u_1^{m_1} \, \partial u_2^{m_2} \cdots \partial u_n^{m_n}} \phi_{\mathbf{X}}(\mathbf{u}) \right]_{\mathbf{u}=0} \tag{2.100}$$

It is clear that, since the joint moments of the first- and second-order completely specify the joint Gaussian distribution, these moments also determine the joint moments of orders higher than two. We can show that, if the r.v.'s $X_1, X_2, \dots, X_n$ have zero means, all odd order moments of these r.v.'s vanish and, for n even,

$$E\{X_1 X_2 \cdots X_n\} = \sum_{m_1, m_2, \dots, m_n} E\{X_{m_1} X_{m_2}\} E\{X_{m_2} X_{m_3}\} \cdots E\{X_{m_{n-1}} X_{m_n}\} \tag{2.101}$$

The sum above is taken over all possible combinations of $n/2$ pairs of n r.v.'s. The number of terms in the summation is $1 \cdot 3 \cdot 5 \cdots (n-3)(n-1)$.

References

1. K. L. Chung, *A Course in Probability Theory*. Harcourt, New York, 1968.
2. H. Cramér, *Mathematical Methods of Statistics*. Princeton Univ. Press, Princeton, New Jersey, 1946.
3. J. L. Doob, *Stochastic Processes*. Wiley, New York, 1953.
4. W. Feller, *An Introduction to Probability Theory and Its Applications*, Vol. I. Wiley, New York, 1958.
5. M. Loéve, *Probability Theory*. Van Nostrand-Reinhold, Princeton, New Jersey, 1963.
6. A. Papoulis, *Probability, Random Variables, and Stochastic Processes*. McGraw-Hill, New York, 1965.
7. J. W. Lindeberg, Eine neue Herleitung des Exponentialgesetzes in der Wahrscheinlichkeitsrechnung. *Math. Z.* **15**, 211–225 (1922).

PROBLEMS

2.1. Listed below are some of the commonly encountered density functions for a random variable. For each of the density functions, find its associated characteristic function and, when they exist, the mean and the variance.

(a)   Point distribution

$$f(x) = \delta(x - a)$$

(b)   Binomial distribution (*n* trials with success probability *p*)

$$\sum_{k=0}^{n} \binom{n}{k} p^k q^{n-k} \, \delta(x - k), \qquad p + q = 1$$

(c)   Geometric distribution (with success probability *p*)

$$\sum_{k=0}^{\infty} pq^k \, \delta(x - k), \qquad p + q = 1$$

(d)   Poisson distribution (with parameter $\lambda > 0$)

$$e^{-\lambda} \sum_{k=0}^{\infty} \frac{\lambda^k}{k!} \, \delta(x - k)$$

(e)   Uniform distribution

$$1/2a, \qquad -a \leq x \leq a$$

(f)   Triangular distribution

$$(a - |x|)/a^2, \qquad -a \leq x \leq a$$

(g)   Exponential distribution ($\lambda > 0$)

$$\lambda e^{-\lambda x}, \qquad x > 0$$

(h)   Gamma distribution ($a, \lambda > 0$)

$$[\lambda/\Gamma(a)](\lambda x)^{a-1} e^{-\lambda x}, \qquad x > 0$$

(i)   Cauchy distribution ($a > 0$)

$$a/\pi(a^2 + x^2)$$

(j) χ^2-distribution (with n degrees of freedom)

$$\frac{1}{2^{n/2}\Gamma(n/2)}\, x^{n/2-1}e^{-x/2}, \qquad x > 0$$

2.2. The joint density function of two r.v.'s X and Y is

(a) $f(x, y) = 1/x^2y^2, \qquad 1 \leq x, y < \infty$

(b) $f(x, y) = 1/\pi, \qquad x^2 + y^2 \leq 1$

(c) $f(x, y) = y^2 e^{-y(1+x)}, \qquad x, y > 0$

For each of the above, determine the marginal density functions of X and Y. Are the random variables uncorrelated? Are they independent?

2.3. Let the r.v.'s X and Y be independent and let each take the values $+1$ and -1 with probability $\frac{1}{2}$. Show that the r.v.'s X and XY are independent but the r.v.'s X, Y, and XY are not mutually independent.

2.4. Let the r.v.'s X and Y be identically Gaussian distributed. Show that the r.v.'s $X + Y$ and $X - Y$ are independent.

2.5. Show that if $E\{|X|^p\} < \infty$, then $E\{X^q\} < \infty$ for $q \leq p$.

2.6. Verify the Tchebycheff inequality (2.55).

2.7. Verify the Hölder inequality (2.57).

2.8. Show that, given a r.v. X, $P\{X = E\{X\}\} = 1$ if $\sigma_X^2 = 0$.

2.9. Show that, if the indicated moments exist,

$$[E\{|X + Y|^p\}]^{1/p} \leq [E\{|X|^p\}]^{1/p} + [E\{|Y|^p\}]^{1/p}, \qquad p > 1$$

$$E\left\{\left|(1/n)\sum_{j=1}^{n} X_j\right|^p\right\} \leq (1/n)\sum_{j=1}^{n} E\{|X_j|^p\}$$

The first inequality is referred to as the *Minkowski inequality*.

2.10. Given the characteristic function $\phi(u)$ of a r.v. X, consider the Mac-Laurin series expansion of the logarithm of $\phi(u)$ in the form

$$\log \phi(u) = \sum_{j=1}^{\infty} (\lambda_j/j!)(iu)^j$$

Show that the coefficients λ_j are functions of only the jth- and lower-order moments of X. Specifically, the first three coefficients are

$$\lambda_1 = m, \qquad \lambda_2 = \mu_2 = \sigma^2, \qquad \lambda_3 = \mu_3$$

The coefficient λ_k is called the kth *cumulant* of X.

2.11. The r.v.'s X and Y are independent with respective density functions

$$f_X(x) = p \, \delta(x) + q \, \delta(x - 1), \qquad p + q = 1$$
$$f_Y(y) = [1/(2\pi)^{1/2}] \exp(-y^2/2)$$

Determine the density functions of the r.v.'s $X + Y$ and XY.

2.12. Show that the distribution of the sum of n independent and identically distributed r.v.'s with a Poisson (Cauchy, Binomial) distribution is Poisson (Cauchy, Binomial).

2.13. Let $X_1, X_2, \ldots, X_n$ be independent and identically distributed Gaussian r.v.'s with means zero and variances σ^2. Determine the density function of

$$S_n = \sum_{j=1}^{n} X_j^2$$

The distribution of S_n is called the χ^2 distribution of n degrees of freedom (Problem 2.1(j)).

2.14. Show that the Gaussian distribution is closed under linear algebraic operations, that is, if the r.v.'s $X_1, X_2, \ldots, X_n$ are Gaussian, the r.v. Y defined by

$$Y = \sum_{j=1}^{n} a_j X_j$$

where a_j are arbitrary constants, is also Gaussian.

2.15. If $X_1, X_2, \ldots, X_n$ are mutually independent r.v.'s with respective central moments μ_{jk} (kth central moment of X_j) and if μ_k is the kth central moment of the sum $\sum_{j=1}^{n} X_j$. Show that

$$\mu_k = \sum_{j=1}^{n} \mu_{jk}$$

for $k \leq 3$ but it does not hold in general for $k > 3$.

2.16. Show that two uncorrelated Gaussian r.v.'s are independent.

2.17. The Poisson distribution with parameter $\lambda > 0$ is given in Problem 2.1.(d). Show that it approaches a Gaussian distribution as $\lambda \to \infty$.

2.18. Let the r.v.'s $X_1, X_2, \ldots, X_n$ be jointly Gaussian with zero means. Show that

$$E\{X_1 X_2 X_3\} = 0$$

$$E\{X_1 X_2 X_3 X_4\} = E\{X_1 X_2\}E\{X_3 X_4\} + E\{X_1 X_3\}E\{X_2 X_4\}$$
$$+ E\{X_1 X_3\}E\{X_2 X_4\}$$

Generalize the above results and verify Eq. (2.101).

Chapter 3

Stochastic Processes and Their Classifications

Random variables or random vectors are adequate for describing results of random experiments which assume scalar or vector values in a given trial. In many physical applications, however, the outcomes of a random experiment are represented by functions depending upon a parameter. These outcomes are then described by a *random function* $X(t)$, where t is a parameter assuming values in a reference set T. A typical example of random experiments giving rise to random functions is found in the theory of Brownian motion, where each coordinate of a particle executing Brownian motion exhibits random behavior as a function of time. There are, of course, many other examples of random functions in physical situations. Thermal noises in electrical circuits, wind load on structures, earthquake motion, turbulence, population fluctuations, epidemics, solar activities, and many other phenomena are conveniently modeled by random functions.

In our discussions, we will use the terms *random function*, *random process*, and *stochastic process* synonymously. For the sake of consistency, $X(t)$, $t \in T$, will be called a stochastic process (s.p.) $X(t)$ on T or, in short, s.p. $X(t)$. The parameter t may denote the time variable or a spatial variable. If the index set T is finite or countably infinite, the s.p. $X(t)$ is said to be a *discrete-parameter s.p.* The s.p. $X(t)$ is called a *continuous-parameter s.p.* when the index set T is a finite or an infinite interval.

In certain situations, there is a need to consider a random quantity depending upon several parameters. For example, the temperature of a point in a turbulent field can be considered as a stochastic process depending upon time and three spatial coordinates. The name *random field* is generally reserved for these processes.

3.1. Basic Concepts of Stochastic Processes

The foregoing description of a stochastic process is based upon the axiomatic approach in probability theory, originally due to Kolmogorov [1]. Let us call each realization of a given random experiment a *sample function* or a *member function*. This description suggests that a s.p. $X(t)$, $t \in T$, is a family of sample functions of the variable t, all defined on the same underlying probability space $\{\Omega, \mathscr{B}, P\}$. Intuitively, this sample theoretic approach to the concept of a stochastic process is physically plausible since a stochastic process describing a physical phenomenon generally represents a family of individual realizations. To each realization there corresponds a definite deterministic function of t. However, a difficult point arises. A stochastic process defined this way is specified by the probability of realizations of various sample functions. This leads to the consideration of probability measures on a functional space and it generally requires advanced mathematics in measure theory and functional analysis.

In order to circumvent this difficulty, we prefer a simpler approach to the concept of a stochastic process. This is considered in Section 3.1.1. We should emphasize that, although the sample theoretic approach is not essential in our subsequent development, it is a fruitful approach to the development of the mathematical foundation of stochastic processes and its importance should not be overlooked. A brief discussion of this approach is given in Appendix A.

3.1.1. Definition and Probability Distributions

We now give a definition of a stochastic process which is fruitful for our development in this book.

At a fixed t, a s.p. $X(t)$, $t \in T$, is a random variable. Hence, another characterization of a stochastic process is to regard it as a family of random variables, say, $X(t_1)$, $X(t_2)$, ... , depending upon a parameter $t \in T$. The totality of all the random variables defines the s.p. $X(t)$. For discrete-parameter stochastic processes, this set of random variables is finite or countably infinite. For continuous-parameter processes, the number is noncountably infinite. Thus, we see that the mathematical description of a stochastic process is considerably more complicated than that of a random variable. It, in fact, is equivalent to the mathematical description of an infinite and generally noncountable number of random variables.

We recall from Chapter 2 that a finite sequence of random variables is completely specified by its joint distribution or joint density functions. It thus follows from the remarks above that a s.p. $X(t)$, $t \in T$, is defined by a family of joint distribution or density functions, taking into account that we are now dealing with infinitely many random variables.

Let us now make this definition more precise. For simplicity, we shall assume in the sequel that a s.p. $X(t)$, $t \in T$, takes on only real values. The index set T usually is the real line or an interval of the real line.

Definition. If to every finite set $\{t_1, t_2, \ldots, t_n\}$ of $t \in T$, there corresponds a set of r.v.'s $X_1 = X(t_1)$, $X_2 = X(t_2)$, $\ldots$, $X_n = X(t_n)$, having a well defined joint probability distribution function

$$F_{X_1 X_2 \ldots X_n}(x_1, t_1; x_2, t_2; \ldots; x_n, t_n)$$
$$= P\{X_1 \leq x_1 \cap X_2 \leq x_2 \cap \cdots \cap X_n \leq x_n\}, \qquad n = 1, 2, \ldots \qquad (3.1)$$

then this family of joint distribution functions defines a s.p. $X(t)$, $t \in T$.

In the theory of stochastic processes, a commonly used notation for the joint distribution function given above is

$$F_n(x_1, t_1; x_2, t_2; \ldots; x_n, t_n) = F_{X_1 X_2 \ldots X_n}(x_1, t_1; x_2, t_2; \ldots; x_n, t_n) \qquad (3.2)$$

It is called the *n*th *distribution function* of the s.p. $X(t)$. Its associated joint density function, assuming it exists,

$$f_n(x_1, t_1; \ldots; x_n, t_n) = \partial^n F_n(x_1, t_1; \ldots; x_n, t_n)/\partial x_1 \cdots \partial x_n \qquad (3.3)$$

is the *n*th *density function* of $X(t)$.

We note that this collection of distribution functions is not arbitrary; they must satisfy

(1) For $m > n$,

$$F_m(x_1, t_1; \ldots; x_n, t_n; +\infty, t_{n+1}; \ldots; +\infty, t_m)$$
$$= F_n(x_1, t_1; \ldots; x_n, t_n) \qquad (3.4)$$

(2) The joint distribution function given by Eq. (3.1) is invariant under

an arbitrary permutation of the indices, $1, 2, \ldots, n$, that is,

$$F_n(x_1, t_1; x_2, t_2; \ldots; x_n, t_n) = F_n(x_{i_1}, t_{i_1}; x_{i_2}, t_{i_2}; \ldots; x_{i_n}, t_{i_n}) \quad (3.5)$$

where $\{i_1, i_2, \ldots, i_n\}$ is an arbitrary permutation of $\{1, 2, \ldots, n\}$.

The two conditions stated above are referred to as *Kolmogorov compatibility conditions*.

Comparing it with the sample theoretic approach, our definition of a s.p. in terms of joint distribution functions has obvious defects. First of all, it does not always lead to a clear description of the s.p. on the sample function level; we generally cannot, for example, picture the sample functions which make up a continuous-parameter s.p. Secondly, it does not provide unique answers to some probabilistic questions concerning a continuous-parameter s.p. $X(t)$ for *all* $t \in T$. Probabilities of physically useful events, such as $\{| X(t) | < 1, 0 \leq t \leq 1\}$, are not well defined by $\{F_n\}$ without assuming regularity properties such as separability and measurability of the process. Any detailed discussion of separability and measurability concepts requires mathematical details which are out of place in this book. Let us simply mention that Doob [2] has shown that, fortunately, it is possible to choose a model of the process possessing these regularity conditions by imposing either no requirements or minimal requirements on the joint distribution functions.

The advantages of adopting this definition, on the other hand, are many. It is mathematically simple. It also has the physical appeal in that it ties probability distributions to the concept of stochastic processes; this is desirable because the modeling of physical phenomena by stochastic processes is often based upon observations of a probability-of-occurrence nature. Lastly, for practical reasons, the drawbacks mentioned above are not serious. Because of various constraints, it is often necessary for us to work with only the first few distribution functions. This point will become more clear as we proceed.

Based upon the definition above, another convenient and often useful method of specifying a stochastic process is to characterize it in terms of an analytic formula as a function of t containing random variables as parameters. Let $A_1, A_2, \ldots, A_n$ be a set of random variables. A s.p. $X(t)$ characterized this way has the general form

$$X(t) = g(t; A_1, A_2, \ldots, A_n) \quad (3.6)$$

where the functional form of g is given.

Example 3.1. Consider a function of t defined by

$$X(t) = At + B, \qquad t \in T \tag{3.7}$$

where A and B are random variables with a known joint distribution function $F(a, b)$. The function $X(t)$ is clearly random; it is a stochastic process in view of its dependence upon the parameter t. Clearly, the distribution functions of $X(t)$ of all orders can be determined in terms of $F(a, b)$. Thus, Eq. (3.7) together with $F(a, b)$ define the s.p. $X(t)$, $t \in T$.

Example 3.2. A characterization of random noise currents in physical devices, due to Rice [3], is

$$X(t) = \sum_{j=1}^{n} A_j \cos(\Omega_j t + \phi_j) \tag{3.8}$$

where the amplitude A_j, the frequency Ω_j, and the phase ϕ_j are random variables. The s.p. $X(t)$ is completely characterized by the joint probability distribution of these $3n$ random variables.

A distinctive advantage associated with this method is that the mathematical description of $X(t)$ is considerably simplified because it is now governed by the behavior of a *finite* set of random variables. Also, it clearly identifies the associated sample functions. We see from Eq. (3.7) that the sample functions for this particular case are all straight lines. Hence, such sample functional properties as continuity and differentiability can be directly examined.

We remark that the sample functions of the s.p. $X(t)$ defined by Eq. (3.6) are completely fixed by a set of n values assumed by the r.v.'s A_1, A_2, $\ldots$, A_n for all $t \in T$. A s.p. $X(t)$ having this property is said to have n *degrees of randomness*. Based upon this definition, the s.p. in Example 3.1 has two degrees of randomness, and, in Example 3.2, $3n$ degrees of randomness.

Following the definition of a random vector, let us now introduce the notion of a *vector stochastic process*. In many physical situations, there is a need to consider not a single stochastic process but a sequence of stochastic processes, say, $X_1(t), X_2(t), \ldots, X_m(t)$, $t \in T$. Let $\mathbf{X}(t)$, $t \in T$, be a vector stochastic process whose components are $X_1(t), X_2(t), \ldots, X_m(t)$. We can give a definition for a sequence $\{X_m(t)\}$ of stochastic processes in much the same way as we did for a single process.

Definition. If to every finite set $\{t_1, t_2, \ldots, t_n\}$ of $t \in T$, there corresponds a set of m-dimensional random vectors $\mathbf{X}_1 = \mathbf{X}(t_1)$, $\mathbf{X}_2 = \mathbf{X}(t_2)$, $\ldots$,

$\mathbf{X}_n = \mathbf{X}(t_n)$, having a well-defined joint distribution function

$$F_{nm}(\mathbf{x}_1, t_1; \mathbf{x}_2, t_2; \ldots; \mathbf{x}_n, t_n)$$
$$= P\{\mathbf{X}_1 \leq \mathbf{x}_1 \cap \mathbf{X}_2 \leq \mathbf{x}_2 \cap \cdots \cap \mathbf{X}_n \leq \mathbf{x}_n\} \tag{3.9}$$

then this family of the joint distribution functions defines the sequence $\{X_m(t)\}, t \in T$, where $X_1(t), \ldots, X_m(t)$ are the components of $\mathbf{X}(t)$.

It is noted that the joint distribution function given above is one associated with nm random variables.

We will occasionally speak of complex-valued stochastic processes. A *complex s.p.* $Z(t)$ can be represented by

$$Z(t) = X(t) + iY(t) \tag{3.10}$$

where $X(t)$ and $Y(t)$ are real s.p.'s. It is clear that $Z(t)$ is completely characterized by a two-dimensional vector stochastic process.

3.1.2. Moments

As in the case of random variables, some of the most important properties of a s.p. are characterized by its moments, particularly those of the first and the second order. In the sequel, the existence of density functions shall be assumed.

In terms of its first density function $f_1(x, t)$, the nth moment of a s.p. $X(t)$ at a given $t \in T$, $\alpha_n(t)$, is defined by

$$\alpha_n(t) = E\{X^n(t)\} = \int_{-\infty}^{\infty} x^n f_1(x, t)\, dx \tag{3.11}$$

The first moment, $\alpha_1(t)$, is the mean of the s.p. $X(t)$ at t. It is sometimes denoted by $m_X(t)$ or simply $m(t)$. The mean square value of $X(t)$ at t is given by $\alpha_2(t)$.

The nth central moment of $X(t)$ at a given t is

$$\mu_n(t) = E\{[X(t) - m(t)]^n\} = \int_{-\infty}^{\infty} (x - m)^n f_1(x, t)\, dx \tag{3.12}$$

Of particular importance is $\mu_2(t)$, the variance of $X(t)$ at t, and it is commonly denoted by $\sigma_X{}^2(t)$ or $\sigma^2(t)$.

The moments of $X(t)$ defined in terms of its second density function $f_2(x_1, t_1; x_2, t_2)$ are, in effect, joint moments of two random variables.

The joint moment $\alpha_{nm}(t_1, t_2)$ of $X(t)$ at t_1 and t_2 is defined by

$$\alpha_{nm}(t_1, t_2) = E\{X^n(t_1) X^m(t_2)\}$$

$$= \int_{-\infty}^{\infty} \int_{-\infty}^{\infty} x_1^n x_2^m f_2(x_1, t_1; x_2, t_2) \, dx_1 \, dx_2 \qquad (3.13)$$

An important measure of linear interdependence between $X(t_1)$ and $X(t_2)$ is contained in $\alpha_{11}(t_1, t_2)$, which plays a central role in the theory of stochastic processes. Moreover, its importance justifies a special notation and name. We shall call $\alpha_{11}(t_1, t_2)$ the *correlation function* of the s.p. $X(t)$ and denoted it by $\Gamma_{XX}(t_1, t_2)$. It is in general a function of t_1 and t_2. The double subscript XX indicates the correlation of $X(t)$ with itself.

A more specific terminology used for $\Gamma_{XX}(t_1, t_2)$ is the *autocorrelation function*. It is intended to distinguishing it from the *cross-correlation function*, defined by

$$\Gamma_{XY}(t_1, t_2) = E\{X(t_1) Y(t_2)\} \qquad (3.14)$$

where the random variables involved belong to two different stochastic processes.

Similarly, the *autocovariance function* of $X(t)$ is given by

$$\mu_{XX}(t_1, t_2) = E\{(X(t_1) - m(t_1))(X(t_2) - m(t_2))\}$$

$$= \int_{-\infty}^{\infty} \int_{-\infty}^{\infty} (x_1 - m_2)(x_2 - m_2) f_2(x_1, t_1; x_2, t_2) \, dx_1 \, dx_2 \qquad (3.15)$$

It becomes $\sigma_X^2(t)$, the variance of $X(t)$ at t, when $t_1 = t_2 = t$. The normalized $\mu_{XX}(t_1, t_2)$ is called the *correlation-coefficient function* and is denoted by

$$\varrho_{XX}(t_1, t_2) = \mu_{XX}(t_1, t_2)/\sigma_X(t_1) \, \sigma_X(t_2) \qquad (3.16)$$

It is analogous to the correlation coefficient ϱ defined in (2.49) with the same "measure of linear interdependence" interpretation.

Associated with two s.p.'s $X(t)$ and $Y(t)$ we have the *cross-covariance function*

$$\mu_{XY}(t_1, t_2) = E\{(X(t_1) - m_X(t_1))(Y(t_2) - m_Y(t_2))\} \qquad (3.17)$$

The extension of the definitions above to the case of vector stochastic processes can be carried out in a straightforward manner. The mean of a vector stochastic process now takes the form of a mean vector. The correlation functions and covariance functions become correlation function matrices and covariance function matrices.

The importance of the correlation functions or, equivalently, the covariance functions rests in part on the fact that their properties define the mean square properties of the stochastic processes associated with them. This will become clear in Chapter 4. In what follows, we note some of the important properties associated with these functions.

(1) $\Gamma_{XX}(t_1, t_2) = \Gamma_{XX}(t_2, t_1)$

$\Gamma_{XY}(t_1, t_2) = \Gamma_{YX}(t_2, t_1)$
$$(3.18)$$

for every $t_1, t_2 \in T$.

(2) $\Gamma_{XX}^2(t_1, t_2) \leq \Gamma_{XX}(t_1, t_1)\, \Gamma_{XX}(t_2, t_2)$

$\Gamma_{XY}^2(t_1, t_2) \leq \Gamma_{XX}(t_1, t_1)\, \Gamma_{YY}(t_2, t_2)$
$$(3.19)$$

for every $t_1, t_2 \in T$.

(3) $\Gamma_{XX}(t_1, t_2)$ is nonnegative definite on $T \times T$; that is, for every n, t_1, $t_2, \ldots, t_n \in T$, and for an arbitrary function $g(t)$ defined on T,

$$\sum_{j,k=1}^{n} \Gamma_{XX}(t_j, t_k)\, g(t_j)\, g(t_k) \geq 0 \qquad (3.20)$$

The first property is clear from definition. Property 2 is a direct consequence of the Schwarz inequality (2.56). In order to show that Property 3 is true, consider a r.v. Y defined by

$$Y = \sum_{j=1}^{n} X(t_j)\, g(t_j) \qquad (3.21)$$

we get

$$E\{Y^2\} = E\left\{ \sum_{j,k=1}^{n} X(t_j)\, X(t_k)\, g(t_j)\, g(t_k) \right\} \geq 0 \qquad (3.22)$$

Eq. (3.20) thus follows immediately.

3.1.3. Characteristic Functions

Following the definition of a characteristic function associated with a sequence of random variables, the characteristic functions of a s.p. $X(t)$, $t \in T$, are defined in an analogous way. There is, of course, a family of characteristic functions since $X(t)$ is characterized by a family of density functions. Again, we assume the existence of the density functions in what follows.

The *n*th *characteristic function* of $X(t)$ is defined by

$$\phi_n(u_1, t_1; u_2, t_2; \ldots; u_n, t_n) = E\left\{\exp\left[i \sum_{j=1}^{n} u_j X(t_j)\right]\right\}$$

$$= \int_{-\infty}^{\infty} \cdots \int_{-\infty}^{\infty} \exp\left[i \sum_{j=1}^{n} x_j u_j\right] f_n(x_1, t_1; \ldots; x_n, t_n) \, dx_1 \cdots dx_n \quad (3.23)$$

It exists for all $u_1, u_2, \ldots, u_n$ and, together with the *n*th density function, they form an *n*-dimensional Fourier transform pair.

Let the joint moment $\alpha_{j_1 j_2 \ldots j_n}(t_1, t_2, \ldots, t_n)$ be defined by

$$\alpha_{j_1 j_2 \ldots j_n}(t_1, t_2, \ldots, t_n) = E\{X^{j_1}(t_1) X^{j_2}(t_2) \cdots X^{j_n}(t_n)\} \quad (3.24)$$

It is related to the *n*th characteristic function by

$$\partial^m \phi_n(0, t_1; \ldots; 0, t_n)/\partial u_1^{j_1} \, \partial u_2^{j_2} \cdots \partial u_n^{j_n}$$

$$= i^m \alpha_{j_1 j_2 \ldots j_n}(t_1, t_2, \ldots, t_n) \quad (3.25)$$

where $m = j_1 + j_2 + \cdots + j_n$.

3.1.4. Classification of Stochastic Processes

We have seen that a s.p. $X(t)$, $t \in T$, is defined by a system of probability distribution functions. This system consists of

$$F_n(x_1, t_1; x_2, t_2; \ldots; x_n, t_n)$$

for every n and for every finite set $t_1, t_2, \ldots, t_n$ in the fixed index set T.

In general, all F_n's are needed to specify a stochastic process completely. It is not uncommon, however, that stochastic processes modeling physical phenomena are statistically simpler processes, simple in the sense that all the statistical information about a given process is contained in a relatively small number of probability distribution functions. The statistical complexity of a stochastic process is determined by the properties of its distribution functions and, in the analysis of stochastic processes, it is important to consider them in terms of these properties.

We consider in the remainder of this chapter classification of stochastic processes according to the properties of their distribution functions. Presented below are two types of classifications that are important to us in the analysis to follow, namely, classification based upon the statistical regularity of a process over the index set T and classification based upon its memory.

We note that these two classifications are not mutually exclusive and, in fact, many physically important processes possess some properties of each. In what follows, it is convenient to think of the parameter t as time.

3.2. Classification Based Upon Regularity

In terms of statistical regularities, stochastic processes can be grouped into two classes: *stationary stochastic processes* and *nonstationary stochastic processes*.

The nth probability distribution function, $F_n(x_1, t_1; x_2, t_2; \ldots, x_n, t_n)$, of a s.p. $X(t)$, $t \in T$, is in general a function of $x_1, x_2, \ldots, x_n$ and $t_1, t_2, \ldots, t_n$. A nonstationary s.p. is one whose distribution functions depend upon the values of time parameters explicitly. The statistical behavior of a nonstationary s.p. thus depends upon the absolute origin of time. Clearly, most stochastic processes modeling physical phenomena are nonstationary. In particular, all physical processes having a transient period or certain damping characteristics are of the nonstationary type. For example, ground motion due to strong motion earthquakes, noise processes in devices with a starting transient, seasonal temperature variations, and epidemic models are described by nonstationary stochastic process.

On the other hand, many stochastic processes occurring in nature have the property that their statistical behavior does not vary significantly with respect to their parameters. The surface of the sea in spatial and time coordinates, noise in time in electric circuits under steady state operation, impurities in engineering materials and media as functions of spatial coordinates all have the appearance that their fluctuations as functions of time or spatial positions stay roughly the same. Because of powerful mathematical tools which exist for treating stationary s.p.'s, this class of s.p.'s is of great practical importance, and we consider it in some detail in the following sections. More exhaustive accounts of stationary stochastic processes can be found in the work of Yaglom [4] and Cramér and Leadbetter [5].

3.2.1. Stationary and Wide-Sense Stationary Processes

A s.p. $X(t)$, $t \in T$, is said to be *stationary* or *strictly stationary* if its collection of probability distributions stay invariant under an arbitrary translation of the time parameter, that is, for each n and for an arbitrary τ,

$$F_n(x_1, t_1; x_2, t_2; \ldots; x_n, t_n) = F_n(x_1, t_1 + \tau; x_2, t_2 + \tau; \ldots; x_n, t_n + \tau)$$

$$t_j + \tau \in T, \ j = 1, 2, \ldots, n, \qquad (3.26)$$

identically.

Let $\tau = -t_1$ in the above. We see that the probability distributions depend upon the time parameters only through their differences. In other words, the statistical properties of a stationary s.p. are independent of the absolute time origin.

We can easily see that a stationary s.p. $X(t)$, $t \in T$, possesses the following important properties for its moments, if they exist.

Since the first distribution function $F_1(x, t)$ is not a function of t, we have

$$E\{X(t)\} = \text{const} \tag{3.27}$$

Since

$$F_2(x_1, t_1; x_2, t_2) = F_2(x_1, x_2; t_2 - t_1) \tag{3.28}$$

we have

$$E\{X(t)\, X(t + \tau)\} = \Gamma(\tau) \tag{3.29}$$

In the above, the double subscript XX for $\Gamma(\tau)$ is omitted for convenience. In view of the general relation (3.18), it follows that

$$\Gamma(\tau) = \Gamma(-\tau) \tag{3.30}$$

the correlation function of a stationary stochastic process is thus an even function of τ.

Properties of higher moments can also be easily derived.

Given a physical problem, it is often quite difficult to ascertain whether the stationary property holds since Eq. (3.26) must hold for all n. For practical purposes, we are often interested in a wider class of stationary stochastic processes.

Definition. A s.p. $X(t)$, $t \in T$, is called a *wide-sense stationary s.p.* if

$$|\, E\{X(t)\}\,| = \text{const} < \infty \tag{3.31}$$

and

$$E\{X^2(t)\} < \infty, \qquad E\{X(t_1)\, X(t_2)\} = \Gamma(t_2 - t_1) \tag{3.32}$$

A wide-sense stationary s.p. is sometimes called *second-order stationary*, *weakly stationary*, or *covariance stationary*. It is clear that a strictly stationary process whose second moment is finite is also wide-sense stationary, but the converse is not true in general. An important exception is the Gaussian s.p. As we shall see, a Gaussian stochastic process is completely specified by its means and covariance functions. Hence, a wide-sense stationary Gaussian stochastic process is also strictly stationary. We give below several examples of stationary and wide-sense stationary processes.

Example 3.3. Let X_n, $n = \ldots, -2, -1, 0, 1, 2, \ldots$, be a sequence of independent and identically distributed r.v.'s with means zero and variances σ^2. Consider a discrete-parameter s.p. $X(t)$, $t = \ldots, -2, -1, 0, 1, 2, \ldots$, defined by

$$X(t) = X_t \tag{3.33}$$

It is easily shown that

$$E\{X(t)\} = 0 \tag{3.34}$$

and

$$E\{X(t + \tau) X(t)\} = \begin{cases} \sigma^2, & \tau = 0 \\ 0, & \tau \neq 0 \end{cases} \tag{3.35}$$

Hence, it is wide-sense stationary. It is clear that the process is also strictly stationary. The sequence $\{X_n\}$ in this case is called a *stationary random sequence*.

Example 3.4. A *constant process* is given by

$$X(t) = Y \tag{3.36}$$

where Y is a random variable. It is clear that $X(t)$ is strictly stationary and thus wide-sense stationary if $E\{Y^2\} < \infty$.

Example 3.5. A *binary noise* $X(t)$, $t \in T$, is one which takes values either $+1$ or -1 throughout successive time intervals of a fixed length Δ; the value it takes in one interval is independent of the values taken in any other interval; and all member functions differing only by a shift along the t-axis are equally likely. It is an example of two-valued processes.

A possible representation of $X(t)$ is

$$X(t) = Y(t - A) \tag{3.37}$$

with

$$Y(t) = Y_n, \qquad (n - 1)\Delta < t < n\Delta \tag{3.38}$$

where the r.v.'s Y_n, $n = \ldots, -2, -1, 0, 1, 2, \ldots$, are independent and identically distributed with the density function

$$f(y) = \tfrac{1}{2}[\delta(y + 1) + \delta(y - 1)] \tag{3.39}$$

The r.v. A in Eq. (3.37) has a uniform distribution in the interval $[0, \Delta]$ and it is independent of Y_n.

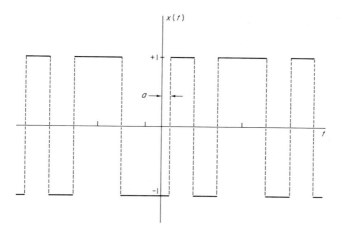

Fig. 3.1. A typical sample function of binary noise.

A typical sample function of a binary noise is shown in Figure 3.1 where the shift a is a sample value of the r.v. A.

Clearly, the mean of $X(t)$ is

$$E\{X(t)\} = 0 \qquad (3.40)$$

Consider now the correlation function $\Gamma(t_1, t_2)$. It is also clear that

$$\Gamma(t_1, t_2) = 0, \qquad |t_2 - t_1| > \Delta \qquad (3.41)$$

due to independence of the values of the process $X(t)$ in different intervals. For $|t_2 - t_1| \leq \Delta$, whether or not $X(t_1)$ and $X(t_2)$ are situated in a same interval depends upon the value A takes. Hence

$$\Gamma(t_1, t_2) = 1 \cdot P\{A < \Delta - |t_2 - t_1|\}$$
$$= 1 - |t_2 - t_1|/\Delta, \qquad |t_2 - t_1| \leq \Delta \qquad (3.42)$$

Combining Eqs. (3.41) and (3.42) and setting $t_2 - t_1 = \tau$, we have

$$\Gamma(\tau) = \begin{cases} 1 - |\tau|/\Delta, & |\tau| \leq \Delta \\ 0, & |\tau| > \Delta \end{cases} \qquad (3.43)$$

It follows from Eqs. (3.40) and (3.43) that a binary noise is wide-sense stationary.

As a more general case, if the r.v.'s Y_n in Eq. (3.38) are independent two-valued random variables with means zero and variances σ^2, the correlation function has the form

$$\Gamma(\tau) = \begin{cases} \sigma^2(1 - |\tau|/\Delta), & |\tau| \le \Delta \\ 0, & |\tau| > \Delta \end{cases} \tag{3.44}$$

Example 3.6. Another important two-valued stochastic process is the *random telegraph signal.* It has the properties that its values are $+1$ and -1 at successive intervals and the time instants at which the values change obey a Poisson distribution. More precisely, the probability of exactly k changes in values from $+1$ to -1 or vice versa in a time interval τ is given by

$$\frac{(\lambda\tau)^k}{k!} e^{-\lambda\tau}$$

A typical sample function of a random telegraph signal is shown in Fig. 3.2.

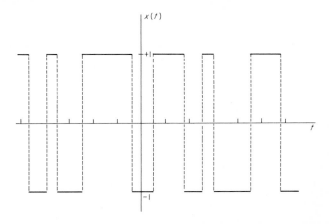

Fig. 3.2. A typical sample function of random telegraph signal.

In mathematical terms, the random telegraph signal can be represented by

$$X(t) = A(-1)^{Y(t)} \tag{3.45}$$

where $Y(t)$ is a s.p. having the properties that, for $t_1 < t_2 < t_3 < t_4$, the random variables $Y(t_2) - Y(t_1)$ and $Y(t_4) - Y(t_3)$ are independent, and

$$P\{[Y(t + \tau) - Y(t)] = k\} = [(\lambda\tau)^k/k!]e^{-\lambda\tau} \tag{3.46}$$

The random variable A is independent of $Y(t)$ and it is distributed according to

$$f(a) = \tfrac{1}{2}[\delta(a + 1) + \delta(a - 1)] \tag{3.47}$$

Without going into details, we can show that this stochastic process is wide-sense stationary with mean zero and correlation function

$$\Gamma(\tau) = e^{-2\lambda|\tau|} \tag{3.48}$$

Example 3.7. Consider a s.p. $X(t)$, $t \in T$, defined by

$$X(t) = a \cos(\omega t + \Phi) \tag{3.49}$$

where a and ω are real constants and Φ is a r.v. uniformly distributed in the interval $[0, 2\pi]$. It is a simplified version of the Rice noise discussed in Example 3.2.

We obtain easily

$$E\{X(t)\} = E\{a \cos(\omega t + \Phi)\} = (a/2\pi) \int_0^{2\pi} \cos(\omega t + \phi) \, d\phi = 0 \tag{3.50}$$

$$\Gamma(t_1, t_2) = a^2 E\{\cos(\omega t_1 + \Phi) \cos(\omega t_2 + \Phi)\} = \tfrac{1}{2}a^2 \cos \omega(t_2 - t_1) \tag{3.51}$$

The s.p. $X(t)$ defined by Eq. (3.49) is thus wide-sense stationary.

3.2.2. Spectral Densities and Correlation Functions

We restrict ourselves here to the study of wide-sense stationary stochastic processes and develop further useful concepts in the analysis of this class of processes. It has been noted that the covariance or correlation function of a stochastic process plays a central role. Let us first develop a number of additional properties associated with the correlation function

$$E\{X(t) X(t + \tau)\} = \Gamma(\tau) \tag{3.52}$$

of a wide-sense stationary s.p. $X(t)$, $t \in T$.

(1) $\Gamma(\tau)$ is an even function of τ, that is

$$\Gamma(\tau) = \Gamma(-\tau) \tag{3.53}$$

This has been derived in Section 3.2.1, Eq. (3.30).

(2) $\Gamma(\tau)$ always exists and is finite. Furthermore

$$|\Gamma(\tau)| \leq \Gamma(0) = E\{X^2(t)\} \tag{3.54}$$

This result is a direct consequence of Eq. (3.19).

(3) If $\Gamma(\tau)$ is continuous at the origin, then it is uniformly continuous in τ. In order to show this, let us consider the difference

$$\Gamma(\tau + \varepsilon) - \Gamma(\tau) = E\{[X(t - \varepsilon) - X(t)] X(t + \tau)\} \qquad (3.55)$$

Using the Schwarz inequality we obtain the inequality

$$| \Gamma(\tau + \varepsilon) - \Gamma(\tau) |^2 \leq 2\Gamma(0)[\Gamma(0) - \Gamma(\varepsilon)] \qquad (3.56)$$

Since $\Gamma(\tau)$ is continuous at the origin, $\Gamma(0) - \Gamma(\varepsilon) \to 0$ as $\varepsilon \to 0$. Hence, the difference $\Gamma(\tau + \varepsilon) - \Gamma(\tau)$ tends to zero with ε uniformly in τ.

(4) As indicated by Eq. (3.20), $\Gamma(\tau)$ is nonnegative definite in τ.

The next important property associated with $\Gamma(\tau)$ is its relation with the so-called power spectral density of $X(t)$. Let us first give a mathematical development.

(5) Let $\Gamma(\tau)$ be continuous at the origin. It can be represented in the form

$$\Gamma(\tau) = \tfrac{1}{2} \int_{-\infty}^{\infty} e^{i\omega\tau} \, d\Phi(\omega) \qquad (3.57)$$

If $\Phi(\omega)$ is absolutely continuous, we have

$$S(\omega) = d\Phi(\omega)/d\omega \qquad (3.58)$$

and

$$\Gamma(\tau) = \tfrac{1}{2} \int_{-\infty}^{\infty} e^{i\omega\tau} \, S(\omega) \, d\omega \qquad (3.59)$$

where $S(\omega)$ is real and nonnegative.

By virtue of Properties (2), (3), and (4) possessed by $\Gamma(\tau)$, this result is a direct consequence of the following important theorem due to Bochner [6].

Theorem. A function $\Gamma(\tau)$ is continuous, real, and nonnegative definite if, and only if, it has the representation (3.57), where $\Phi(\omega)$ is real, non-decreasing, and bounded. Cramér [7] also gives a proof of this theorem.

We see from Eq. (3.59) that the functions $\Gamma(\tau)$ and $S(\omega)$ form a Fourier transform pair. The relations

$$S(\omega) = \frac{1}{\pi} \int_{-\infty}^{\infty} e^{-i\omega\tau} \, \Gamma(\tau) \, d\tau, \qquad \Gamma(\tau) = \tfrac{1}{2} \int_{-\infty}^{\infty} e^{i\omega\tau} \, S(\omega) \, d\omega \qquad (3.60)$$

are called the *Wiener–Khintchine formulas*.

Equations (3.60) lead to the definition of the power spectral density function of a wide-sense stationary process whose correlation function is continuous.

Definition. The power spectral density function of a wider-sense stationary s.p. $X(t)$ is defined by the first equation of (3.60), that is,

$$S(\omega) = \frac{1}{\pi} \int_{-\infty}^{\infty} e^{-i\omega\tau} \, \Gamma(\tau) \, d\tau \qquad (3.61)$$

When clarification warrants, the notation $S_{XX}(\omega)$ will be used to indicate that it is associated with the s.p. $X(t)$.

We have already noted that the power spectral density function $S(\omega)$ is real and nonnegative. Furthermore, since $\Gamma(\tau)$ is an even function of τ, the Wiener–Khintchine formulas may be written as

$$S(\omega) = \frac{2}{\pi} \int_0^{\infty} \Gamma(\tau) \cos \omega\tau \, d\tau, \qquad \Gamma(\tau) = \int_0^{\infty} S(\omega) \cos \omega\tau \, d\omega \qquad (3.62)$$

Hence, $S(\omega)$ is also an even function of ω.

When $\tau = 0$, the second of Eqs. (3.62) reduces to

$$\int_0^{\infty} S(\omega) \, d\omega = \Gamma(0) = E\{X^2(t)\} \qquad (3.63)$$

It is well known that, in nonrandom time series analysis, the Fourier analysis in the frequency domain is a powerful tool from both mathematical and physical points of view. The notion of power spectral density functions plays such a role in the analysis of stochastic processes. Its usefulness in physical applications will be discussed in detail in Chapters 7–9. In what follows we give the correlation functions and the spectral density functions of several well-known wide-sense stationary processes.

(1) Binary Noise (Fig. 3.3):

$$\Gamma(\tau) = \begin{cases} \sigma^2(1 - |\tau|/\Delta), & |\tau| \leq \Delta \\ 0, & |\tau| > \Delta \end{cases} \qquad (3.64)$$

$$S(\omega) = \frac{\sigma^2 \Delta}{\pi} \frac{\sin^2(\omega\Delta/2)}{(\omega\Delta/2)^2} \qquad (3.65)$$

(2) Random Telegraph Signal (Fig. 3.4):

$$\Gamma(\tau) = e^{-\lambda|\tau|} \qquad (3.66)$$

$$S(\omega) = 2\lambda/\pi(\omega^2 + \lambda^2) \qquad (3.67)$$

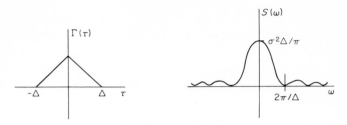

Fig. 3.3. Binary noise.

(3) White Noise (Fig. 3.5). This seems to be an appropriate place to give the definition of one of the most used, and perhaps most abused, stochastic processes, namely, the white noise. The term 'white' is introduced in connection with the 'white light,' which has the property that its power spectral density is flat over the visible portion of the electromagnetic

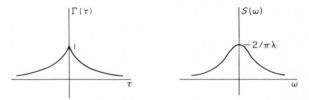

Fig. 3.4. Random telegraph signal.

spectrum. We thus define a white noise process to mean one whose power spectral density is a constant over the whole spectrum, that is

$$S(\omega) = S_0, \qquad \omega \in (-\infty, \infty) \tag{3.68}$$

Its correlation function has the form

$$\Gamma(\tau) = \pi S_0 \, \delta(\tau) \tag{3.69}$$

where $\delta(\tau)$ is the Dirac delta function.

Fig. 3.5. White noise.

We see that, for any $t_1 \neq t_2$, $X(t_1)$ and $X(t_2)$ are uncorrelated. Furthermore, the variance of $X(t)$ at any t approaches infinity. It is clear that such a process is an abstraction or a limiting process. For example, we can formally show that the white noise is the limiting process of the binary noise when $\sigma^2 \to \infty$ and $\Delta \to 0$ in such a way that $\sigma^2 \Delta \to \pi S_0$ in Eqs. (3.64) and (3.65).

However, because of its mathematical simplicity, we often use white noise as an approximation to a great number of physical phenomena. Much more will be said of it when we encounter it in the theory of random differential equations.

In closing this section, we shall discuss briefly the notion of *cross-power spectral density function* associated with two s.p.'s $X(t)$ and $Y(t)$, $t \in T$.

Definition. Let $X(t)$ and $Y(t)$ be two wide-sense stationary s.p.'s whose cross-correlation function $\Gamma_{XY}(t_1, t_2)$ is stationary and continuous. The *cross-power spectral density function* $S_{XY}(\omega)$ of $X(t)$ and $Y(t)$ is defined by

$$S_{XY}(\omega) = (1/\pi) \int_{-\infty}^{\infty} \Gamma_{XY}(\tau) e^{-i\omega\tau} \, d\tau \tag{3.70}$$

The other member of the Fourier transform pair is

$$\Gamma_{XY}(\tau) = \tfrac{1}{2} \int_{-\infty}^{\infty} S_{XY}(\omega) e^{i\omega\tau} \, d\omega \tag{3.71}$$

The function $S_{XY}(\omega)$ is in general complex. In view of the property

$$\Gamma_{XY}(\tau) = \Gamma_{YX}(-\tau) \tag{3.72}$$

we have

$$S_{XY}(\omega) = S_{YX}^*(\omega) \tag{3.73}$$

where the superscript * denotes complex conjugate.

We frequently use cross-power spectral density functions when there are more than one stochastic process involved in a given problem. A common application of this concept is in the consideration of a sum of two or more wide-sense stationary processes. Let

$$Z(t) = X(t) + Y(t) \tag{3.74}$$

where $X(t)$, $Y(t)$, and $Z(t)$ are wide-sense stationary and $\Gamma_{XY}(t_1, t_2)$ is stationary. We can easily show that

$$S_{ZZ}(\omega) = S_{XX}(\omega) + S_{YY}(\omega) + S_{XY}(\omega) + S_{YX}(\omega)$$
$$= S_{XX}(\omega) + S_{YY}(\omega) + 2\text{Re}[S_{XY}(\omega)] \qquad (3.75)$$

where Re[ ] indicates the real part.

3.2.3. Ergodicity

The study of ergodic theory was first introduced in classical statistical mechanics and kinetic theory in order to, for example, relate the average properties of a particular system of molecules in a closed space to the ensemble behavior of all molecules at any given time. In our context, ergodicity deals with the specific question of relating statistical or ensemble averages of a stationary stochastic process to time averages of its individual sample functions. The interchangeability of ensemble and time averages has considerable appeal in practice because, when statistical averages of a stochastic process need to be computed, what is generally available is not a representative collection of sample functions but rather certain pieces of sample functions or a long single observation of one sample function. One naturally asks if certain statistical averages of a stochastic process can be determined from appropriate time averages associated with a single sample function.

Let $x(t)$ be a sample function of a stationary s.p. $X(t)$, $t \in T$. The time average of a given function of $x(t)$, $g[x(t)]$, denoted by $\overline{g[x(t)]}$, is defined by

$$\overline{g[x(t)]} = \lim_{T \to \infty} (1/2T) \int_{-T}^{T} g[x(t + \tau)] \, d\tau \qquad (3.76)$$

if the limit exists. In order to define the ergodic property of a stationary stochastic process, the sample function $x(t)$ is allowed to range over all possible sample functions and, for a fixed T, over all possible time translations; $\overline{g[x(t)]}$ thus defines a random variable. For the sake of consistency in notation, we may replace $x(t)$ by $X(t)$ in Eq. (3.76) without confusion.

Definition. A stationary process $X(t)$, $t \in T$, is said to be ergodic relative to G if, for every $g[X(t)] \in G$, G being an appropriate domain of $g[\ ]$,

$$\overline{g[X(t)]} = E\{g[X(t)]\} \qquad (3.77)$$

with probability one, that is, with a possible exception of a subset of sample functions $g[X(t)]$ with zero probability of occurring.

It is, of course, entirely possible that, given a s.p. $X(t)$, the ergodicity condition stated above is satisfied for certain functions of $X(t)$ and is not satisfied for others. In physical applications we are primarily interested in the following:

(1) Ergodicity in the Mean: $g[X(t)] = X(t)$. The ergodicity condition becomes

$$\overline{X(t)} = \lim_{T \to \infty} (1/2T) \int_{-T}^{T} X(t)\, dt = E\{X(t)\} \tag{3.78}$$

with probability one.

(2) Ergodicity in the Mean Square: $g[X(t)] = X^2(t)$. The ergodicity condition is

$$\overline{X^2(t)} = \lim_{T \to \infty} (1/2T) \int_{-T}^{T} X^2(t)\, dt = E\{X^2(t)\} \tag{3.79}$$

with probability one.

(3) Ergodicity in the Correlation Function: $g[X(t)] = X(t)\,X(t+\tau)$. The ergodicity condition is

$$\overline{X(t)\,X(t+\tau)} = \lim_{T \to \infty} (1/2T) \int_{-T}^{T} X(t)\,X(t+\tau)\, dt = \Gamma(\tau) \tag{3.80}$$

with probability one.

For these simple ergodicity criteria, mathematical conditions can be derived for the purpose of verifying the ergodicity conditions. The procedure follows the fundamental result (see Problem 2.8) that, given a random variable Y,

$$Y = E\{Y\} \tag{3.81}$$

with probability one if

$$\sigma_Y{}^2 = 0 \tag{3.82}$$

Consider, for example, the ergodicity in the mean criterion. We can easily show that

$$E\{\overline{X(t)}\} = \lim_{T \to \infty} (1/2T) \int_{-T}^{T} E[X(t)]\, dt = E[X(t)] = m \tag{3.83}$$

since $X(t)$ is stationary. The variance of $X(t)$ is given by

$$
\begin{aligned}
\operatorname{var}\{\overline{X(t)}\} &= E\{(\overline{X(t)} - m)^2\} \\
&= \lim_{T \to \infty} (1/4T^2) \int_{-T}^{T} \int_{-T}^{T} E\{X(t_1)\, X(t_2)\}\, dt_1\, dt_2 - m^2 \\
&= \lim_{T \to \infty} (1/4T^2) \int_{-T}^{T} \int_{-T}^{T} \Gamma(t_2 - t_1)\, dt_1\, dt_2 - m^2 \qquad (3.84)
\end{aligned}
$$

To put this result in a simpler form, consider the change of variables according to $\tau_1 = t_2 + t_1$ and $\tau_2 = t_2 - t_1$. The Jacobian of this transformation is

$$
|\, \partial(t_1, t_2)/\partial(\tau_1, \tau_2)\, | = \tfrac{1}{2}
$$

Equation (3.84) becomes, in terms of the new variables,

$$
\operatorname{var}\{\overline{X(t)}\} = \lim_{T \to \infty} (1/4T^2) \iint \tfrac{1}{2}\Gamma(\tau_1)\, d\tau_1\, d\tau_2 - m^2
$$

where the domain of integration is the square shown in Fig. 3.6. It is seen that the integrand is an even function of τ_1 and is not a function of τ_2.

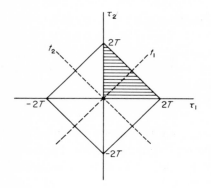

Fig. 3.6. Domain of integration.

Hence, the value of the integral is four times the value of the integral over the shaded area. Thus,

$$
\begin{aligned}
\operatorname{var}\{\overline{X(t)}\} &= \lim_{T \to \infty} (1/2T^2) \int_{0}^{2T} \int_{0}^{2T - \tau_1} \Gamma(\tau_1)\, d\tau_2\, d\tau_1 - m^2 \\
&= \lim_{T \to \infty} (1/T) \int_{0}^{2T} \Gamma(\tau_1)(1 - \tau_1/2T)\, d\tau_1 - m^2 \\
&= \lim_{T \to \infty} (1/T) \int_{0}^{2T} (1 - \tau_1/2T)[\Gamma(\tau_1) - m^2]\, d\tau_1 \qquad (3.85)
\end{aligned}
$$

The result obtained above may be stated as a theorem given below.

Theorem 3.2.1 (Ergodic Theorem in the Mean). A s.p. $X(t)$, $t \in T$, is ergodic in the mean if, and only if,

$$\lim_{T \to \infty} (1/T) \int_0^{2T} (1 - \tau/2T)[\Gamma(\tau) - m^2]\, d\tau = 0 \qquad (3.86)$$

Another important ergodic theorem is one with respect to the correlation function.

Theorem 3.2.2 (Ergodic Theorem in the Correlation Function). A s.p. $X(t)$, $t \in T$, is ergodic in the correlation function if, and only if,

$$\lim_{T \to \infty} (1/T) \int_0^{2T} (1 - \tau_1/2T)[\xi(\tau_1) - \Gamma^2(\tau)]\, d\tau_1 = 0 \qquad (3.87)$$

where

$$\xi(\tau_1) = E\{X(t + \tau + \tau_1)\, X(t + \tau_1)\, X(t + \tau)\, X(t)\} \qquad (3.88)$$

The proof follows essentially the same steps in the establishment of Theorem 3.2.1, and it is left as an exercise.

Similar conditions can be derived for other ergodicity criteria. But they become more complex as the function $g[X(t)]$ becomes more complex and it is difficult to verify these ergodicity properties. In practice, therefore, ergodic properties are generally regarded as hypotheses. Because of their great utility, ergodic conditions are often assumed to be valid in physical situations where we "expect" them to be true.

Example 3.8. Let us first consider a trivial example in order to show that stationarity does not imply ergodicity. Hence, ergodic stationary processes are a subclass of stationary stochastic processes.

Let

$$X(t) = Y \qquad (3.89)$$

where Y is a random variable with mean zero and variance σ^2. It is easily shown that $X(t)$ is not ergodic in the mean, since

$$\lim_{T \to \infty} (1/T) \int_0^{2T} (1 - \tau/2T)[\Gamma(\tau) - m^2]\, d\tau = \sigma^2 \qquad (3.90)$$

and Eq. (3.86) is not satisfied.

The above conclusion can also be drawn simply by inspection, as the time average of a sample function simply takes a sample value of Y which in general is different from other sample values and is different from zero.

Example 3.9. Let us investigate the ergodic property in the mean of the random telegraph signal. From Eq. (3.66) the correlation function is given by

$$\Gamma(\tau) = e^{-\lambda|\tau|} \tag{3.91}$$

Upon substituting Eq. (3.91) into Eq. (3.86) we obtain

$$\lim_{T\to\infty} \frac{1}{T} \int_0^{2T} \left(1 - \frac{\tau}{2T}\right) e^{-\lambda\tau}\, d\tau = \lim_{T\to\infty} \frac{1}{\lambda T}\left(1 - \frac{1 - e^{-\lambda T}}{2\lambda T}\right) = 0 \tag{3.92}$$

The random telegraph signal is therefore ergodic in the mean.

Example 3.10. Consider the stochastic process

$$X(t) = A \sin t + B \cos t, \qquad t \in T \tag{3.93}$$

where A and B are independent Gaussian random variables with means zero and variances σ^2. The correlation function of $X(t)$ is

$$\Gamma(\tau) = \sigma^2 \cos \tau \tag{3.94}$$

It is easily shown that Eq. (3.86) is satisfied and thus the stochastic process is ergodic in the mean. Again, this property can be easily verified by inspection.

Let us consider its ergodic property in the mean square. We shall do this by direct evaluation of the time average indicated in Eq. (3.79).

Let

$$x(t) = a_1 \sin t + b_1 \cos t \tag{3.95}$$

be a representative sample function of $X(t)$, a_1 and b_1 being, respectively, specific sample values that the random variables A and B can assume. The time average in mean square with respect to this particular sample function takes the form

$$\lim_{T\to\infty} (1/2T) \int_{-T}^{T} x^2(t)\, dt = \lim_{T\to\infty} (1/2T) \int_{-T}^{T} (a_1 \sin t + b_1 \cos t)^2\, dt$$

$$= \tfrac{1}{2}(a_1^2 + b_1^2) \tag{3.96}$$

It is seen that the time average in mean square depends upon the particular sample function selected. Thus the process is not ergodic in the mean square.

3.3. Classification Based Upon Memory

In this classification, a s.p. $X(t)$, $t \in T$, is classified according to the manner in which its present state is dependent upon its past history. It is centered around one of the most important classes of stochastic processes, that is, the Markov processes.

3.3.1. The Purely Stochastic Process

According to memory properties, the simplest s.p. is one without memory. More precisely, a s.p. $X(t)$, $t \in T$, has *no memory* or is *purely stochastic* when a random variable defined by $X(t)$ at a given t is independent of the random variables defined by $X(t)$ at all other t. The nth distribution function of a purely stochastic process is thus given by

$$F_n(x_1, t_1; x_2, t_2; \ldots; x_n, t_n) = \prod_{j=1}^{n} F_1(x_j, t_j) \tag{3.97}$$

for all n. We see that all the distribution functions in this case can be generated from the knowledge of F_1. The first distribution functions of a purely stochastic process, therefore, contain all the statistical information about the process.

It is clear that, although mathematically simple, a continuous-parameter purely stochastic process is not physically realizable because it implies absolute independence between the present and the past no matter how closely they are spaced. In physical situations we certainly expect $X(t_1)$ to be dependent on $X(t_2)$ when t_1 is sufficiently close to t_2. In practical applications, therefore, purely stochastic processes with a continuous parameter are considered to be limiting processes. An important example here is the "white noise"; much more will be said about it later.

On the other hand, discrete-parameter purely stochastic processes are not uncommon. Throwing a die at a sequence of time instants, for example, constitute a discrete-parameter purely stochastic process. Along the same line, discrete stochastic models of this type are used in many physical situations.

3.3.2. The Markov Process

Logically, the next in order of complexity are stochastic processes whose statistical information is completely contained in their second probability distribution functions. An important class of stochastic processes possessing this property is the class of *Markov processes*, named after A. A. Markov, who in 1906 initiated the study of stochastic processes of this type. Rigorous treatment of Markov dependence was given by Kolmogorov [8] in his fundamental paper on the subject, and later in a series of basic papers by Feller [9–11]. Markov properties of stochastic processes are discussed extensively in the books of Doob [2] and Loéve [12]. Books devoted entirely to Markov processes and their applications include Bharucha–Reid [13], Stratonovich [14], and, from a more abstract measure-theoretic point of view, Dynkin [15].

We give below a brief introduction of Markov processes. The important role they play in the theory of random differential equations will unfold as we proceed. In this book, emphasis will be placed upon continuous-parameter Markov processes, particularly continuous Markov processes known as diffusion processes.

Definition. A s.p. $X(t)$, $t \in T$, is called a *Markov process* if for every n and for $t_1 < t_2 < \cdots < t_n$ in T we have

$$F\{x_n, t_n \mid x_{n-1}, t_{n-1}; x_{n-2}, t_{n-2}; \ldots; x_1, t_1\} = F\{x_n, t_n \mid x_{n-1}, t_{n-1}\} \quad (3.98)$$

Equation (3.98) is equivalent to

$$f(x_n, t_n \mid x_{n-1}, t_{n-1}; x_{n-2}, t_{n-2}; \ldots; x_1, t_1) = f(x_n, t_n \mid x_{n-1}, t_{n-1}) \quad (3.99)$$

if the indicated density functions exist.

To see that a Markov process is completely specified by its first and second distribution functions, we start with the general relation

$$
\begin{aligned}
f_n(x_1, t_1; x_2, t_2; \ldots; x_n, t_n) \\
= f(x_n, t_n \mid x_{n-1}, t_{n-1}; \ldots; x_1, t_1) f(x_{n-1}, t_{n-1} \mid x_{n-2}, t_{n-2}; \ldots; x_1, t_1) \\
\cdots f(x_2, t_2 \mid x_1, t_1) f_1(x_1, t_1) \quad (3.100)
\end{aligned}
$$

Using Eq. (3.99) we immediately have, for $n \geq 3$ and $t_1 < t_2 < \cdots < t_n$,

$$
\begin{aligned}
f_n(x_1, t_1; x_2, t_2; \ldots; x_n, t_n) &= f_1(x_1, t_1) \prod_{k=1}^{n-1} f(x_{k+1}, t_{k+1} \mid x_k, t_k) \\
&= \prod_{k=1}^{n-1} f_2(x_k, t_k; x_{k+1}, t_{k+1}) \Big/ \prod_{k=2}^{n-1} f_1(x_k, t_k) \quad (3.101)
\end{aligned}
$$

Equation (3.98) or (3.99) implies that a Markov process represents a collection of trajectories whose conditional probability distributions at a given instant given all past observations depends only upon the latest past. It is the probabilistic analog of the deterministic trajectory of a particle in mechanics. We recall in classical mechanics that a trajectory at a given time t_2 is completely determined by its state at some time $t_1 < t_2$, requiring no knowledge of its states at times prior to t_1.

Example 3.11. Let $Y_1, Y_2, \ldots$ be mutually independent random variables and let a discrete-parameter s.p. $X(t)$ be defined by

$$X(t_j) = \sum_{k=1}^{j} Y_k \qquad (3.102)$$

In this case, $\{t_j\} = T$. A simple way to show that $X(t)$ is a Markov process is to rewrite it in the recursive form

$$X(t_j) = X(t_{j-1}) + Y_j, \qquad j = 2, 3, \ldots \qquad (3.103)$$

Since Y_j are independent random variables, it follows that the properties of $X(t)$ at t_j are functions of those at t_{j-1} only. Hence Eq. (3.102) defines a Markov process.

The example above considers a simple and yet physically important process. The situation describes, for example, an elementary one-dimensional random walk problem.

We will see in the chapters to follow that Markov processes play an important role in physical problems. A great number of physical situations are modeled or can be closely approximated by processes of this type. The Brownian motion process, a fundamental process in physical sciences, is Markovian. Noise and signal processes in engineering systems, communication networks, and transport phenomena are also frequently modeled by Markov processes. Some important Markov processes will be considered in some detail in the next two sections.

An important equation known as the *Smoluchowski–Chapman–Kolmogorov* equation for Markov processes can be established from Eq. (3.101). Consider the general relation

$$\int_{-\infty}^{\infty} f_3(x_1, t_1; x_2, t_2; x_3, t_3)\, dx_2 = f_2(x_1, t_1; x_3, t_3) \qquad (3.104)$$

Let $t_1 < t_2 < t_3$. In view of Eq. (3.101), the left-hand side of Eq. (3.104) becomes

$$\int_{-\infty}^{\infty} f(x_3, t_3 \mid x_2, t_2) f(x_2, t_2 \mid x_1, t_1) f_1(x_1, t_1)\, dx_2$$

and the right-hand side gives

$$f(x_3, t_3 \mid x_1, t_1) f_1(x_1, t_1)$$

We thus have the Smoluchowski–Chapman–Kolmogorov equation

$$\int_{-\infty}^{\infty} f(x_3, t_3 \mid x_2, t_2) f(x_2, t_2 \mid x_1, t_1) \, dx_2 = f(x_3, t_3 \mid x_1, t_1) \qquad (3.105)$$

This relation essentially describes the flow or transition of probability densities from an instant t_1 to another instant t_3 as a function of the probability density transitions between consecutive random variables. Accordingly, the conditional density functions in Eq. (3.105) are commonly interpreted as *transition probability densities* and they occupy a central place in the theory of Markov processes.

We have indicated that, roughly speaking, if the present value of a Markov process is known, the distribution of any future values is independent of the manner in which the present value was reached. Markov processes can be made more complex by giving "the present value" a broader interpretation. For example, a more complex Markov process can be defined by the relation

$$F(x_n, t_n \mid x_{n-1}, t_{n-1}; \ldots; x_1, t_1) = F(x_n, t_n \mid x_{n-1}, t_{n-1}; x_{n-2}, t_{n-2}) \qquad (3.106)$$

Processes of this type are completely specified by its third distribution functions.

Markov processes defined by Eq. (3.106) and so on are sometimes called *Markov processes of higher order*. In contrast, we sometimes use the name *simple Markov processes* to denote those defined by Eq. (3.98). Based upon this terminology, a purely stochastic process is also called a *Markov process of zeroth order*.

3.3.3. Independent-Increment Processes

In this and the next sections we consider this physically important class of Markov processes, and briefly describe two processes belonging to it.

Definition. Consider a s.p. $X(t)$, $t \geq 0$. The random variable $X(t_2) - X(t_1)$, $0 \leq t_1 < t_2$, is denoted by $X(t_1, t_2)$ and is called an *increment* of $X(t)$ on $[t_1, t_2]$. If for all $t_1 < t_2 < \cdots < t_n$ the increments $X(t_1, t_2)$, $X(t_2, t_3), \ldots, X(t_{n-1}, t_n)$ are mutually independent, the s.p. $X(t)$, $t \geq 0$, is called an *independent-increment stochastic process*. In practice, this definition is used only in the case of continuous-parameter stochastic processes.

We see that if $X(t)$, $t \geq 0$, has independent increments, a new process $Y(t)$, defined by $Y(t) = X(t) - X(0)$, $t \geq 0$, is also an independent increment process, has the same increments as $X(t)$ does, and has the property that $P\{Y(0) = 0\} = 1$. It is therefore not at all restrictive to add the property $P\{X(0) = 0\} = 1$ to $X(t)$ itself.

We now give a proof that a continuous-parameter independent increment stochastic process is Markovian. Consider a continuous s.p. $X(t)$, $t \geq 0$, with independent increments and with $P\{X(0) = 0\} = 1$ or, equivalently, consider the process $X(t) - X(0)$. We see that $X(t_n) - X(0)$ is a partial sum of the series

$$\sum_{j=1}^{n} [X(t_j) - X(t_{j-1})]$$

of mutually independent random variables. The s.p. $X(t)$ is therefore a continuous version of a discrete-parameter stochastic process whose random variables at discrete time instants are partial sums of a series of independent random variables. We have indicated in Example 3.11 that this discrete process in Markovian; it thus follows that the process $X(t)$ is a continuous-parameter Markov process.

Definition. Let a s.p. $X(t)$, $t \geq 0$, be an independent-increment process. If the probability distributions of its increments $X(t_1, t_2)$, $X(t_2, t_3)$, $\ldots$, $X(t_{n-1}, t_n)$ depend only on the parameter differences $t_2 - t_1$, $t_3 - t_2$, $\ldots$, $t_n - t_{n-1}$, the process $X(t)$ is said to have *stationary independent increments*.

For a stationary independent increment process $X(t)$, $t \geq 0$, we are able to give a general form for its nth characteristic function. We first note that the sum

$$\sum_{j=1}^{n} u_j X(t_j)$$

can be written in the form

$$\sum_{j=1}^{n} u_j X(t_j)$$

$$= X(t_1) \left[\sum_{j=1}^{n} u_j - \sum_{j=2}^{n} u_j \right] + X(t_2) \left[\sum_{j=2}^{n} u_j - \sum_{j=3}^{n} u_j \right] + \cdots + X(t_n) u_n$$

$$= X(0, t_1) \sum_{j=1}^{n} u_j + X(t_1, t_2) \sum_{j=2}^{n} u_j + \cdots + X(t_{n-1}, t_n) u_n \qquad (3.107)$$

where the condition $P\{X(0) = 0\} = 1$ is invoked. Let $0 < t_1 < \cdots < t_n$. The nth characteristic function of $X(t)$ becomes

$$
\begin{aligned}
\phi_n(u_1, t_1; u_2, t_2; \ldots; u_n, t_n) &= E\left\{\exp\left[i \sum_{j=1}^{n} u_j X(t_j)\right]\right\} \\
&= \prod_{k=1}^{n} E\left\{\exp\left[iX(t_{k-1}, t_k) \sum_{j=k}^{n} u_j\right]\right\}
\end{aligned} \tag{3.108}
$$

To see what form the last expectation in Eq. (3.108) must take, let us consider the expectation

$$
\begin{aligned}
E\{\exp[iuX(t_1, t_3)]\} &= E\{\exp(iu[X(t_2, t_3) + X(t_1, t_2)])\} \\
&= E\{\exp[iuX(t_2, t_3)]\} \, E\{\exp[iuX(t_1, t_2)]\} \tag{3.109}
\end{aligned}
$$

Due to stationarity of the increments, Eq. (3.109) is in the form of the functional equation

$$
\phi(u, t_3 - t_1) = \phi(u, t_3 - t_2) \, \phi(u, t_2 - t_1) \tag{3.110}
$$

or

$$
\phi(u, t + t') = \phi(u, t) \, \phi(u, t') \tag{3.111}
$$

Its solution has the general form

$$
\phi(u, t) = e^{tg(u)}, \qquad g(0) = 0 \tag{3.112}
$$

where the function $g(u)$ must be such that $\phi(u, t)$ is a characteristic function but it is otherwise arbitrary.

Going back to Eq. (3.108), the nth characteristic function of $X(t)$ thus takes the general form

$$
\phi_n(u_1, t_1; u_2, t_2; \ldots; u_n, t_n) = \prod_{k=1}^{n} \exp\left[(t_k - t_{k-1}) g\left(\sum_{j=k}^{n} u_j\right)\right] \tag{3.113}
$$

In the next section we give two important examples of independent-increment stochastic processes.

3.3.4. The Wiener Process and the Poisson Process

The investigation of Brownian motion marked the beginning of the study of random differential equations, and the mathematical theory evolved in this intensive research certainly represents one of the major developments in the theory of stochastic processes.

In 1828, Robert Brown, a botanist, observed that small particles immersed in a liquid execute irregular movements. This phenomenon, correctly described by Brown as the result of impacts of the molecules in the liquid, is called *Brownian motion*. A first satisfactory theory of Brownian motion was advanced independently by Einstein and Smoluchowski at the beginning of this century. A complete summary of these results can be found in the review papers by Uhlenbeck and Ornstein [16] and by Wang and Uhlenbeck [17].

Let $X(t)$ be the coordinate of a free particle on a straight line. Einstein was able to show that, assuming $P\{X(0) = 0\} = 1$, the probability distribution of $X(t) - X(t')$ is Gaussian with

$$E\{X(t) - X(t')\} = 0 \qquad (3.114)$$

$$E\{[X(t) - X(t')]^2\} = 2D \,|\, t - t' \,|, \qquad t, t' \geq 0 \qquad (3.115)$$

The quantity D is a physical constant.

Furthermore, it was shown that, if $0 \leq t_1 < t_2 < \cdots < t_n$, the increments of $X(t)$, $X(t_1, t_2)$, $X(t_2, t_3)$, $\ldots$, $X(t_{n-1}, t_n)$, are mutually independent random variables. It follows that, in view of Eqs. (3.114) and (3.115), the s.p. $X(t)$, $t \geq 0$, is a stationary independent-increment process.

To honor Wiener [18] and Lévy [19] who gave the first rigorous treatment of Brownian motion, the stochastic process $X(t)$ is also referred to as the *Wiener process* or the *Wiener–Lévy process*. The most significant contribution of Wiener is his study of the sample function behavior of the Brownian motion process. Specifically, he was able to show that its sample functions or trajectories are continuous but almost all nondifferential functions. We are not equipped to discuss his method of approach at this time but this topic will be revisited again in Appendix A.

The second important example of an independent-increment process is the *Poisson process*. Shot effect, thermal noise, and a large class of impulse noises are examples of physical situations modeled mathematically by the Poisson process. A Poisson process $X(t)$, $t \geq 0$, has independent integer-valued increments. The construction of this mathematical model is based upon the following observations.

Let an increment $X(t_1, t_2)$, $0 \leq t_1 < t_2$, of $X(t)$ take integer values each of which is the number of some specific events occurring in the interval $[t_1, t_2)$.

Let

$$P_k(t_1, t_2) = P\{X(t_1, t_2) = k\} \qquad (3.116)$$

The probability $P_k(t_1, t_2)$ is assumed to satisfy

(1) The random variables $X(t_1, t_2)$, $X(t_2, t_3)$, . . . , $X(t_{n-1}, t_n)$, $t_1 < t_2 <$ $\cdots < t_n$, are mutually independent.

(2) For sufficiently small Δt,

$$P_1(t, t + \Delta t) = \lambda \Delta t + o(\Delta t) \tag{3.117}$$

where $o(\Delta t)$ stands for functions such that

$$\lim_{\Delta t \to 0} (o(\Delta t)/\Delta t) = 0 \tag{3.118}$$

The parameter λ is called the intensity of the process; it is assumed to be a constant for simplicity.

(3) For sufficiently small Δt,

$$\sum_{j=2}^{\infty} P_j(t, t + \Delta t) = o(\Delta t) \tag{3.119}$$

It follows from Eqs. (3.117) and (3.119) that

$$P_0(t, t + \Delta t) = 1 - \sum_{j=1}^{\infty} P_j(t, t + \Delta t)$$
$$= 1 - \lambda \Delta t + o(\Delta t) \tag{3.120}$$

Let us first determine $P_0(t_0, t)$ on the basis of the assumptions stated above. It is observed that

$$P_0(t_0, t + \Delta t) = P_0(t_0, t) P_0(t, t + \Delta t)$$
$$= P_0(t_0, t)[1 - \lambda \Delta t + o(\Delta t)], \qquad t > t_0 \tag{3.121}$$

Dividing Eq. (3.121) by Δt and letting $\Delta t \to 0$ we obtain a differential equation satisfied by $P_0(t_0, t)$. It has the form

$$dP_0(t_0, t)/dt = -\lambda P_0(t_0, t) \tag{3.122}$$

The solution of Eq. (3.122) satisfying the initial condition $P_0(t_0, t_0) = 1$ is

$$P_0(t_0, t) = \exp[-\lambda(t - t_0)], \qquad t > t_0 \tag{3.123}$$

The determination of $P_1(t_0, t)$ is similar. We first observe that

$$P_1(t_0, t + \Delta t) = P_0(t_0, t)P_1(t, t + \Delta t) + P_1(t_0, t)P_0(t, t + \Delta t) \tag{3.124}$$

Substituting Eqs. (3.117), (3.120), and (3.123) into Eq. (3.124) and letting $\Delta t \to 0$ gives

$$dP_1/d\tau = -P_1 + e^{-\tau}, \qquad \tau = \lambda(t - t_0) \tag{3.125}$$

which yields

$$P_1(t_0, t) = \lambda(t - t_0) \exp[-\lambda(t - t_0)], \qquad t > t_0 \tag{3.126}$$

Continuing in this way we find

$$P_k(t_0, t) = \frac{[\lambda(t - t_0)]^k}{k!} \exp[-\lambda(t - t_0)], \qquad t > t_0 \tag{3.127}$$

Joint probabilities can be written down easily in view of the independent increment property of $X(t)$. For example, we have

$$P[X(0, t_1) = j, X(0, t_2) = k]$$
$$= \frac{(\lambda t_1)^j [\lambda(t_2 - t_1)]^{k-j}}{j!(k - j)!} e^{-\lambda t_2}, \qquad k \geq j, t_2 > t_1 \tag{3.128}$$

The mean and the variance of the increment $X(t_1, t_2)$, $0 < t_1 < t_2$, are given by

$$E\{X(t_1, t_2)\} = E\{X(t_2) - X(t_1)\} = \lambda(t_2 - t_1) \tag{3.129}$$

$$\text{var}\{X(t_1, t_2)\} = \text{var}\{X(t_2) - X(t_1)\} = \lambda(t_2 - t_1) \tag{3.130}$$

and the covariance has the form

$$\text{cov}\{X(t_1) X(t_2)\} = \lambda \min(t_1, t_2), \qquad t_1, t_2 > 0 \tag{3.131}$$

It is easy to see that the sample functions of Poisson processes are continuous and differentiable everywhere except at a finite number of points

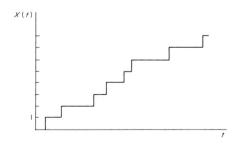

Fig. 3.7. A sample function of the Poisson process.

in a given finite interval. The shape of each function is a staircase of step height one, with steps occurring at random time instants. A typical form of the sample functions is shown in Figure 3.7.

3.4. The Gaussian Stochastic Process

A discussion of stochastic processes is incomplete without giving considerations to Gaussian stochastic processes.

Classifications of stochastic processes based upon regularity and memory do not make specific references to the detailed probability distributions associated with a stochastic process. Gaussian stochastic processes are characterized by a distinct probabilistic law and they are extremely useful and most often encountered in practice. Moreover, we will give repeated considerations to Gaussian processes coupled with stationarity or Markovian properties, or both.

Definition. A s.p. $X(t)$, $t \in T$, is called a *Gaussian stochastic process* if, for every finite set $t_1, t_2, \ldots, t_n$, the random variables $X(t_1)$, $X(t_2)$, $\ldots$, $X(t_n)$ have the joint characteristic function

$$\phi_n(u_1, t_1; u_2, t_2; \ldots; u_n, t_n) = \exp[i\, \mathbf{m}^T(t_i)\, \mathbf{u} - \tfrac{1}{2}\mathbf{u}^T \Lambda(t_i, t_j)\, \mathbf{u}] \qquad (3.132)$$

where $\mathbf{u}^T = [u_1\ u_2\ \ldots\ u_n]$, $\mathbf{m}^T(t_i) = [E\{X(t_1)\}\ E\{X(t_2)\}\ \ldots\ E\{X(t_n)\}]$, and $\Lambda(t_i, t_j) = [\mu_{ij}]$ is the $n \times n$ covariance matrix associated with $X(t_1)$, $X(t_2)$, $\ldots$, $X(t_n)$ with

$$\mu_{ij} = E\{[X(t_i) - m(t_i)][X(t_j) - m(t_j)]\} \qquad (3.133)$$

the superscript T denotes matrix transpose.

We observe that Eq. (3.132) is essentially the same as the joint characteristic function in Eq. (2.99) associated with a Gaussian random vector. Other properties of a Gaussian random vector also apply here.

A Gaussian stochastic process can, of course, be defined in terms of its joint density functions; however, as we have noted in the discussion of Gaussian random vectors in Chapter 2, the definition given above is more desirable because it avoids the difficulties when the covariance matrix becomes singular.

In the stationary case, the mean vector $\mathbf{m}(t)$ is a constant and the covariance matrix is a function of the time instants only through their differences.

Based upon the central limit theorem, a Gaussian stochastic process can be expected to occur whenever it represents a sum of very large number of small independent random effects at each instant. We thus expect that Gaussian processes are models or at least idealizations of many physical phenomena. The Wiener process, for example, is a Gaussian process. Noise currents and voltages in electronic devices, atmospheric disturbances, limiting shot noise, and thermo noise processes are other examples of Gaussian processes.

Gaussian processes have a number of important properties which lead to many mathematical advantages. An important property has been mentioned in Section 3.2.1, namely, since a Gaussian process is completely specified by its means and covariance functions, wide-sense stationarity implies strict stationarity in this case.

Another important property is that the Gaussianity of a Gaussian stochastic process is preserved under linear transformations. This property will be explored in detail in Chapter 4 and we shall see that it is extremely useful in treating random differential equations involving Gaussian stochastic processes.

References

1. A. N. Kolmogorov, *Foundation of the Theory of Probability*, 2nd Engl. ed., Chelsea, Bronx, New York, 1956.
2. J. L. Doob, *Stochastic Processes*. Wiley, New York, 1953.
3. S. O. Rice, Mathematical analysis of random noise. *Bell System Tech. J.* **23**, 282–332 (1944); **24**, 46–156 (1945). Reprinted in *Selected Papers on Noise and Stochastic Processes* (N. Wax, ed.). Dover, New York, 1954.
4. A. M. Yaglom, *An Introduction to the Theory of Stationary Random Functions*. Prentice-Hall, Englewood Cliffs, New Jersey, 1962.
5. H. Cramér and M. R. Leadbetter, *Stationary and Related Stochastic Processes*. Wiley, New York, 1967.
6. S. Bochner, Monotone Funktionen Stieltjessche Integrale und harmonische Analyse. *Math. Ann.* **108**, 376–385 (1933).
7. H. Cramér, On the representation of a function by certain Fourier integrals. *Trans. Amer. Math. Soc.* **46**, 191–201 (1939).
8. A. N. Kolmogorov, "Über die analytischen Methoden in der Wahrscheinlichkeitsrechnung. *Math. Ann.* **104**, 415–458 (1931).
9. W. Feller, Zur Theorie der Stochastishen Prozesse (Existenz und Eindentigkeitssatze). *Math. Ann.* **113**, 113–160 (1936).
10. W. Feller, On the integro-differential equations of purely discontinuous Markov processes. *Trans. Amer. Math. Soc.* **48**, 488–518 (1940); Errata *Trans. Amer. Math. Soc.* **58**, 474 (1945).
11. W. Feller, *An Introduction to Probability Theory and Its Applications*, Vol. 1. Wiley, New York, 1957.

12. M. Loéve, *Probability Theory*. Van Nostrand-Reinhold, Princeton, New Jersey, 1963.
13. A. T. Bharucha-Reid, *Elements of the Theory of Markov Processes and Their Applications*. McGraw-Hill, New York, 1960.
14. R. L. Stratonovich, *Conditional Markov Processes and Their Applications to the Theory of Optimal Control*, Amer. Elsevier, New York, 1968.
15. E. B. Dynkin, *Markov Processes*, Vols. 1 and 2. Springer-Verlag, Berlin and New York, 1965.
16. G. E. Uhlenbeck and L. S. Ornstein, On the theory of the Brownian motion. *Phys. Rev.* **36**, 823–841 (1930). Reprinted in *Selected Papers on Noise and Stochastic Processes* (N. Wax, ed.). Dover, New York, 1954.
17. M. C. Wang and G. E. Uhlenbeck, On the theory of the Brownian motion II. *Rev. Modern Phys.* **17**, 323–342 (1945). Reprinted in *Selected Papers on Noise and Stochastic Processes* (N. Wax, ed.). Dover, New York, 1954.
18. N. Wiener, Differential space. *J. Math. and Phys.* **2**, 131–174 (1923).
19. P. Lévy, *Processus stochastiques et mouvement brownien*. Gauthier-Villars, Paris, 1948.

PROBLEMS

3.1. Determine the first and the second density functions of the stochastic processes given below.

(a) $X(t) = At + B$, where A and B are independent and identically distributed r.v.'s, each uniformly distributed in the interval $[0, 1]$.

(b)   Binary noise defined in Example 3.5.

(c)   Random telegraph signal defined in Example 3.6.

3.2. A s.p. $X(t)$, $t \in T$, consists of two member functions each occurring with probability $\frac{1}{2}$. They are

$$X(t) = \begin{cases} \cos \pi t \\ 2t \end{cases}$$

Determine its first distribution functions $F_1(x, \frac{1}{2})$ and $F_1(x, 1)$, and its second distribution function $F_2(x_1, \frac{1}{2}; x_2, 1)$.

   For each of the s.p. $X(t)$, $t \in T$, defined in Problems 3.3 to 3.8, determine

(a)   Whether $X(t)$ is wide-sense stationary.

(b)   If it is, determine its spectral density function.

3.3. A discrete-parameter s.p. defined by

$$X(t) = \sin \Omega t, \qquad t = 1, 2, \ldots$$

where Ω is a random variable uniformly distributed in the interval $[0, 2\pi]$.

3.4.
$$X(t) = A \cos(\omega_0 t + \Theta)$$

where A is Rayleigh distributed with density function

$$f(a) = \frac{a}{\sigma^2} \exp\left[-\frac{a^2}{2\sigma^2}\right], \qquad a \geq 0$$

and Θ is uniformly distributed in the interval $[0, 2\pi]$, independent of A. The quantity ω_0 is a deterministic constant. This situation represents the noise voltage when a Gaussian noise is passed through a narrow-band filter with center frequency ω_0.

3.5.
$$X(t) = A \cos(\Omega t + \Theta)$$

The random variables A, Ω, Θ are mutually independent. A and Θ are as defined above, and Ω takes only positive values with the density function $f(\omega)$, $\omega > 0$.

3.6. $X(t)$, $t \geq 0$, is defined by

$$X(t) = Y(t + \Delta) - Y(t)$$

where $Y(t)$, $t \geq 0$, is the Poisson process with intensity λ, and Δ is a positive constant. The s.p. $X(t)$ is called an *increment Poisson process*.

3.7.
$$X(t) = \exp[Y(t)]$$

where $Y(t)$ is a stationary Gaussian process with mean zero and correlation function $\Gamma_{YY}(\tau)$.

3.8.
$$X(t) = X_1(t) X_2(t)$$

where $X_1(t)$ and $X_2(t)$ are independent wide-sense stationary stochastic processes.

3.9. Consider the first four properties associated with $\Gamma(\tau)$ of a wide-sense stationary s.p. $X(t)$ as listed in Section 3.2.2. What are the corresponding properties associated with a correlation function matrix of a vector wide-sense stationary process?

3.10. Let $X(t)$ be a stationary Gaussian stochastic process with zero mean. Let a new stochastic process $Y(t)$ be defined by

$$Y(t) = X^2(t)$$

Show that

$$\Gamma_{YY}(\tau) = \Gamma_{XX}^2(0) + 2\Gamma_{XX}^2(\tau)$$

3.11. Given a stationary stochastic process $X(t)$ with zero mean, covariance $\mu(\tau)$ and second density function $f_2(x_1, t; x_2, t + \tau) = f(x_1, x_2, \tau)$

(a)   Show that

$$P\{|\ X(t + \tau) - X(t)\ | \geq a\} \leq 2[\mu(0) - \mu(\tau)]/a^2$$

(b)   Express $P\{|\ X(t + \tau) - X(t)\ | \geq a\}$ in terms of $f(x_1, x_2, \tau)$.

3.12. Let a s.p. $X(t)$, $t \in T$, be defined by

$$X(t) = Ah(t)$$

where A is a r.v. and $h(t)$ is a deterministic function of t. Show that $X(t)$ is wide-sense stationary if, and only if,

$$h(t) = e^{i\omega t}$$

where ω is a real constant.

3.13. Supply a proof for Theorem 3.2.2.

3.14. Using the result of Theorem 3.2.2 and show that the stochastic process considered in Example 3.10 is not ergodic in the correlation function.

3.15. Is the Wiener process ergodic in the mean?

3.16. Show that a Markov process is also Markovian in reverse, that is, for $t_1 < t_2 < \cdots < t_n$, we have

$$F(x_1, t_1 \mid x_2, t_2; x_3, t_3; \ldots; x_n, t_n) = F(x_1, t_1 \mid x_2, t_2)$$

3.17. Let $X(t)$, $t \in T$, be a Markov process of second-order defined by Eq. (3.106). Show that

$$f_n(x_1, t_1; x_2, t_2; \ldots; x_n, t_n) = \frac{\displaystyle\prod_{k=1}^{n-2} f_3(x_k, t_k; x_{k+1}, t_{k+1}; x_{k+2}, t_{k+2})}{\displaystyle\prod_{k=2}^{n-2} f_2(x_k, t_k; x_{k+1}, t_{k+1})}, \quad n \geq 4,$$

and that the Smoluchowski–Chapman–Kolmogorov equation takes the form

$$f(x_3, t_3 \mid x_1, t_1; x_2, t_2)$$

$$= \int_{-\infty}^{\infty} f(x_3, t_3 \mid x_2, t_2; x', t') f(x', t' \mid x_1, t_1; x_2, t_2) \, dx'$$

with $t_1 \leq t_2 \leq t_3$ for all t' such that $t_1 \leq t' \leq t_3$.

3.18. Consider a one-dimensional random walk. A walker, starting at the origin, takes a step of length λ every Δ seconds at each tossing of a coin, a step to the right if heads occur and left if tails occur. If $X(t)$ denotes the position of the walker to the left from the origin at the end of n steps, show that

(a) $$E\{X(t = n\Delta)\} = 0, \qquad E\{X^2(t = n\Delta)\} = t\lambda^2/\Delta$$

(b) In the limiting case when Δ tends to zero, λ tends to zero as $\Delta^{1/2}$, and $n \to \infty$ such that $n\Delta$ tends to a constant, show that the limiting process is a Wiener process.

3.19. Let $X(t)$, $t \geq 0$ and $P\{X(0) = 0\} = 1$, be the Wiener process. Show that the covariance of $X(t)$ is

$$\mu(t_1, t_2) = 2D \min(t_1, t_2), \qquad t_1, t_2 \geq 0$$

Is it continuous and differentiable in the first quadrant of the (t_1, t_2) plane?

3.20. Show that a Gaussian process can be obtained as a limiting case of the Poisson process as the intensity $\lambda \to \infty$.

Chapter 4

Stochastic Limits and Calculus in Mean Square

In Chapter 3, we introduced several important classes of stochastic processes together with examples illustrating the role stochastic processes play in modeling physical phenomena. There is, however, still a long way to go in accomplishing the modest goal we set out to reach in this book. Many random phenomena in nature which directly interest us are expressed mathematically in terms of limiting sums, derivatives, integrals, and differential and integral equations. Hence, the next step in our discussion is that of developing a calculus associated with a class of stochastic processes.

Let us remark that when we speak of limiting operations such as sums and derivatives, we clearly see the need for having a Borel field $\mathscr{B}$ in the development of probability theory. If we did not have closure under countable set operations, such limits would lead to quantities to which probabilities might not be readily assigned.

We have chosen to study stochastic processes in terms of a system of probability distribution functions. In this chapter we shall develop a calculus relevant to this approach, namely, the calculus in mean square or m.s. calculus. The m.s. calculus is important for several practical reasons. First of all, its importance lies in the fact that simple yet powerful and well-developed methods exist. Secondly, the development of m.s. calculus and its application to physical problems follows in broad outline the same steps used in considering calculus of ordinary (deterministic) functions. It is thus easier to grasp for engineers and scientists who have had a solid background in the analysis of ordinary functions. Furthermore, the m.s. calculus is attractive because it is defined in terms of distributions and moments which are our chief concern.

It was indicated in Chapter 3 that stochastic processes can also be studied in terms of its sample functions. Indeed, stochastic processes in nature generally represent collections of ordinary deterministic functions. In dealing with problems in the real world, we certainly prefer a calculus which makes sense on the sample level. Unfortunately, sample behavior of stochastic processes in physical situations is generally difficult to ascertain. In Appendix A we will give a discussion of this type of calculus (sample calculus) and explore its connection with the m.s. calculus. Let us simply point out here that another important reason for studying m.s. calculus is that, for the important case of Gaussian processes, m.s. properties lead to properties on the sample function level (Loéve [1, p. 485]).

For additional reading on the subject, the reader is referred to Loéve [1] and Moyal [2]. Some of the examples studied in this chapter were first discussed by Bartlett [3]. The proofs of some theorems in the development of m.s. calculus follow Ruymgaart [4].

4.1. Preliminary Remarks

We give below some definitions and discussions relevant to the development of m.s. calculus.

Let us consider the properties of a class of real r.v.'s $X_1, X_2, \ldots$, whose second moments, $E\{X_1{}^2\}, E\{X_2{}^2\}, \ldots$, are finite. They are called *second-order random variables*.

(1) From the Schwarz inequality (Eq. (2.56))

$$E^2\{|\ X_1 X_2\ |\} \leq E\{X_1{}^2\}\, E\{X_2{}^2\} \tag{4.1}$$

it follows directly that

$$E\{(X_1 + X_2)^2\} < \infty \tag{4.2}$$

$$E\{(cX_1)^2\} = c^2\, E\{X_1{}^2\} \tag{4.3}$$

where c is real and finite. Hence, the class of all second order r.v.'s on a probability space constitute a *linear vector space* if all equivalent r.v.'s are identified. Two r.v.'s X and Y are called *equivalent* if $P\{X \neq Y\} = 0$.

(2) Let us use the notation

$$E\{X_1 X_2\} = \langle X_1, X_2 \rangle \tag{4.4}$$

The Schwarz inequality (4.1) leads to

$$|\langle X_1, X_2 \rangle| < \infty \tag{4.5}$$

It is easily shown that $\langle X_1, X_2 \rangle$ satisfies the following inner product properties:

$$\langle X, X \rangle \geq 0; \quad \langle X, X \rangle = 0 \text{ if, and only if, } X = 0 \tag{4.6}$$

with probability one

$$\langle X_1, X_2 \rangle = \langle X_2, X_1 \rangle \tag{4.7}$$

$$\langle c X_1, X_2 \rangle = \langle c X_2, X_1 \rangle = c \langle X_1, X_2 \rangle, \quad c \text{ is real and finite} \tag{4.8}$$

$$\langle X_1 + X_2, X_3 \rangle = \langle X_1, X_3 \rangle + \langle X_2, X_3 \rangle \tag{4.9}$$

(3) Define

$$\| X \| = \langle X, X \rangle^{1/2} \tag{4.10}$$

It follows directly from Eqs. (4.6) to (4.9) that $\| X \|$ possesses the norm properties:

$$\| X \| \geq 0; \quad \| X \| = 0 \text{ if, and only if, } X = 0 \text{ with probability one} \tag{4.11}$$

$$\| c X \| = | c | \cdot \| X \|, \quad c \text{ is real and finite} \tag{4.12}$$

$$\| X_1 + X_2 \| \leq \| X_1 \| + \| X_2 \| \tag{4.13}$$

Inequality (4.13) holds because

$$
\begin{aligned}
\| X_1 + X_2 \|^2 &= \langle X_1 + X_2, X_1 + X_2 \rangle \\
&= \langle X_1, X_1 \rangle + 2 \langle X_1, X_2 \rangle + \langle X_2, X_2 \rangle \\
&\leq \langle X_1, X_1 \rangle + 2 \| X_1 \| \cdot \| X_2 \| + \langle X_2, X_2 \rangle \\
&= \{ \| X_1 \| + \| X_2 \| \}^2
\end{aligned}
\tag{4.14}
$$

where we have again made use of Eq. (4.1).

(4) Define the distance between X_1 and X_2 by

$$d(X_1, X_2) = \| X_1 - X_2 \| \tag{4.15}$$

The distance $d(X_1, X_2)$ possesses the usual distance properties:

$$d(X_1, X_2) \geq 0; \quad d(X_1, X_2) = 0 \text{ if, and only if, } X_1 = X_2 \tag{4.16}$$

with probability one

$$d(X_1, X_2) = d(X_2, X_1) \tag{4.17}$$

$$d(X_1, X_2) \leq d(X_1, X_3) + d(X_3, X_2) \tag{4.18}$$

It is seen from Eq. (4.16) that the equivalence of two random variables can also be defined using the distance. That is, $X = Y$ if, and only if, $d(X, Y) = 0$.

The linear vector space of second-order random variables with the inner product, the norm, and the distance defined above is called a L_2-*space*.

We state without proof an important theorem associated with L_2-spaces (for proof, see Loéve [1, p. 161]).

L_2-Completeness Theorem. L_2-spaces are complete in the sense that any Cauchy sequence in L_2 has a unique limit in L_2. That is, let the sequence $\{X_n\}$ be defined on the set of natural numbers. There is a unique element $X \in L_2$ such that

$$\| X_n - X \| \to 0 \qquad \text{as} \quad n \to \infty \tag{4.19}$$

if, and only if,

$$d(X_n, X_m) = \| X_n - X_m \| \to 0 \tag{4.20}$$

as $n, m \to \infty$ in any manner whatever.

We thus see that L_2-spaces are complete normed linear spaces (Banach spaces) and complete inner product spaces (Hilbert spaces).

Now consider a s.p. $X(t)$, $t \in T$. $X(t)$ is called a *second-order stochastic process* if, for every set $t_1, t_2, \ldots, t_n$, the r.v.'s $X(t_1), X(t_2), \ldots, X(t_n)$ are elements of L_2-space. A second order s.p. $X(t)$ is characterized by

$$\| X(t) \|^2 = E\{X^2(t)\} < \infty, \qquad t \in T \tag{4.21}$$

In the development of m.s. calculus we shall deal exclusively with second-order random variables and stochastic processes.

4.2. Convergence of a Sequence of Random Variables

As in the case of developing a calculus for deterministic functions, the first step in the development of a calculus for stochastic processes is to define the convergence of a sequence of random variables. In this development, more than one mode of convergence can be given, and we discuss in this section four modes of convergence and their relations with each other. These four modes are convergence in distribution, convergence in probability, convergence in mean square, and almost sure convergence.

The development of the m.s. calculus is based upon the concept of convergence in mean square or m.s. convergence. This concept shall be explored first. It is followed by the discussions of the other three modes of convergence and of how all four are related.

4.2.1. Convergence in Mean Square

Definition. A sequence of r.v.'s $\{X_n\}$ *converges in m.s.* to a r.v. X as $n \to \infty$ if

$$\lim_{n \to \infty} \| X_n - X \| = 0 \tag{4.22}$$

This type of convergence is often expressed by

$$X_n \xrightarrow{\text{m.s.}} X \quad \text{or} \quad \underset{n \to \infty}{\text{l.i.m.}}\ X_n = X \tag{4.23}$$

The symbol l.i.m. denotes the *limit in the mean*. Other commonly used names are *convergence in quadratic mean* and *second-order convergence*.

Let us consider some properties of the m.s. convergence.

Theorem 4.2.1. Let $\{X_n\}$ be a sequence of second-order r.v.'s. If

$$X_n \xrightarrow{\text{m.s.}} X \tag{4.24}$$

then

$$\lim_{n \to \infty} E\{X_n\} \to E\{X\} \tag{4.25}$$

In words, "l.i.m." and "expectation" commute. This is certainly a desirable result.

Proof. First, it is easy to show that $E\{X_n\}$ exists. This is seen from (or directly from Problem 2.5)

$$| E\{X_n\} | \leq E | X_n | \leq \| X_n \| < \infty \tag{4.26}$$

The second inequality follows from the Schwarz inequality (2.56) with Y being identically one.

Since L_2-spaces are complete, the limit X is of second order and

$$| E\{X_n\} - E\{X\} | \leq E\{| X_n - X |\} \leq \| X_n - X \| \tag{4.27}$$

Since $\| X_n - X \| \to 0$ as $n \to \infty$, we have

$$\lim_{n \to \infty} E\{X_n\} \to E\{X\} \tag{4.28}$$

The L_2-completeness theorem also leads to the following theorems.

Theorem 4.2.2. Limits in m.s. convergence are unique, that is, if

$$X_n \xrightarrow{\text{m.s.}} X \qquad \text{and} \qquad X_n \xrightarrow{\text{m.s.}} Y \tag{4.29}$$

then X and Y are equivalent, that is, $P\{X \neq Y\} = 0$. This result also follows directly from the distance property (4.16).

Definition. A sequence $\{X_n\}$ of r.v.'s is *fundamental in mean square* or *m.s. fundamental* if

$$\| X_m - X_n \| \to 0 \tag{4.30}$$

as $n, m \to \infty$ is any manner whatever.

Theorem 4.2.3. A sequence $\{X_n\}$ of r.v.'s converges in mean square to a r.v. X as $n \to \infty$ if, and only if, the sequence is m.s. fundamental.

This theorem is simply a restatement of the L_2-completeness theorem. It is clearly useful in testing the m.s. convergence since the criterion (4.30) does not involve the limiting r.v. X, which is frequently unknown. We see that this theorem is the counterpart of the Cauchy's mutual convergence criterion in the analysis of ordinary functions.

Let us now give some examples.

Example 4.1. Consider a sequence $\{X_n\}$ of r.v.'s defined by

$$X_n = (1/n) \sum_{j=1}^{n} Y_j \tag{4.31}$$

where the r.v.'s Y_j are independent and identically distributed with means m and variances σ^2. We first investigate whether the sequence $\{X_n\}$ converges in mean square to some random variable.

Using Eq. (4.30) we form

$$
\begin{aligned}
\| X_m - X_n \|^2 &= E\{[(X_m - m) - (X_n - m)]^2\} \\
&= E\{(X_m - m)^2 - 2(X_m - m)(X_n - m) + (X_n - m)^2\} \\
&= \sigma^2\{1/m + 1/n - 2[\max(n, m)]^{-1}\} \to 0
\end{aligned}
\tag{4.32}
$$

as $n, m \to \infty$. Hence, it follows from Theorem 4.2.3 that the sequence is m.s. convergent.

Let us consider next whether the sequence converges in mean square to $E\{X\} = m$. From Eq. (4.22) we observe that

$$\lim_{n \to \infty} \| X_n - X \| = \lim_{n \to \infty} \| X_n - m \| = \lim_{n \to \infty} \sigma/\sqrt{n} = 0 \qquad (4.33)$$

The sequence thus converges in mean square to m identically.

The result of this example is one form of the law of large numbers. It represents the same situation as that in sampling theory where we take independent observations of a r.v. $Y, y_1, y_2, \ldots, y_n$, and ask the question whether the ensemble mean of Y, $E\{Y\}$, can be closely approximated by the formula

$$\lim_{n \to \infty} (1/n) \sum_{j=1}^{n} y_j \qquad (4.34)$$

We have shown in this example that it converges to the mean in the mean square sense.

If we define a discrete-parameter s.p. $X(t)$ by

$$X(t_j) = Y_j, \qquad j = 1, 2, \ldots, n, \qquad (4.35)$$

this is also a good example in showing that $X(t)$ is ergodic in the mean (Section 3.2.3).

Example 4.2. Let $\{X_n\}$ be a sequence of r.v.'s, mutually independent and distributed according to

$$X_n = \begin{cases} 1, & P(1) = 1/n \\ 0, & P(0) = 1 - 1/n \end{cases} \qquad (4.36)$$

Does it converge in mean square to zero identically? We have

$$\lim_{n \to \infty} \| X_n - X \| = \lim_{n \to \infty} \| X_n \| = \lim_{n \to \infty} 1/\sqrt{n} = 0 \qquad (4.37)$$

Thus

$$X_n \xrightarrow{\text{m.s.}} 0 \qquad (4.38)$$

Example 4.3. If we keep all statements given in Example 4.2 unchanged except changing Eq. (4.36) to

$$X_n = \begin{cases} n, & P(n) = 1/n^2 \\ 0, & P(0) = 1 - 1/n^2 \end{cases} \qquad (4.39)$$

Does the new sequence converge in mean square? Consider

$$\| X_m - X_n \|^2 = E\{X_m{}^2 - 2X_nX_m + X_n{}^2\} = 2[1 - 1/nm] \to 2 \quad (4.40)$$

as $n, m \to \infty$. Hence, the sequence $\{X_n\}$ is not m.s. convergent.

4.2.2. Convergence in Probability

Definition. A sequence of r.v.'s $\{X_n\}$ *converges in probability* or *converges i.p.* to a r.v. X as $n \to \infty$ if, for every $\varepsilon > 0$,

$$\lim_{n \to \infty} P\{| X_n - X | > \varepsilon\} = 0 \quad (4.41)$$

This type of convergence is often expressed by

$$X_n \xrightarrow{\text{i.p.}} X, \qquad \underset{n \to \infty}{\text{l.i.p.}} X_n = X \quad (4.42)$$

where "l.i.p." reads *"limit in probability."* Convergence in probability is sometimes called *stochastic convergence* or *convergence in measure.*

Definition. A sequence $\{X_n\}$ of r.v.'s is *fundamental in probability* or *fundamental i.p.* if, for every $\varepsilon > 0$,

$$P\{| X_m - X_n | > \varepsilon\} \to 0 \quad (4.43)$$

as $n, m \to \infty$ in any manner whatever.

Theorem 4.2.4. A sequence $\{X_n\}$ converges i.p. to a r.v. X as $n \to \infty$ if, and only if, that $\{X_n\}$ is fundamental i.p.

Proof. The event $| X_m - X_n | > \varepsilon$ can be written as

$$\{| X_m - X_n | > \varepsilon\} \subset \{| X_m - X | > \tfrac{1}{2}\varepsilon\} \cup \{| X_n - X | > \tfrac{1}{2}\varepsilon\} \quad (4.44)$$

we thus have

$$P\{| X_m - X_n | > \varepsilon\} \leq P\{| X_m - X | > \tfrac{1}{2}\varepsilon\} + P\{| X_n - X | > \tfrac{1}{2}\varepsilon\} \quad (4.45)$$

Hence, if $X_n \xrightarrow{\text{i.p.}} X$,

$$P\{| X_m - X_n | > \varepsilon\} \to 0 \quad (4.46)$$

as $n, m \to \infty$.

We have proved the "only if" part. A discussion of the "if" part of the proof is found in Problem 4.8, which we omit here because it requires some additional work. As in the case of m.s. convergence, Eq. (4.46) is useful in testing the convergence i.p.

Theorem 4.2.5. Limits of convergence in probability are unique in the sense that, if a sequence $\{X_n\}$ of r.v.'s converges i.p. to r.v.'s X and Y, then X and Y are equivalent.

Proof. Consider the event $|X - Y| > \varepsilon$. We have

$$\{|X - Y| > \varepsilon\} \subset \{|X - X_n| > \tfrac{1}{2}\varepsilon\} \cup \{|Y - X_n| > \tfrac{1}{2}\varepsilon\} \qquad (4.47)$$

and thus

$$P\{|X - Y| > \varepsilon\} \leq P\{|X - X_n| > \tfrac{1}{2}\varepsilon\} + P\{|Y - X_n| > \tfrac{1}{2}\varepsilon\} \qquad (4.48)$$

By hypothesis, the right-hand side of Eq. (4.48) approaches zero as $n \to \infty$. Hence, for every $\varepsilon > 0$,

$$P\{|X - Y| > \varepsilon\} \to 0 \qquad (4.49)$$

which is equivalent to

$$P\{X \neq Y\} = 0 \qquad (4.50)$$

which completes the proof.

As shown in the theorem below, convergence in probability is weaker than m.s. convergence.

Theorem 4.2.6. Convergence in mean square implies convergence in probability.

Proof. We write the Tchebycheff inequality (2.55) in the form

$$P\{|X_n - X| > \varepsilon\} \leq \|X_n - X\|^2/\varepsilon^2 \qquad (4.51)$$

which holds for every $\varepsilon > 0$. As $n \to \infty$

$$\|X_n - X\| \to 0 \qquad (4.52)$$

by hypothesis. Thus

$$\lim_{n \to \infty} P\{|X_n - X| > \varepsilon\} = 0 \qquad (4.53)$$

or

$$X_n \xrightarrow{\text{i.p.}} X \qquad (4.54)$$

It is clear that the limiting r.v. X is the same for both modes of convergence.

In view of Theorem 4.2.6, the two sequences of random variables considered in Examples 4.1 and 4.2 also converge i.p. The result of Example 4.1 in this mode of convergence is called the *weak law of large numbers*.

Example 4.4. It is of interest to see whether the sequence of r.v.'s in Example 4.3 converges i.p. to zero. We have, for every $\varepsilon > 0$,

$$\lim_{n\to\infty} P\{|X_n - X| > \varepsilon\} = \lim_{n\to\infty} P\{|X_n| > \varepsilon\}$$

$$= \lim_{n\to\infty} P\{X_n = n\} = \lim_{n\to\infty} 1/n^2 = 0 \qquad (4.55)$$

Hence, the sequence converges i.p. to zero identically.

This example points out an important feature which separates m.s. convergence from convergence i.p. Mean square convergence is dependent upon both the values that the random variables can take and the probabilities associated with them. Convergence in probability, on the other hand, is only concerned with the probability of an event. Example 4.4 also shows that the converse of Theorem 4.2.6 is not true.

The next example illustrates a difficulty one encounters in dealing with convergence i.p.

Example 4.5. Consider a sequence $\{X_n\}$ of r.v.'s defined by

$$X_n = \begin{cases} n, & P(n) = 1/n \\ -1, & P(-1) = 1 - 1/n \end{cases} \qquad (4.56)$$

Clearly,

$$X_n \xrightarrow{\text{i.p.}} X \equiv -1 \qquad (4.57)$$

Now consider the expectations. We have $E\{X\} = -1$ and

$$\lim_{n\to\infty} E\{X_n\} = \lim_{n\to\infty} [n(1/n) - (1 - 1/n)] = \lim_{n\to\infty} 1/n = 0 \qquad (4.58)$$

Hence

$$\lim_{n\to\infty} E\{X_n\} \neq E\{X\} \qquad (4.59)$$

The limit of expectation is not equal to expectation of the limit in this case. We thus expect difficulties, for example, in interchanging expectations and integrals when the integrals are defined in the sense of i.p.

4.2.3. Convergence Almost Surely

Referred to as *"strong"* convergence in probability theory, convergence almost surely plays an important role in the limit theorems concerning laws of large numbers.

Definition. A sequence $\{X_n\}$ of r.v.'s is said to *converge almost surely* or *converge a.s.* to a r.v. X as $n \to \infty$ if

$$P\left\{\lim_{n \to \infty} X_n = X\right\} = 1 \qquad (4.60)$$

This type of convergence is sometimes called *almost certain convergence*, *convergence almost everywhere*, or *convergence with probability one*. It is often expressed by

$$X_n \xrightarrow{\text{a.s.}} X \qquad (4.61)$$

Definition. A sequence $\{X_n\}$ of r.v.'s is *fundamental almost surely* or *fundamental a.s.* if, for any $\varepsilon > 0$ and $\delta > 0$, there is an $N > 0$ such that

$$P\left\{\bigcup_{m \geq n}^{\infty} |X_n - X_m| > \varepsilon\right\} < \delta \qquad (4.62)$$

for all $n \geq N$.

The counterpart of Theorems 4.2.3 and 4.2.4 for this type of convergence is stated below without proof.

Theorem 4.2.7. A sequence $\{X_n\}$ of r.v.'s converges a.s. to a r.v. X as $n \to \infty$ if, and only if, the sequence is fundamental a.s.

Although the criterion (4.62) does not involve the limiting r.v. X, it is still difficult to apply for the convergence a.s. test. This is because that this type of convergence is dependent upon the *joint* probabilistic behavior of the r.v.'s $X_n - X_m$ for all $n, m \geq N$. As convergence i.p. is only concerned with the *marginal* probabilistic behavior of the r.v.'s $X_n - X_m$, $n, m \geq N$, we see that convergence a.s. is a stronger criterion than convergence i.p. The fact that convergence a.s. implies convergence i.p. will be seen more clearly later.

In order to test the convergence a.s., we give below three equivalent statements which are often useful.

Consider the statement (4.60). It is equivalent to

$$P\left\{\lim_{j \to \infty} X_j \neq X\right\} = 0 \qquad (4.63)$$

or, for any $\varepsilon > 0$, there is an $n > 0$ such that

$$P\{|X_j - X| > \varepsilon\} = 0 \tag{4.64}$$

for *all* $j \geq n$.

Let the event $|X_j - X| > \varepsilon$ be denoted by $A_j(\varepsilon)$. The statement (4.64) is equivalent to

$$\lim_{n \to \infty} P\left\{\bigcup_{j \geq n} A_j(\varepsilon)\right\} = 0 \tag{4.65}$$

Now the event $\bigcup_{j \geq n} A_j(\varepsilon)$ is one that, at least for one $j \geq n$,

$$|X_j - X| > \varepsilon \tag{4.66}$$

then the event

$$\sup_{j \geq n} |X_j - X| > \varepsilon \tag{4.67}$$

is true. Conversely, if Eq. (4.67) holds, there exists one j at least as large as n such that Eq. (4.66) is true. Hence, the statement (4.65) is also equivalent to

$$\lim_{n \to \infty} P\left\{\sup_{j \geq n} |X_j - X| > \varepsilon\right\} = 0 \tag{4.68}$$

Lastly, by means of the DeMorgan law of sets,

$$\left\{\bigcup_{j=n}^{\infty} A_j(\varepsilon)\right\}' = \bigcap_{j=n}^{\infty} A_j'(\varepsilon) \tag{4.69}$$

where the primes indicate complements of sets, Eq. (4.65) is equivalent to

$$\lim_{n \to \infty} P\left\{\bigcap_{j \geq n} E_j'(\varepsilon)\right\} = 1 \tag{4.70}$$

or

$$\lim_{n \to \infty} P\left\{\bigcap_{j \geq n} |X_j - X| \leq \varepsilon\right\} = 1 \tag{4.71}$$

The results given above are summarized in the following theorem.

Theorem 4.2.8. Equations (4.65), (4.68), and (4.71) are all necessary and sufficient conditions for the convergence a.s. of the sequence $\{X_j\}$ to the r.v. X as $j \to \infty$.

We can now easily give the following results relating convergence a.s. to convergence i.p.

Theorem 4.2.9. Convergence almost surely implies convergence in probability.

Proof. Clearly, for every $\varepsilon > 0$,

$$\{| X_n - X | > \varepsilon\} \subset \left\{\sup_{j \geq n} | X_j - X | > \varepsilon\right\} \tag{4.72}$$

Hence

$$P\{| X_n - X | > \varepsilon\} \leq P\left\{\sup_{j \geq n} | X_j - X | > \varepsilon\right\} \tag{4.73}$$

Equation (4.68) immediately leads to

$$\lim_{n \to \infty} P\{| X_n - X | > \varepsilon\} = 0 \tag{4.74}$$

which completes the proof.

Theorem 4.2.10. Limits in convergence a.s. are unique in the sense of equivalence.

Proof. We have shown that convergence a.s. implies convergence i.p. and that limits of convergence i.p. are unique. It follows that limits in convergence a.s. are also unique in the same sense. Again we point out that the limiting r.v. X, if it exists, is the same for both modes of convergence.

We note that Theorem 4.2.2 on the uniqueness of limits in m.s. convergence can also be proven in the same manner.

The relationship between convergence a.s. and convergence i.p. has been explored. The next question one naturally asks is whether convergence a.s. can be related to m.s. convergence in any way. We shall show by means of examples that convergence a.s. does not imply m.s. convergence, nor vice versa.

Example 4.6. Consider again the sequence $\{X_n\}$ in Example 4.2. We have shown that the $\{X_n\}$ converges m.s. to zero identically. Let us now show that $\{X_n\}$ does not converge a.s. to zero.

We see that

$$\begin{aligned}
\lim_{n \to \infty} P\left\{\bigcap_{j \geq n} | X_j - X | \leq \varepsilon\right\} &= \lim_{n \to \infty} P\left\{\bigcap_{j \geq n} X_j = 0\right\} \\
&= \lim_{n \to \infty} (1 - 1/n)(1 - 1/(n+1)) \cdots \\
&= \lim_{n \to \infty} \prod_{j=0}^{\infty} (1 - 1/(n+j)) \\
&= \lim_{n \to \infty} \exp\left[- \sum_{j=0}^{\infty} 1/(n+j)\right] = 0 \tag{4.75}
\end{aligned}$$

According to Eq. (4.71), the sequence thus fails to converge a.s. to zero. Hence, m.s. convergence *does not* imply convergence a.s. This example also serves to show that the converse of Theorem 4.2.9 is not true.

Example 4.7. Let us take another look at the sequence $\{X_n\}$ in Example 4.3.

Using Eq. (4.65) as the convergence a.s. criterion, we have, for $X \equiv 0$,

$$\lim_{n \to \infty} P\left\{\bigcup_{j \geq n} |X_j - X| > \varepsilon\right\} = \lim_{n \to \infty} P\left\{\bigcup_{j \geq n} X_j = j\right\} \leq \lim_{n \to \infty} \sum_{j \geq n} 1/j^2 = 0 \quad (4.76)$$

It thus converges to zero almost surely. As the sequence fails to convergence in m.s., we have shown also that convergence a.s. *does not* imply m.s. convergence.

The reason that the sequence in Example 4.7 converges a.s. and the one in Example 4.6 does not is quite simple. In the first instance the convergence depends upon the property of the sum $\sum 1/n$, which is divergent, where the second, a convergent sum $\sum 1/n^2$.

Finally, we mention that the sequence in Example 4.1 is also convergent a.s. The proof is lengthy and will be omitted here (see Doob [5, pp. 122–128]). This result is referred to as the *"strong" law of large numbers.*

4.2.4. Convergence in Distribution

Let us digress from our treatment of the first three modes of convergence and discuss a related situation. Consider a sequence $\{X_n\}$ of r.v.'s which converges to a r.v. X as $n \to \infty$ in some sense. Since all three modes of convergence deal with probability statements concerning the sequence $\{X_n\}$ and X, one may ask if they are related in any way to a condition related to their associated distribution functions in the form

$$\lim_{n \to \infty} F_{X_n}(x) = F_X(x) \quad (4.77)$$

at every continuity point of $F_X(x)$. An application of this mode of convergence is found in the statement of the important central limit theorem in Section 2.5.

The next theorem shall show that the convergence criterion (4.77) or *convergence in distribution* is weaker than convergence i.p., and hence the weakest of all four convergence criteria.

Theorem 4.2.11. Convergence in probability implies convergence in distribution.

Proof. We start from the probability relation

$$F_{X_n}(x) = P\{X_n \leq x\}$$
$$= P\{(X_n \leq x \cap X \leq x + \varepsilon) \cup (X_n \leq x \cap X > x + \varepsilon)\}$$
$$= P\{X_n \leq x \cap X \leq x + \varepsilon\} + P\{X_n \leq x \cap X > x + \varepsilon\} \quad (4.78)$$

$$F_X(x + \varepsilon) = P\{X \leq x + \varepsilon\}$$
$$= P\{(X_n \leq x \cap X \leq x + \varepsilon) \cup (X_n > x \cap X \leq x + \varepsilon)\}$$
$$= P\{X_n \leq x \cap X \leq x + \varepsilon\} + P\{X_n > x \cap X \leq x + \varepsilon\} \quad (4.79)$$

for every $\varepsilon > 0$. Subtracting Eq. (4.79) from Eq. (4.78) gives

$$F_{X_n}(x) - F_X(x + \varepsilon) = P\{X_n \leq x \cap X > x + \varepsilon\} - P\{X_n > x \cap X \leq x + \varepsilon\}$$
$$\leq P\{X_n \leq x \cap X > x + \varepsilon\} \quad (4.80)$$

Since

$$\{X_n \leq x \cap X > x + \varepsilon\} \subset \{|X_n - X| > \varepsilon\} \quad (4.81)$$

We have

$$F_{X_n}(x) \leq F_X(x + \varepsilon) + P\{|X_n - X| > \varepsilon\} \quad (4.82)$$

Following the similar steps, we also have

$$F_{X_n}(x) \geq F_x(x - \varepsilon) - P\{|X_n - X| > \varepsilon\} \quad (4.83)$$

By the hypothesis,

$$\lim_{n \to \infty} P\{|X_n - X| > \varepsilon\} = 0 \quad (4.84)$$

The substitution of Eq. (4.84) into Eqs. (4.82) and (4.83) thus yields

$$F_X(x - \varepsilon) \leq \lim_{n \to \infty} F_{X_n}(x) \leq F_X(x + \varepsilon) \quad (4.85)$$

for every $\varepsilon > 0$. Hence, we have

$$\lim_{n \to \infty} F_{X_n}(x) = F_X(x) \quad (4.86)$$

at every continuity point of $F_X(x)$.

It is easily shown that the converse of the above theorem is not true. A sequence of independent and identically distributed random variables, for example, converges in distribution but fails to converge in probability.

The next theorem exhibits a stronger convergence criterion in terms of distribution functions.

Theorem 4.2.12. A sequence $\{X_n\}$ of r.v.'s convergences i.p. to a random variable X as $n \to \infty$ if, and only if, the probability distribution function $F_{Y_n}(y)$ of $Y_n = X_n - X$ tends to that of $Y \equiv 0$ as $n \to \infty$.

Proof. For every $\varepsilon > 0$, we have

$$P\{|X_n - X| > \varepsilon\} = 1 - P\{|X_n - X| \le \varepsilon\}$$
$$= 1 - [F_{Y_n}(\varepsilon) - F_{Y_n}(-\varepsilon)] \qquad (4.87)$$

Hence, $X_n \xrightarrow{\text{i.p.}} X$ if, and only if,

$$\lim_{n \to \infty} [F_{Y_n}(\varepsilon) - F_{Y_n}(-\varepsilon)] = 1 \qquad (4.88)$$

for every $\varepsilon > 0$. In view of the properties of probability distribution functions, we must have

$$\lim_{n \to \infty} F_{Y_n}(y) = u(y) \qquad (4.89)$$

where $u(y)$ is a unit step function, and the theorem is proved.

Fig. 4.1. Relationships of four modes of convergence.

In summary, the way in which the four modes of convergence relate to each other is indicated in Fig. 4.1. This ranking can be justified in part by the distinctive features that, given a sequence of random variables, convergence in distribution needs only a knowledge of the *univariate* distributions, convergence m.s. and i.p. need only a knowledge of the *bivariate* distributions, and convergence a.s. needs a knowledge of all *multivariate* distributions.

4.3. Mean Square Continuity

The concept of mean square convergence can be extended from a random sequence to a second-order s.p. $X(t)$ where t is continuous over a finite interval. As in the analysis of ordinary functions, this extension leads to the notion of continuity and, in this case, *continuity in mean square*. Before doing so, let us recapitulate some of the properties of second-order stochastic processes which are useful in what follows. These properties can be deduced directly from that of second-order random variables outlined in Section 4.1.

(1) If $X(t)$, $t \in T$, is second-order, its expectation $E\{X(t)\}$ is finite on T. Consider a new s.p. $Y(t)$, $t \in T$, defined by

$$Y(t) = X(t) - E\{X(t)\} \tag{4.90}$$

It follows that $Y(t)$ is also second-order and $E\{Y(t)\} = 0$. Hence, there is no loss of generality in assuming that the mean values of second-order s.p.'s are zero. This we shall do in this section unless the contrary is stated.

(2) $X(t)$, $t \in T$, is a second-order s.p. if, and only if, its correlation function $\Gamma(t, s)$ exists and is finite on $T \times T$. The "only if" part of this statement follows from Eq. (4.5). For the "if" part, the existence and finiteness of $\Gamma(t, s)$ on $T \times T$ imply

$$\Gamma(t, t) = E\{X^2(t)\} < \infty \tag{4.91}$$

on T.

Hence, second-order stochastic processes possess correlation functions. The correlation or covariance function determines the second-order properties of a second-order s.p.

We need the following two theorems for the development of mean square continuity properties.

Theorem 4.3.1. Assume that n and n' vary over some index set N, and that n_0 and n_0' are limit points of N. Let X_n and $X_{n'}'$ be two sequences of second-order r.v.'s. If

$$\underset{n \to n_0}{\text{l.i.m.}} \ X_n = X \qquad \text{and} \qquad \underset{n' \to n_0'}{\text{l.i.m.}} \ X_{n'}' = X' \tag{4.92}$$

then

$$\underset{\substack{n \to n_0 \\ n' \to n_0'}}{\lim} \ E\{X_n X_{n'}'\} = E\{XX'\} \tag{4.93}$$

Proof. Since L_2-spaces are linear, the r.v.'s $\Delta X = X_n - X$ and $\Delta X' = X_{n'}' - X'$ are second-order. We write

$$| E\{X_n X_{n'}' - XX'\} | = | E\{X \Delta X'\} + E\{X' \Delta X\} + E\{\Delta X \Delta X'\} |$$

$$\leq \| X \| \cdot \| \Delta X' \| + \| X' \| \cdot \| \Delta X \| + \| \Delta X \| \cdot \| \Delta X' \| \tag{4.94}$$

where the inequality follows from the Schwarz inequality. By hypothesis,

$$\lim_{n \to n_0} \| \varDelta X \| = 0 \quad \text{and} \quad \lim_{n' \to n_0'} \| \varDelta X' \| = 0 \tag{4.95}$$

we thus have

$$\lim_{\substack{n \to n_0 \\ n' \to n_0'}} | E\{X_n X_{n'}' - XX'\} | = 0 \tag{4.96}$$

and the proof is complete.

Theorem 4.3.2 (Convergence in Mean Square Criterion). Let $\{X_n(t)\}$, $t \in T$, be a sequence of second-order s.p.'s. $\{X_n(t)\}$ converges to a second-order process $X(t)$ on T if, and only if, the functions $E\{X_n(t) X_{n'}(t)\}$ converge to a finite function on T as $n, n' \to n_0$ in any manner whatever. Then

$$\varGamma_{X_n X_n}(t, s) \to \varGamma_{XX}(t, s), \qquad n \to n_0 \tag{4.97}$$

on $T \times T$.

Proof. To establish the "if" part, consider the Cauchy mutual convergence criterion in mean square. We have, with $\varGamma(t, s) = \varGamma_{XX}(t, s)$,

$$\| X_n(t) - X_{n'}(t) \|^2 = E\{X_n{}^2(t)\} - 2E\{X_n(t) X_{n'}(t)\} + E\{X_{n'}^2(t)\}$$
$$\to \varGamma(t, t) - 2\varGamma(t, t) + \varGamma(t, t) = 0 \tag{4.98}$$

as $n, n' \to n_0$ in any manner whatever. By means of Theorem 4.2.3, the result

$$\| X_n(t) - X_{n'}(t) \| \to 0, \qquad n, n' \to n_0 \tag{4.99}$$

implies

$$\underset{n \to n_0}{\text{l.i.m.}} \; X_n(t) = X(t) \tag{4.100}$$

To establish the "only if" part, we make use of Theorem 4.3.1. Replacing X_n by $X_n(t)$, $X_{n'}'$ by $X_{n'}(s)$, X by $X(t)$, and X' by $X(s)$, Eq. (4.93) then becomes

$$E\{X_n(t) X_{n'}(s)\} \to E\{X(t) X(s)\} = \varGamma(t, s) \tag{4.101}$$

as $n, n' \to n_0$ in any manner whatever. On letting $s = t$ we have

$$E\{X_n(t) X_{n'}(t)\} \to \varGamma(t, t), \qquad n, n' \to n_0 \tag{4.102}$$

which has been shown to be finite. The "only if" part is thus established. Equation (4.97) is established by letting $n = n'$ in Eq. (4.101).

Definition. A second-order s.p. $X(t)$, $t \in T$, is *continuous in mean square* or *m.s. continuous* at a fixed t if

$$\underset{\tau \to 0}{\text{l.i.m.}}\ X(t + \tau) = X(t) \qquad (4.103)$$

for $t + \tau \in T$, or

$$\lim_{\tau \to 0} \| X(t + \tau) - X(t) \| = 0 \qquad (4.104)$$

Theorem 4.3.3 (Continuity in Mean Square Criterion). A second-order s.p. $X(t)$, $t \in T$, is m.s. continuous at t if, and only if, $\Gamma(t, s)$ is continuous at (t, t).

Proof. It follows from Theorem 4.3.2 that

$$\underset{\tau \to 0}{\text{l.i.m.}}\ X(t + \tau) = X(t) \qquad (4.105)$$

if, and only if

$$\underset{\substack{\tau \to 0 \\ \tau' \to 0}}{\lim}\ E\{X(t + \tau)\, X(t + \tau')\} = E\{X^2(t)\} \qquad (4.106)$$

This condition, that is

$$\Gamma(t + \tau, t + \tau') \to \Gamma(t, t) \qquad (4.107)$$

as $\tau, \tau' \to 0$ in any manner whatever, is just the requirement that $\Gamma(t, s)$ be continuous at (t, t).

It is important to bear in mind that, in the discussion of continuity of $\Gamma(t, s)$, it is the *joint* continuity of the pair (t, s) that is of interest.

In the case where $X(t)$ is wide-sense stationary, we have

$$\Gamma(t, s) = \Gamma(s - t) = \Gamma(\tau) \qquad (4.108)$$

Hence, a wide-sense second-order s.p. $X(t)$ in m.s. continuous at *all* $t \in T$ if, and only if, $\Gamma(\tau)$ is continuous at $\tau = 0$.

The continuity in mean square criterion is of fundamental importance because it shows that the m.s. continuity properties of a second order s.p. are determined by the *ordinary* continuity properties of its associated correlation function. This in part justifies the statement made at the beginning of this chapter, namely, the development of m.s. calculus follows in broad outline that of ordinary calculus. For example, we can easily see that the Wiener process and the Poisson process discussed in Section 3.3.4 are m.s. continuous at every finite $t \in T$. Similarly, the binary noise and the random telegraph signal considered in Section 3.2.1 are also m.s. continuous at every $t \in T$. Purely stochastic processes discussed in Section 3.3.1, however, does not enjoy this property.

The following corollary points out an interesting property associated with the correlation function.

Corollary. If a correlation function $\Gamma(t, s)$ on $T \times T$ is continuous at every $(t, t) \in T \times T$, then it is continuous on $T \times T$; in other words, let the square in Fig. 4.2 represent the region $T \times T$, it states that $\Gamma(t, s)$ is continuous everywhere on the square if it is continuous on the diagonal line $0a$.

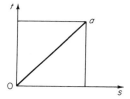

Fig. 4.2. The continuity region of $\Gamma(t, s)$.

Proof. If $\Gamma(t, s)$ is continuous at every $(t, t) \in T \times T$, Theorem 4.3.3 states that the associated s.p. $X(t)$ is m.s. continuous at every $t \in T$. Hence, for $t, t + \tau, s, s + \tau' \in T$,

$$\underset{\tau \to 0}{\text{l.i.m.}} \ X(t + \tau) = X(t) \tag{4.109}$$

$$\underset{\tau' \to 0}{\text{l.i.m.}} \ X(s + \tau') = X(s) \tag{4.110}$$

It is thus seen from Theorem 4.3.1 that

$$\lim_{\tau, \tau' \to 0} E\{X(t + \tau) X(s + \tau')\} = E\{X(t) X(s)\} \tag{4.111}$$

Equation (4.111) is the requirement that $\Gamma(t, s)$ be continuous at all $(t, s) \in T \times T$.

Definition. If a second-order s.p. $X(t)$, $t \in T$ is m.s. continuous at *every* $t \in [t_1, t_2] \subset T$, then $X(t)$ is *m.s. continuous on the interval* $[t_1, t_2]$.

We clearly have the following corollary.

Corollary. A second-order s.p. $X(t)$, $t \in T$, is m.s. continuous on an interval $[t_1, t_2] \subset T$ if, and only if, $\Gamma(t, s)$ is continuous at (t, t) for *every* $t \in [t_1, t_2]$.

In closing, let us remark that m.s. continuity does not imply continuity properties at the sample function level. We mentioned, for example, that the Poisson process is m.s. continuous for every $t \in T$, but we see from Fig. 3.7 that almost all of its sample functions have discontinuities over a finite interval.

4.4. Mean Square Differentiation

The concept of mean square differentiation follows naturally from that of mean square continuity.

Definition. A second-order s.p. $X(t)$, $t \in T$, has a *mean square derivative* or *m.s. derivative* $\dot{X}(t)$ at t if

$$\text{l.i.m.} \atop \tau \to 0 \ [X(t + \tau) - X(t)]/\tau = \dot{X}(t) \tag{4.112}$$

Higher order m.s. derivatives are defined analogously.

Theorem 4.4.1 (Differentiation in Mean Square Criterion). A second-order s.p. $X(t)$, $t \in T$, is m.s. differentiable at t if, and only if, the second generalized derivative

$$\lim_{\tau, \tau' \to 0} (1/\tau\tau') \, \Delta_\tau \, \Delta_{\tau'} \, \Gamma(t, s) = \lim_{\tau, \tau' \to 0} (1/\tau\tau')[\Gamma(t + \tau, s + \tau') - \Gamma(t + \tau, s)$$
$$-\Gamma(t, s + \tau') + \Gamma(t, s)] \tag{4.113}$$

exists at (t, t) and is finite.

Proof. This theorem is the immediate result of the convergence in mean square criterion if we let

$$X_n(t) = [X(t + n) - X(t)]/n, \qquad n = \tau, \quad n' = \tau', \quad n_0 = 0 \tag{4.114}$$

in Theorem 4.3.2.

We remark that the existence of the second generalized derivative is not equivalent to the existence of the second partial derivative as partial derivatives do not require *joint* continuity of $\Gamma(t, s)$ in (t, s). However, if the second generalized derivative exists, then the first and second partial derivatives exist. On the other hand, if the partial derivatives up to the second order do not exist, then the second generalized derivative does not exist.

The following example shows that the existence of the second partial derivative does not ensure m.s. differentiability.

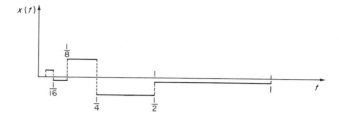

Fig. 4.3. A sample function of $X(t)$.

Example 4.8. Let a second-order s.p. $X(t)$, $t \in T = [0, 1]$, be defined by

$$X(0) = 0 \quad \text{with probability one} \tag{4.115}$$

and

$$X(t) = Y_j, \quad 1/2^j < t \le 1/(2^{j-1}), \quad j = 1, 2, \ldots \tag{4.116}$$

where the r.v.'s Y_j are mutually independent and identically distributed with means zero and variances one. A typical sample function has the appearance shown in Fig. 4.3.

The correlation function $\Gamma(t, s)$ of $X(t)$ has the value one on the semi-closed squares as shown in Fig. 4.4 and zero elsewhere.

Fig. 4.4. The correlation function $\Gamma(t, s)$.

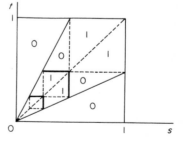

It is easily seen that

$$\lim_{\tau \to 0}(1/\tau^2) \, \Delta_\tau \Delta_\tau \Gamma(0, 0) = \lim_{\tau \to 0} 1/\tau^2 \to \infty \tag{4.117}$$

Hence, $X(t)$ *does not* have a m.s. derivative at $t = 0$ (it is not even m.s. continuous at $t = 0$). However, the second order partial derivative exists at $(0, 0)$ and is equal to zero.

Example 4.9. Let a s.p. $X(t)$, $t \in T$, be defined by

$$X(t) = At \tag{4.118}$$

where A is a second-order r.v. with mean zero and variance σ^2. The s.p. $X(t)$ is second-order for finite t with

$$\Gamma(t, s) = \sigma^2 ts \tag{4.119}$$

The second generalized derivative of $\Gamma(t, s)$ is

$$\lim_{\tau, \tau' \to 0} \Delta_\tau \Delta_{\tau'} \Gamma(t, s)$$
$$= \lim_{\tau, \tau' \to 0} (\sigma^2/\tau\tau')[(t + \tau)(s + \tau') - (t + \tau)s - t(s + \tau') + ts]$$
$$= \lim_{\tau, \tau' \to 0} (\sigma^2/\tau\tau')[\tau\tau'] = \sigma^2 < \infty \tag{4.120}$$

Hence, $X(t)$ is m.s. differentiable at every finite t.

In terms of mean square differentiability, we see from Eq. (3.131) that the covariance function of the Poisson process does not possess second-order partial derivatives. It is thus not m.s. differentiable anywhere. Similarly, we see that the Wiener process is also nondifferentiable in the mean square sense. It is instructive to note that, while both the Poisson and the Wiener processes are not m.s. differentiable, the sample functions of the Poisson process are differentiable in the ordinary sense in any finite interval except at isolated jumps; whereas the sample functions of the Wiener process are nondifferentiable anywhere with probability one.

Definition. If a second-order s.p. $X(t)$, $t \in T$, is m.s. differentiable at *every* $t \in [t_1, t_2] \subset T$, then $X(t)$ is *m.s. differentiable on the interval* $[t_1, t_2]$.

We see from Theorem 4.4.1 that $X(t)$ is m.s. differentiable on $[t_1, t_2]$ if, and only if, the second generalized derivative (4.113) exists and is finite at *every* $(t, t) \in [t_1, t_2]$.

It is noted that, if the s.p. $X(t)$ is also wide-sense stationary, $\Gamma(t, s)$ is only a function of $\tau = s - t$ and the process $X(t)$, $t \in T$, is m.s. differentiable everywhere if, and only if, $X(t)$ is m.s. differentiable at any t.

As a direct consequence of Theorem 4.4.1, the differentiation in mean square criterion for wide sense stationary processes can be stated as follows.

Corollary. A wide-sense stationary second-order process $X(t)$, $t \in T$, is m.s. differentiable if, and only if, the first- and the second-order derivatives of $\Gamma(\tau)$ exist and are finite at $\tau = 0$.

Example 4.10. Consider the s.p.

$$X(t) = a \cos(\omega t + \Phi) \tag{4.121}$$

where Φ is uniformly distributed in the interval $[0, 2\pi]$. The correlation function of $X(t)$ is

$$\Gamma(\tau) = \tfrac{1}{2}a^2 \cos \omega\tau \qquad (4.122)$$

which is differentiable indefinitely. Hence, $X(t)$ has m.s. derivatives of all orders.

Example 4.11. The correlation functions of binary noise, random telegraph signal, and white noise are given by, respectively, Eqs. (3.64), (3.66), and (3.69). It is clear that their derivatives of any order do not exist at $\tau = 0$. Hence, these stochastic processes are not m.s. differentiable anywhere.

4.4.1. Properties of Mean Square Derivatives

We give below several properties associated with the mean square derivatives of a second-order s.p. $X(t)$. They are seen to be consistent with those associated with ordinary derivatives of deterministic functions. In what follows we shall use the symbols $\dot{X}(t)$ and $dX(t)/dt$ interchangeably to denote the m.s. derivative of $X(t)$. The same symbols, when used in connection with ordinary functions, are to be interpreted as ordinary derivatives.

(1) Mean square differentiability of $X(t)$ at $t \in T$ implies m.s. continuity of $X(t)$ at t, since if $t + \tau \in T$

$$\lim_{\tau \to 0} \| X(t + \tau) - X(t) \|^2 = \lim_{\tau \to 0} | \tau |^2 \cdot \| [X(t + \tau) - X(t)]/\tau \|^2$$

$$= 0 \cdot \lim_{\tau \to 0}(1/\tau^2)\, \Delta\Delta\Gamma(t, t) \qquad (4.123)$$

The last quantity is zero because the second generalized derivative is finite by hypothesis.

(2) The m.s. derivative $\dot{X}(t)$ of $X(t)$ at $t \in T$, if it exists, is unique. This property follows directly from the m.s. convergence uniqueness property.

(3) If $X(t)$ and $Y(t)$ are m.s. differentiable at $t \in T$, then the m.s. derivative of $aX(t) + bY(t)$ exists at t and

$$\frac{d}{dt}\,[aX(t) + bY(t)] = a\dot{X}(t) + b\dot{Y}(t) \qquad (4.124)$$

where a and b are constants. The proof is immediate as, by the norm property (4.13),

$$\left\| \frac{aX(t+\tau) + bY(t+\tau) - aX(t) - bY(t)}{\tau} - a\dot{X}(t) - b\dot{Y}(t) \right\|$$

$$\leq \left\| a\left[\frac{X(t+\tau) - X(t)}{\tau} - \dot{X}(t) \right] \right\| + \left\| b\left[\frac{Y(t+\tau) - Y(t)}{\tau} - \dot{Y}(t) \right] \right\|$$

(4.125)

By hypothesis, the last two terms tend to zero as $\tau \to 0$.

(4) If an ordinary function $f(t)$ is differentiable at $t \in T$ and $X(t)$ is m.s. differentiable at $t \in T$, then $f(t) X(t)$ is m.s. differentiable at t and

$$\frac{d}{dt}\left[f(t) X(t) \right] = \frac{df(t)}{dt} X(t) + f(t) \frac{dX(t)}{dt}$$

(4.126)

To show that this property is true, we again use the norm property (4.13) and consider

$$\left\| \frac{f(t+\tau) X(t+\tau) - f(t) X(t)}{\tau} - \frac{df(t)}{dt} X(t) - f(t) \frac{dX(t)}{dt} \right\|$$

$$\leq \left\| \frac{f(t+\tau) X(t+\tau) - f(t) X(t+\tau)}{\tau} - \frac{df(t)}{dt} X(t) \right\|$$

$$+ \left\| \frac{f(t) X(t+\tau) - f(t) X(t)}{\tau} - f(t) \frac{dX(t)}{dt} \right\|$$

$$\leq \left\| \left[\frac{f(t+\tau) - f(t)}{\tau} - \frac{df(t)}{dt} \right] X(t+\tau) \right\|$$

$$+ \left\| \frac{df(t)}{dt} \left[X(t+\tau) - X(t) \right] \right\|$$

$$+ \left\| f(t)\left[\frac{X(t+\tau) - X(t)}{\tau} - \frac{dX(t)}{dt} \right] \right\|$$

$$\leq \left| \frac{f(t+\tau) - f(t)}{\tau} - \frac{df(t)}{dt} \right| \cdot \| X(t+\tau) \|$$

$$+ \left| \frac{df(t)}{dt} \right| \cdot \| X(t+\tau) - X(t) \|$$

$$+ | f(t) | \cdot \left\| \frac{X(t+\tau) - X(t)}{\tau} - \frac{dX(t)}{dt} \right\|$$

(4.127)

The first and third terms of the last expression in Eq. (4.127) tend to zero as $\tau \to 0$ by the hypothesis. The second term tends to zero as $\tau \to 0$ because of Property (1). Equation (4.126) is thus established. Same procedure can be employed to establish various extensions of this property.

4.4.2. Means and Correlation Functions of Mean Square Derivatives

We have seen that the m.s. derivative of $X(t)$ and higher order ones if they exist are stochastic processes. We shall show below that their correlation functions and cross-correlation functions are determined simply in terms of the correlation function of $X(t)$. Let us first give an elementary result concerning the means of the m.s. derivatives of $X(t)$.

It is easily shown that, if $X(t)$, $t \in T$, is n times m.s. differentiable on T, the means of these m.s. derivatives of $X(t)$ exist on T and they are given by

$$E\left\{\frac{d^n X(t)}{dt^n}\right\} = \frac{d^n}{dt^n} E\{X(t)\} \tag{4.128}$$

To establish Eq. (4.128), let us first consider the case when $n = 1$. By definition, we see that

$$E\{\dot{X}(t)\} = E\left\{\underset{\tau \to 0}{\text{l.i.m.}} \left[\frac{X(t + \tau) - X(t)}{\tau}\right]\right\}$$
$$= \lim_{\tau \to 0}\left[\frac{E\{X(t + \tau)\} - E\{X(t)\}}{\tau}\right] = \frac{d}{dt} E\{X(t)\} \tag{4.129}$$

The second step in the derivation of Eq. (4.129) follows from the interchangeability of the symbols "E" and "l.i.m." as established in Theorem 4.2.1.

The relation (4.128) for higher-order derivatives can be established by the repeated use of Eq. (4.129).

Theorem 4.4.2. If the second generalized derivative (4.113) exists at (t, t) for *every* $t \in T$, then the partial derivatives $\partial \Gamma(t, s)/\partial t$, $\partial \Gamma(t, s)/\partial s$, $\partial^2 \Gamma(t, s)/\partial t \, \partial s$ exist and are finite on $T \times T$.

Proof. According to Theorem 4.4.1, the m.s. derivative $\dot{X}(t)$ of $X(t)$ exists at all t. Hence, $E\{\dot{X}(t) X(s)\}$ exists and is finite on $T \times T$ with

$$E\{\dot{X}(t) X(s)\} = E\left\{\underset{\tau \to 0}{\text{l.i.m.}} \left[\frac{X(t + \tau) - X(t)}{\tau}\right] X(s)\right\}$$
$$= \lim_{\tau \to 0}(1/\tau) E\{X(t + \tau) X(s) - X(t) X(s)\}$$
$$= \lim_{\tau \to 0}(1/\tau)[\Gamma(t + \tau, s) - \Gamma(t, s)] = \partial \Gamma(t, s)/\partial t \tag{4.130}$$

where again we have invoked Theorem 4.2.1.

Similarly, we establish that

$$E\{X(t)\,\dot{X}(s)\} = \partial\Gamma(t, s)/\partial s \qquad (4.131)$$

exists and is finite on $T \times T$.

Finally, the existence and finiteness of $E\{\dot{X}(t)\,\dot{X}(s)\}$ imply that

$$
\begin{aligned}
E\{\dot{X}(t)\,\dot{X}(s)\} &= E\left\{\operatorname*{l.i.m.}_{\tau,\tau'\to 0}\left[\frac{X(t+\tau)-X(t)}{\tau}\right]\left[\frac{X(s+\tau')-X(s)}{\tau'}\right]\right\} \\
&= \lim_{\tau,\tau'\to 0}\frac{1}{\tau}\,E\left\{\left[\frac{X(t+\tau)\,X(s+\tau')-X(t+\tau)\,X(s)}{\tau'}\right]\right. \\
&\qquad\qquad \left. -\left[\frac{X(t)\,X(s+\tau')-X(t)\,X(s)}{\tau'}\right]\right\} \\
&= \lim_{\tau\to 0}\frac{1}{\tau}\,\lim_{\tau'\to 0}\left\{\frac{\Gamma(t+\tau,s+\tau')-\Gamma(t+\tau,s)}{\tau'}\right. \\
&\qquad\qquad \left. -\frac{\Gamma(t,s+\tau')-\Gamma(t,s)}{\tau'}\right\} \\
&= \lim_{\tau\to 0}\frac{1}{\tau}\left[\frac{\partial\Gamma(t+\tau,s)}{\partial s}-\frac{\partial\Gamma(t,s)}{\partial s}\right] = \frac{\partial^2\Gamma(t,s)}{\partial t\,\partial s} \qquad (4.132)
\end{aligned}
$$

exists and is finite on $T \times T$.

Generalizations of the results given above are straightforward. Let the cross-correlation function of $d^n X(t)/dt^n$ and $d^m X(t)/dt^m$ be denoted by $\Gamma_{X(n)X(m)}(t, s)$, we have the general formula

$$\Gamma_{X(n)X(m)}(t, s) = \frac{\partial^{n+m}\Gamma(t, s)}{\partial t^n\,\partial s^m} \qquad (4.133)$$

if the indicated m.s. derivatives exist.

In the case where $X(t)$ is also wide-sense stationary, Eq. (4.133) reduces to

$$\Gamma_{X(n)X(m)}(t, s) = \Gamma_{X(n)X(m)}(\tau) = (-1)^n\frac{d^{n+m}\Gamma(\tau)}{d\tau^{n+m}}, \qquad \tau = s - t \qquad (4.134)$$

Furthermore, using Eq. (3.61), Eq. (4.134) leads to

$$S_{X(n)X(m)}(\omega) = (i\omega)^n(i\omega)^m S(\omega) \qquad (4.135)$$

It relates the cross-power spectral density of $d^n X(t)/dt^n$ and $d^m X(t)/dt^m$ to the power spectral density $S(\omega)$ of $X(t)$.

4.4.3. Mean Square Analyticity

We remark that the concept of mean square differentiation leads to the notion of mean square Taylor series expansion of a s.p. and the definition of mean square analyticity.

Definition. A second-order s.p. $X(t)$, $t \in T$, is *m.s. analytic* on T if it can be expanded in the m.s. convergent Taylor series

$$X(t) = \sum_{n=0}^{\infty} (t - t_0)^n X^{(n)}(t_0)/n!, \qquad t, t_0 \in T \qquad (4.136)$$

In connection with the m.s. analyticity, we state without proof the following theorem (Loéve [1, p. 471]).

Theorem 4.4.3. A second-order s.p. $X(t)$, $t \in T$ is m.s. analytic on T if, and only if, its correlation function $\Gamma(t, s)$ is analytic at (t, t) for *every* $t \in T$, and then $\Gamma(t, s)$ is analytic on $T \times T$.

We thus see that the analytic properties in the mean square sense of a stochastic process are also determined by the ordinary analytic properties of its correlation function.

4.5. Mean Square Integration

The development of mean square integrals of stochastic processes is more involved. We shall consider mainly one type of integrals which directly interests us, namely, the mean square Riemann integral. Mean square Riemann–Stieltjes integrals will be treated in Section 4.5.3 in a somewhat cursory manner.

Consider a collection of all finite partitions $\{p_n\}$ of an interval $[a, b]$. The partition p_n is defined by the subdivision points t_k, $k = 0, 1, \ldots, n$, such that

$$a = t_0 < t_1 < t_2 < \cdots < t_n = b \qquad (4.137)$$

Let

$$\Delta_n = \max_k (t_k - t_{k-1})$$

and let t_k' be an arbitrary point in the interval $[t_{k-1}, t_k)$.

Let $X(t)$ be a second-order s.p. defined on $[a, b] \subset T$. Let $f(t, u)$ be an ordinary function defined on the same interval for t and Riemann integrable for every $u \in U$. We form the random variable

$$Y_n(u) = \sum_{k=1}^{n} f(t_k', u) X(t_k')(t_k - t_{k-1}) \tag{4.138}$$

Since L_2-space is linear, $Y_n(u)$ is an element of the L_2-space. It is a random variable defined for each partition p_n and for each $u \in U$.

Definition. If, for $u \in U$,

$$\underset{\substack{n \to \infty \\ \Delta_n \to 0}}{\text{l.i.m.}}\, Y_n(u) = Y(u) \tag{4.139}$$

exists for some sequence of subdivisions p_n, the s.p. $Y(u)$, $u \in U$, is called the *mean square Riemann integral* or *m.s. Riemann integral* of $f(t, u) X(t)$ over the interval $[a, b]$, and it is denoted by

$$Y(u) = \int_a^b f(t, u) X(t)\, dt \tag{4.140}$$

It is independent of the sequence of subdivisions as well as the positions of $t_k' \in [t_{k-1}, t_k)$.

Theorem 4.5.1 (Integration in Mean Square Criterion). The s.p. $Y(u)$, $u \in U$, defined by Eq. (4.140) exists if, and only if, the ordinary double Riemann integral

$$\int_a^b \int_a^b f(t, u) f(s, u)\, \Gamma_{XX}(t, s)\, dt\, ds \tag{4.141}$$

exists and is finite.

Proof. This theorem is an immediate result of the convergence in mean square criterion if we let

$$X_n(t) = \sum_{k=1}^{n} f(t_k', t) X(t_k')(t_k - t_{k-1}), \qquad t = u, n_0 = \infty \tag{4.142}$$

in Theorem 4.3.2.

Improper m.s. Riemann integrals are defined in the same manner. An improper m.s. Riemann integral, say, $\int_a^\infty f(t, u) X(t)\, dt$, is defined by

$$\int_a^\infty f(t, u) X(t)\, dt = \underset{b \to \infty}{\text{l.i.m.}} \int_a^b f(t, u) X(t)\, dt \tag{4.143}$$

It is easily seen that it exists if, and only if, the improper ordinary double

Riemann integral

$$\int_a^\infty \int_a^\infty f(t, u) f(s, u) \, \Gamma_{XX}(t, s) \, dt \, ds \qquad (4.144)$$

exists and is finite.

We note again the desirable features associated with the m.s. calculus. The m.s. Riemann integral properties of a stochastic process are determined by the ordinary Riemann integral properties of its correlation function. For example, on the basis of Theorem 4.5.1, it is easily shown that the Wiener process and the Poisson process are m.s. integrable over any finite interval. The binary noise and the random telegraph signal are also m.s. integrable. As an illustration, let us carry out the specific steps for the Wiener process.

Example 4.12. Consider the m.s. integral

$$Y(u) = \int_0^u X(t) \, dt \qquad (4.145)$$

where $X(t), t \geq 0$, is the Wiener process with (Prob. 3.18)

$$\Gamma_{XX}(t, s) = 2D \min(t, s) \qquad (4.146)$$

We have

$$\int_0^u \int_0^u \Gamma_{XX}(t, s) \, dt \, ds = \int_0^u \int_0^u 2D \min(t, s) \, dt \, ds$$

$$= 2D \int_0^u ds \left[\int_0^s t \, dt + \int_s^u s \, dt \right] = 2Du^3/3 \qquad (4.147)$$

The double integral thus exists and is finite for all $u < \infty$. Hence, $Y(u)$ of Eq. (4.145) exists for all finite u.

4.5.1. Properties of Mean Square Riemann Integrals

It is shown below that, as in the case of m.s. differentiation, many useful rules in ordinary integration are valid within the framework of m.s. integration. In what follows, it is understood that the integral sign stands for either a m.s. integral or an ordinary integral depending upon whether the integrand is random or deterministic.

(1) Mean square continuity of $X(t)$ on $[a, b]$ implies m.s. Riemann integrability of $X(t)$ on $[a, b]$. We see from Theorem 4.3.3 and its corollary that, if $X(t)$ is m.s. continuous on $[a, b]$, its correlation function $\Gamma_{XX}(t, s)$

is continuous on $[a, b] \times [a, b]$. Hence, existence theorems on ordinary integrals state that

$$\int_a^b \int_a^b \Gamma_{XX}(t, s) \, dt \, ds \qquad (4.148)$$

exists. It is also finite as $X(t)$ is of second order. Hence, in view of Theorem 4.5.1, property (1) is established.

(2) The m.s. integral of $X(t)$ on an interval $[a, b]$, if it exists, is unique. This property follows immediately from the m.s. convergence uniqueness property.

(3) If $X(t)$ is m.s. continuous on $[a, b]$, then

$$\left\| \int_a^b X(t) \, dt \right\| \leq \int_a^b \| X(t) \| \, dt \leq M(b - a) \qquad (4.149)$$

where

$$M = \max_{t \in [a,b]} \| X(t) \| \qquad (4.150)$$

Proof. It follows from Property (1) above that the first integral in Eq. (4.149) exists. The second integral exists because $\| X(t) \|$ is a real-valued, continuous function of $t \in [a, b]$ (Theorem 4.3.3).

Now

$$\| Y_n \| = \left\| \sum_{k=1}^{n} X(t_k')(t_k - t_{k-1}) \right\| \rightarrow \left\| \int_a^b X(t) \, dt \right\| \qquad (4.151)$$

as $n \rightarrow \infty$ and $\Delta_n \rightarrow 0$. We have, from the norm property (4.13),

$$\| Y_n \| \leq \sum_{k=1}^{n} \| X(t_k') \| (t_k - t_{k-1}) \rightarrow \int_a^b \| X(t) \| \, dt, \qquad (4.152)$$

The above sum also gives

$$\sum_{k=1}^{n} \| X(t_k') \| (t_k - t_{k-1}) \leq M \sum_{k=1}^{n} (t_k - t_{k-1}) = M(b - a) \qquad (4.153)$$

Equations (4.151) and (4.152) yield the first inequality; Equations (4.152) and (4.153) give the second.

(4) If the m.s. integrals of $X(t)$ and $Y(t)$ exist on $[a, c]$, then

$$\int_a^c [\alpha X(t) + \beta Y(t)] \, dt = \alpha \int_a^c X(t) \, dt + \beta \int_a^c Y(t) \, dt \qquad (4.154)$$

$$\int_a^c X(t) \, dt = \int_a^b X(t) \, dt + \int_b^c X(t) \, dt, \qquad a \leq b \leq c \qquad (4.155)$$

The procedure for establishing these two relations follows the same steps given in Eq. (4.125) in establishing the corresponding m.s. derivative property.

(5) If $X(t)$ is m.s. continuous on $[a, t] \subset T$, then

$$Y(t) = \int_a^t X(s) \, ds \tag{4.156}$$

is m.s. continuous on T; it is also m.s. differentiable on T with

$$\dot{Y}(t) = X(t) \tag{4.157}$$

Proof. We only need to prove the second part on differentiability since, according to Property (1) in Section 4.4.1, it implies m.s. continuity.

Consider

$$\left\| (1/\tau) \left[\int_a^{t+\tau} X(s) \, ds - \int_a^t X(s) \, ds \right] - X(t) \right\|$$

$$= \left\| (1/\tau) \int_t^{t+\tau} [X(s) - X(t)] \, ds \right\|$$

$$\leq | \, 1/\tau \, | \cdot \int_t^{t+\tau} \| X(s) - X(t) \| \, ds$$

$$\leq \max_{s \in [t, t+\tau]} \| X(s) - X(t) \| \tag{4.158}$$

The last inequality follows from Property (3). The last quantity approaches zero as $\tau \to 0$ from hypothesis, and the proof is complete.

We state without proof two important corollaries to the above result. They are, respectively, the m.s. counterparts of the Leibniz rule and integration by parts in ordinary integral calculus.

Corollary (Leibniz Rule). If $X(t)$ is m.s. integrable on T and if the ordinary function $f(t, s)$ is continuous on $T \times T$ with a finite first partial derivative $\partial f(t, s)/\partial t$, then the m.s. derivative of

$$Y(t) = \int_a^t f(t, s) \, X(s) \, ds \tag{4.159}$$

exists at all $t \in T$, and

$$\dot{Y}(t) = \int_a^t \frac{\partial f(t, s)}{\partial t} \, X(s) \, ds + f(t, t) \, X(t) \tag{4.160}$$

Corollary (Integration by Parts). Let $X(t)$ be m.s. differentiable on T and let the ordinary function $f(t, s)$ be continuous on $T \times T$ whose partial derivative $\partial f(t, s)/\partial s$ exists. If

$$Y(t) = \int_a^t f(t, s) \, \dot{X}(s) \, ds, \tag{4.161}$$

then

$$Y(t) = f(t, s) \, X(s) \, \Big|_a^t - \int_a^t \frac{\partial f(t, s)}{\partial s} \, X(s) \, ds \tag{4.162}$$

(6) Let $f(t, s) \equiv 1$ in Eqs. (4.161) and (4.162). We have the useful result that, if $\dot{X}(t)$ is m.s. Riemann integrable on T, then

$$X(t) - X(a) = \int_a^t \dot{X}(s) \, ds, \qquad [a, t] \in T \tag{4.163}$$

This property is seen to be m.s. counterpart of the fundamental theorem of ordinary calculus. It may be properly called the *fundamental theorem of mean square calculus.*

4.5.2. Means and Correlation Functions of Mean Square Riemann Integrals

Consider first the mean of a m.s. Riemann integral. If

$$Y(u) = \int_a^b f(t, u) \, X(t) \, dt \tag{4.164}$$

exists, it is easily shown that

$$E\{Y(u)\} = \int_a^b f(t, u) \, E\{X(t)\} \, dt \tag{4.165}$$

as, with the aid of Theorem 4.2.1,

$$E\{Y(u)\} = E\Big\{ \underset{\substack{n \to \infty \\ \Delta_n \to 0}}{\text{l.i.m.}} \sum_{k=1}^n f(t_k', u) \, X(t_k')(t_k - t_{k-1}) \Big\}$$

$$= \lim_{\substack{n \to \infty \\ \Delta_n \to 0}} \sum_{k=1}^n f(t_k', u) \, E\{X(t_k')\}(t_k - t_{k-1})$$

$$= \int_a^b f(t, u) \, E\{X(t)\} \, dt \tag{4.166}$$

The determination of the correlation function of $Y(u)$ in terms of that of $X(t)$ is also simple. In fact, if we wrote out the proof of Theorem 4.5.1, it would result that the correlation function $\Gamma_{YY}(u, v)$ of $Y(u)$ in Eq. (4.164) is given by

$$\Gamma_{YY}(u, v) = \int_a^b \int_a^b f(t, u) f(s, v) \, \Gamma_{XX}(t, s) \, dt \, ds \qquad (4.167)$$

Appropriate changes should be made when the integration limits a and b are functions of u. If a and b in Eq. (4.164) are replaced by, respectively, $a(u)$ and $b(u)$, Eq. (4.167) then takes the form

$$\Gamma_{YY}(u, v) = \int_{a(v)}^{b(v)} \int_{a(u)}^{b(u)} f(t, u) f(s, v) \, \Gamma_{XX}(t, s) \, dt \, ds \qquad (4.168)$$

Example 4.13. Let us consider again the "integrated Wiener process" in Example 4.12, and compute its correlation function. In this case,

$$Y(u) = \int_0^u X(t) \, dt \qquad (4.169)$$

and

$$\Gamma_{XX}(t, s) = 2D \min(t, s) \qquad (4.170)$$

Hence, from Eq. (4.168) we have, for $t \geq s \geq 0$,

$$\Gamma_{YY}(t, s) = 2D \int_0^s \int_0^t \min(\tau, \tau') \, d\tau \, d\tau'$$

$$= 2D \int_0^s (\tau' t - \tau'^2/2) \, d\tau' = (Ds^2/3)(3t - s) \qquad (4.171)$$

4.5.3. *Mean Square Riemann–Stieltjes Integrals*

In the development of the following chapters, there will be the need of considering stochastic Riemann–Stieltjes integrals of the types

$$V_1 = \int_a^b f(t) \, dX(t) \qquad (4.172)$$

and

$$V_2 = \int_a^b X(t) \, df(t) \qquad (4.173)$$

Riemann–Stieltjes integrals in the mean square sense can be developed parallel to the development of m.s. Riemann integrals. As expected, the existence of the m.s. integrals V_1 and V_2 hinges upon the existence of ordinary double Riemann–Stieltjes integrals involving the ordinary function $f(t)$ and the correlation function of $X(t)$. They also possess the formal properties of ordinary Riemann–Stieltjes integrals. Finally, the means and the correlation functions of these integrals can also be determined in a straightforward manner.

Because of these similarities, we will bypass many of the details and only point out some pertinent features associated with the integrals of the types represented by V_1 and V_2.

Definition. Keeping all the definitions given in the beginning of Section 4.5, let us form the random variables

$$V_{1n} = \sum_{k=1}^{n} f(t_k')[X(t_k) - X(t_{k-1})] \qquad (4.174)$$

and

$$V_{2n} = \sum_{k=1}^{n} X(t_k')[f(t_k) - f(t_{k-1})] \qquad (4.175)$$

If

$$\underset{\substack{n\to\infty \\ \Delta_n\to 0}}{\text{l.i.m.}} V_{1n} = V_1 \qquad \left[\underset{\substack{n\to\infty \\ \Delta_n\to 0}}{\text{l.i.m.}} V_{2n} = V_2\right] \qquad (4.176)$$

exists for some sequences of subdivisions p_n, the r.v. V_1 $[V_2]$ is called the *m.s. Riemann–Stieltjes integral* of $f(t)$ $[X(t)]$ on the interval $[a, b]$ with respect to $X(t)$ $[f(t)]$. It is denoted by Eq. (4.172) [Eq. (4.173)].

Theorem 4.5.2. The m.s. Riemann–Stieltjes integral V_1 $[V_2]$ exists if, and only if, the ordinary double Riemann–Stieltjes integral

$$\int_a^b \int_a^b f(t) f(s) \, dd\Gamma_{XX}(t, s) \qquad (4.177)$$

$$\left[\int_a^b \int_a^b \Gamma_{XX}(t, s) \, df(t) \, df(s)\right] \qquad (4.178)$$

exists and is finite.

Again, the proof follows directly from the convergence in mean square criterion after appropriate substitutions.

We thus see that the existence of V_1 and V_2 is determined by the existence of appropriate ordinary integrals. From the existence theorems of ordinary calculus, the double integral (4.177) exists if $\Gamma_{XX}(t, s)$ is of bounded variation on $[a, b] \times [a, b]$ and if $f(t)$ is continuous on $[a, b]$; the double integral (4.178) exists if $\Gamma_{XX}(t, s)$ is continuous on $[a, b] \times [a, b]$ and if $f(t)$ is of bounded variation on $[a, b]$.

An important as well as practical property of m.s. Riemann–Stieltjes integrals is contained in the following theorem.

Theorem 4.5.3. If either V_1 or V_2 exists, then both integrals exist, and

$$\int_a^b X(t)\,df(t) = [f(t)\,X(t)]_a^b - \int_a^b f(t)\,dX(t) \qquad (4.179)$$

Proof. Consider two partitions p_n and p'_{n+1} of $[a, b]$. The partition p_n is defined by subdivision points t_k, $k = 0, 1, \ldots, n$, such that

$$a = t_0 < t_1 < t_2 < \cdots < t_n = b$$

and p'_{n+1} by t'_j, $j = 0, 1, \ldots, n + 1$, such that

$$a = t_0' < t_1' < t_2' < \cdots < t'_{n+1} = b$$

Furthermore, let $t_k' \in [t_{k-1}, t_k)$ as indicated in Fig. 4.5. We observe that, if we let

$$\Delta_n = \max_k (t_k - t_{k-1}) \qquad \text{and} \qquad \Delta_{n+1} = \max_j (t_j' - t_{j-1}') \qquad (4.180)$$

we can write

$$2\Delta'_{n+1} \geq \Delta_n \qquad (4.181)$$

Fig. 4.5. The partitions p_n and p'_{n+1}.

Next, define

$$V'_{1n} = \sum_{j=0}^n f(t_j)[X(t'_{j+1}) - X(t_j')] \qquad (4.182)$$

and

$$V_{2n} = \sum_{k=1}^n X(t_k')[f(t_k) - f(t_{k-1})] \qquad (4.183)$$

The sum of Eqs. (4.182) and (4.183) gives

$$V'_{1n} + V_{2n} = f(t_n)\,X(t'_{n+1}) - f(t_0)\,X(t_0') \qquad (4.184)$$

or we have

$$V_{2n} = f(b)\,X(b) - f(a)\,X(a) - V'_{1n} \qquad (4.185)$$

Now, let us assume that V_1 exists, that is,

$$\underset{\substack{n\to\infty \\ \Delta'_{n+1}\to 0}}{\text{l.i.m.}}\ V'_{1n} = V_1 \qquad (4.186)$$

In view of Eq. (4.181),

$$\underset{\substack{n\to\infty \\ \Delta'_{n+1}\to 0}}{\text{l.i.m.}}\ V'_{1n} = \underset{\substack{n\to\infty \\ \Delta_n\to 0}}{\text{l.i.m.}}\ V'_{1n} \tag{4.187}$$

and, from Eqs. (4.185) and (4.186),

$$\underset{\substack{n\to\infty \\ \Delta_n\to 0}}{\text{l.i.m.}}\ V_{2n} = f(b)\, X(b) - f(a)\, X(a) - V_1 \tag{4.188}$$

which exists. Thus,

$$V_2 = f(b)\, X(b) - f(a)\, X(a) - V_1 \tag{4.189}$$

It is clear that the same relation results if we assume V_2 exists first. Hence, the proof is complete.

Equation (4.179) shows that integration by parts is valid for m.s. Riemann–Stieltjes integration, a very useful result.

Without going into details, we mention that, as expected, other properties of ordinary Riemann–Stieltjes integrals are also valid in this setting.

If V_1 and V_2 exist, the following formulas for the means and the second moments of V_1 and V_2 can be easily verified.

$$E\{V_1\} = \int_a^b f(t)\, dE\{X(t)\} \tag{4.190}$$

$$E\{V_1^2\} = \int_a^b \int_a^b f(t)\, f(s)\, dd\Gamma(t, s) \tag{4.191}$$

$$E\{V_2\} = \int_a^b E\{X(t)\}\, df(t) \tag{4.192}$$

$$E\{V_2^2\} = \int_a^b \int_a^b \Gamma(t, s)\, df(t)\, df(s) \tag{4.193}$$

4.6. Distributions of Mean Square Derivatives and Integrals

Our development of the mean square calculus is now fairly complete. Mean square derivatives and integrals of a stochastic process are defined based upon its second order properties, which in turn determine the second-order properties of its m.s. derivatives and integrals.

For practical reasons, however, it is often desirable that we develop a technique for determining the joint probability distributions of the m.s.

derivatives or of the m.s. integrals of $X(t)$ when the same information about $X(t)$ is known.

Assuming that the m.s. derivative $\dot{X}(t)$ of $X(t)$ exists, we shall give a formal relation between the joint characteristic functions of $X(t)$ and that of $\dot{X}(t)$.

Let the nth characteristic function of $\dot{X}(t)$ be denoted by $\phi_{n\dot{X}}(u_1, t_1; u_2, t_2; \ldots; u_n, t_n)$.

Theorem 4.6.1. If $\dot{X}(t)$ of $X(t)$ exists at all $t \in T$, then for every finite set $t_1, t_2, \ldots, t_n \in T$,

$$\phi_{n\dot{X}}(u_1, t_1; \ldots; u_n, t_n)$$
$$= \lim_{\tau_1, \tau_2, \ldots, \tau_n \to 0} \phi_{2nX}\left(\frac{u_1}{\tau_1}, t_1 + \tau_1; -\frac{u_1}{\tau_1}, t_1; \ldots; \frac{u_n}{\tau_n}, t_n + \tau_n; -\frac{u_n}{\tau_n}, t_n\right)$$

(4.194)

for $t_1 + \tau_1, t_2 + \tau_2, \ldots, t_n + \tau_n \in T$.

Proof. Let $n = 1$. We have

$$\phi_{\dot{X}}(u_1, t_1) = E\{\exp[iu_1 \dot{X}(t_1)]\}$$
$$= E\left\{\text{l.i.m.}_{\tau_1 \to 0} \exp\left[\frac{iu_1 X(t_1 + \tau_1)}{\tau_1} - \frac{iu_1 X(t_1)}{\tau_1}\right]\right\}$$
$$= \lim_{\tau_1 \to 0} E\left\{\exp\left[\frac{iu_1 X(t_1 + \tau_1)}{\tau_1} - \frac{iu_1 X(t_1)}{\tau_1}\right]\right\}$$
$$= \lim_{\tau_1 \to 0} \phi_{2X}\left(\frac{u_1}{\tau_1}, t_1 + \tau_1; -\frac{u_1}{\tau_1}, t_1\right)$$

(4.195)

Similarly, for $n = 2$ we have

$$\phi_{2\dot{X}}(u_1, t_1; u_2, t_2) = E\{\exp[iu_1 \dot{X}(t_1) + iu_2 \dot{X}(t_2)]\}$$
$$= E\left\{\text{l.i.m.}_{\tau_1, \tau_2 \to 0} \exp\left[iu_1\left(\frac{X(t_1 + \tau_1) - X(t_1)}{\tau_1}\right)\right.\right.$$
$$\left.\left. + iu_2\left(\frac{X(t_2 + \tau_2) - X(t_2)}{\tau_2}\right)\right]\right\}$$
$$= \lim_{\tau_1, \tau_2 \to 0} \phi_{4X}\left(\frac{u_1}{\tau_1}, t_1 + \tau_1; -\frac{u_1}{\tau_1}, t_1; \frac{u_2}{\tau_2}, t_2 + \tau_2; -\frac{u_2}{\tau_2}, t_2\right)$$

(4.196)

It is clear that Eq. (4.194) can be obtained by proceeding as above.

We observe that the nth characteristic function of $\dot{X}(t)$ is determined by the $2n$th characteristic function of $X(t)$. The limiting operations in Eq.

(4.194) are difficult to carry out in general. However, it leads to the following important result.

Theorem 4.6.2. If the m.s. derivative $\dot{X}(t)$ of a Gaussian process $X(t)$ exists, then $\dot{X}(t)$ is a Gaussian process.

Proof. Let us make the proof a little simpler by assuming that the mean of $X(t)$ is zero. Then, as seen from Eq. (3.132),

$$\phi_{nX}(u_1, t_1; u_2, t_2; \ldots; u_n, t_n) = \exp(-\tfrac{1}{2}\mathbf{u}^{\mathrm{T}}\Lambda\mathbf{u})$$

$$= \exp\left[-\tfrac{1}{2} \sum_{i,j=1}^{n} u_i u_j \Gamma_{XX}(t_i, t_j)\right] \quad (4.197)$$

The substitution of Eq. (4.197) into Eq. (4.195) gives

$$\phi_{\dot{X}}(u_1, t_1) = \lim_{\tau_1 \to 0} \phi_{2X}\left(\frac{u_1}{\tau_1}, t_1 + \tau_1; -\frac{u_1}{\tau_1}, t_1\right)$$

$$= \lim_{\tau_1 \to 0} \exp\left[-\frac{u_1^2}{2\tau_1^2} \{\Gamma_{XX}(t_1 + \tau_1, t_1 + \tau_1) - \Gamma_{XX}(t_1 + \tau_1, t_1)\right.$$

$$\left. -\Gamma_{XX}(t_1, t_1 + \tau_1) + \Gamma_{XX}(t_1, t_1)\}\right]$$

$$= \exp\left[-\tfrac{1}{2}u_1^2 \frac{\partial^2 \Gamma_{XX}(t, s)}{\partial t \, \partial s}\bigg|_{t=s=t_1}\right]$$

$$= \exp[-\tfrac{1}{2}u_1^2 \Gamma_{\dot{X}\dot{X}}(t_1, t_1)] \quad (4.198)$$

From Eq. (4.196), we obtain

$$\phi_{2\dot{X}}(u_1, t_1; u_2, t_2)$$

$$= \lim_{\tau_1, \tau_2 \to 0} \phi_{4X}\left(\frac{u_1}{\tau_1}, t_1 + \tau_1; -\frac{u_1}{\tau_1}, t_1; \frac{u_2}{\tau_2}, t_2 + \tau_2; -\frac{u_2}{\tau_2}, t_2\right)$$

$$= \exp\left[-\tfrac{1}{2} \sum_{i,j=1}^{2} u_i u_j \Gamma_{\dot{X}\dot{X}}(t_i, t_j)\right] \quad (4.199)$$

Proceeding in this fashion we find that

$$\phi_{n\dot{X}}(u_1, t_1; \ldots; u_n, t_n) = \exp\left[-\tfrac{1}{2} \sum_{i,j=1}^{n} u_i u_j \Gamma_{\dot{X}\dot{X}}(t_i, t_j)\right] \quad (4.200)$$

and the theorem is proved.

Turning now to mean square integrals and consider the simple case

$$Y(t) = \int_a^t X(\tau) \, d\tau, \qquad [a, t] \subset T \quad (4.201)$$

Theorem 4.6.3. If the m.s. integral $Y(t)$, $t \in T$, exists, then, for every $t_1, t_2, \ldots, t_m \in T$,

$$\phi_{mY}(u_1, t_1; \ldots; u_m, t_m)$$

$$= \lim_{\substack{n \to \infty \\ \Delta_n \to 0}} \phi_{mnX}(u_1(\tau_1 - \tau_0), \tau_1'; \ldots; u_1(\tau_n - \tau_{n-1}), \tau_n';$$

$$\ldots; u_m(\tau_1 - \tau_0), \tau_1'; \ldots; u_m(\tau_n - \tau_{n-1}), \tau_n'), \qquad \tau_j' \in [\tau_{j-1}, \tau_j) \subset T$$
(4.202)

Proof. Let $m = 1$. We have

$$\phi_Y(u_1, t_1) = E\{\exp[iu_1 Y(t_1)]\}$$

$$= E\left\{ \underset{\substack{n \to \infty \\ \Delta_n \to 0}}{\text{l.i.m.}} \exp\left[iu_1 \sum_{k=1}^{n} (\tau_k - \tau_{k-1}) X(\tau_k') \right] \right\}$$

$$= \lim_{\substack{n \to \infty \\ \Delta_n \to 0}} \phi_{nX}(u_1(\tau_1 - \tau_0), \tau_1'; \ldots; u_1(\tau_n - \tau_{n-1}), \tau_n') \qquad (4.203)$$

Now let $m = 2$. We find that

$$\phi_{2Y}(u_1, t_1; u_2, t_2) = E\{\exp(i[u_1 Y(t_1) + u_2 Y(t_2)])\}$$

$$= E\left\{ \underset{\substack{n_1, n_2 \to \infty \\ \Delta_{n_1}, \Delta_{n_2} \to 0}}{\text{l.i.m.}} \exp\left[iu_1 \sum_{j=1}^{n_1} (\tau_j - \tau_{j-1}) X(\tau_j') + iu_2 \sum_{k=1}^{n_2} (\sigma_k - \sigma_{k-1}) X(\sigma_k') \right] \right\}$$

$$= \lim_{\substack{n_1, n_2 \to \infty \\ \Delta_{n_1}, \Delta_{n_2} \to 0}} \phi_{(n_1+n_2)X}(u_1(\tau_1 - \tau_0), \tau_1'; \ldots; u_1(\tau_{n_1} - \tau_{n_1-1}), \tau_{n_1}';$$

$$u_2(\sigma_1 - \sigma_0), \sigma_1'; \ldots; u_2(\sigma_{n_2} - \sigma_{n_2-1}), \sigma_{n_2}') \qquad (4.204)$$

Following this procedure we easily obtain the general result for any arbitrary m. Equation (4.202) gives the space-saving version of the general result when $n_1 = n_2 = \cdots = n$ and $\tau_j = \sigma_j = \cdots, j = 1, 2, \ldots, n$.

Consider the case where $X(t)$ is a Gaussian process with its joint characteristic function given by Eq. (4.197). It follows from Eq. (4.203) that

$$\phi_Y(u_1, t_1) = \lim_{\substack{n \to \infty \\ \Delta_n \to 0}} \exp\left[-\tfrac{1}{2} u_1^2 \sum_{j,k=1}^{n} \Gamma_{XX}(\tau_j', \tau_k')(\tau_j - \tau_{j-1})(\tau_k - \tau_{k-1}) \right]$$

$$= \exp\left[-\tfrac{1}{2} u_1^2 \int_a^{t_1} \int_a^{t_1} \Gamma_{XX}(\tau, \sigma) \, d\tau \, d\sigma \right]$$

$$= \exp[-\tfrac{1}{2} u_1^2 \Gamma_{YY}(t_1, t_1)] \qquad (4.205)$$

Proceeding as above and generalizing, we establish the following general result.

Theorem 4.6.4. If the m.s. integral $Y(t)$ of a Gaussian process $X(t)$, defined by

$$Y(t) = \int_a^t f(t, \tau) X(\tau) \, d\tau \qquad (4.206)$$

exists, then $Y(t)$, $t \in T$, is a Gaussian process.

Theorems 4.6.2 and 4.6.4 establish the fact that the Gaussianity of a Gaussian stochastic process is preserved under m.s. integration and differentiation. In view of the derivations given above, it is not difficult to see that this property of Gaussian processes holds under *all* linear transformations.

References

1. M. Loéve, *Probability Theory*. Van Nostrand-Reinhold, Princeton, New Jersey, 1963.
2. J. E. Moyal, Stochastic processes and statistical physics. *J. Roy. Statist. Soc. Ser. B* **11**, 150–210 (1949).
3. M. S. Bartlett, *An Introduction to Stochastic Processes*, 2nd ed. Cambridge Univ. Press, London and New York, 1966.
4. P. A. Ruymgaart, *Notes on Ordinary Stochastic Differential Equations with Random Initial Conditions*. Dept. of Math. Technol. Univ. Delft, Delft, The Netherlands, 1968.
5. J. L. Doob, *Stochastic Processes*. Wiley, New York, 1953.

PROBLEMS

All random variables and stochastic processes considered in the problems below are assumed to be of second order.

4.1. The law of large numbers states that, if the probability of an event A occurring in a given experiment equals p and the experiment is repeated n times, then, for any $\varepsilon > 0$,

$$\lim_{n \to \infty} P\{| \, k/n - p \, | \leq \varepsilon\} = 1$$

where k is the number of times the event A is realized. Establish this result as a limit of a sequence of random variables.

4.2. Consider a sequence $\{X_n\}$ of r.v.'s defined by

$$X_n = \begin{cases} n, & p(n) = 1/n^m \\ 0, & p(0) = 1 - 2/n^m \\ -n, & p(-n) = 1/n^m \end{cases}$$

where m is an arbitrary positive integer. Discuss the convergence properties of this sequence based upon all four modes of convergence.

4.3. Repeat Problem 4.2 when the probabilities are $p(n) = p(-n) = 1/mn$ and $p(0) = 1 - 2/mn$.

4.4. Consider a sequence $\{X_n\}$ and define

$$S_n = (1/n) \sum_{j=1}^{n} X_j$$

Show that

$$X_n \xrightarrow{\text{a.s.}} 0 \Rightarrow S_n \xrightarrow{\text{a.s.}} 0$$

$$X_n \xrightarrow{\text{m.s.}} 0 \Rightarrow S_n \xrightarrow{\text{m.s.}} 0$$

$$X_n \xrightarrow{\text{i.p.}} 0 \not\Rightarrow S_n \xrightarrow{\text{i.p.}} 0$$

4.5. Prove that, if $X_n \xrightarrow{\text{m.s.}} X$, then

$$\lim_{n,m \to \infty} E\{X_n X_m\}$$

exists. Is the converse true?

4.6. Show that the sequence in Example 4.7 converges a.s. by means of Eq. (4.60) and Theorem 4.2.7.

4.7. The Central Limit Theorem stated in Section 2.5 holds based upon convergence in distribution. Show that the sequence of random variables defined by Eq. (2.92) does not converge in probability.

4.8. We state without proof the following theorem (see Loéve [1, p. 116]).

Theorem. If a sequence $\{X_n\}$ is fundamental i.p., then there exists a subsequence $\{X_{n_k}\}$ of $\{X_n\}$ which converges a.s.

With the aid of the theorem above, prove the "if" part of Theorem 4.2.4.

4.9. Investigate the properties of m.s. continuity, m.s. differentiation, and m.s. integration of the following processes. Whenever defined, determine

$\Gamma_{\dot{X}\dot{X}}(t, s)$ and $\Gamma_{YY}(t, s)$ where

$$Y(t) = (1/t) \int_0^t X(\tau) \, d\tau \qquad .$$

(a) $X(t) = At + B$, where A and B are uncorrelated r.v.'s with means m_a and m_b, and with variances σ_a^2 and σ_b^2. Furthermore, show that

$$\dot{X}(t) = A \qquad \text{and} \qquad Y(t) = At/2 + B$$

with probability one at every finite t.

(b) $X(t)$ is a purely stochastic process. At every t,

$$E\{X(t)\} = 0 \qquad \text{and} \qquad E\{X^2(t)\} = 1.$$

(c) $X(t)$, $t \geq 0$, is the Poisson process defined in Section 3.3.4.

(d) $X(t)$ is the binary noise discussed in Example 3.5.

(e) $X(t)$ is stationary with $\Gamma_{XX}(\tau) = e^{-a|\tau|}$, $a > 0$.

(f) $X(t)$ is stationary with

$$\Gamma_{XX}(\tau) = \exp(-a\tau^2) \cos b\tau, \qquad a, b > 0$$

(g) $X(t)$ is stationary with $\Gamma_{XX}(\tau) = 1/(a^2 + \tau^2)$.

4.10. Consider m.s. differentiation. Is the relation (4.126) valid if $f(t)$ is replaced by a m.s. differentiable s.p. $Y(t)$, independent of $X(t)$? What are other generalizations of this chain rule of m.s. differentiation?

4.11. Let $f(t, u) \equiv 1$ in Eq. (4.140). Simplify integration in mean square criterion (Theorem 4.5.1) when $X(t)$ is wide-sense stationary.

4.12. Let $X(t)$, $t \in T$, be wide-sense stationary and m.s. differentiable on T. Show that $X(t)$ and $\dot{X}(t)$ are orthogonal on T.

4.13. Consider the case where $X(t)$ is a zero-mean, stationary, Gaussian, and twice m.s. differentiable process on T. Show that, at every $t \in T$, $X(t)$ is independent of $\dot{X}(t)$ but not independent of $\ddot{X}(t)$. Compute $\Gamma_{X\ddot{X}}(\tau)$.

4.14. Following the proof of Theorem 4.5.3, supply a proof for the corollary on m.s. Riemann integration by parts [Eq. (4.162)].

4.15. Theorem 4.6.1 gives a relation between $\phi_{n\dot{X}}$ and ϕ_{mX}. Give an explicit relation between $f_{n\dot{X}}$ and f_{mX}. (Use the fact that m.s. convergence implies convergence i.p.).

Chapter 5

Random Ordinary Differential Equations

In modeling, analyzing, and predicting behaviors of physical and natural phenomena, greater and greater emphasis has been placed upon probabilistic methods. This is due to combinations of complexity, uncertainties, and ignorance which are present in the formulation of a great number of these problems. As was indicated in Chapter 1, probabilistic concepts have been used in the study of a wide variety of subjects in science and engineering. There is no doubt that we will see an increasingly heavy use of stochastic formulations in most scientific disciplines.

Using the statement of Newton's second law as a typical example, a large class of physically important problems is described by ordinary differential equations. In taking various random effects into account, it is natural that ordinary differential equations involving random elements, that is, random ordinary differential equations, will play an increasingly prominent role.

In this and the following chapters, the concept of mean square calculus is applied to the study of systems of random differential equations of the form

$$\dot{X}_i(t) = f_i(X_1(t), \ldots, X_n(t); Y_1(t), \ldots, Y_m(t); t), \quad t \in T, \quad i = 1, 2, \ldots, n, \tag{5.1}$$

with initial conditions

$$X_i(t_0) = X_{i0} \qquad (5.2)$$

where $Y_j(t)$, $j = 1, 2, \ldots, m$, and X_{i0}, $i = 1, 2, \ldots, n$, are specified only in a stochastic sense. In the above, $T = [t_0, a]$ is an interval (finite or infinite) of the real line.

It is clear that this system of equations also covers the case of random differential equations of higher order. It is always possible, by introducing more variables, to convert higher-order differential equations to a set of coupled first-order equations.

In vector–matrix notation, our basic system of differential equations is represented by

$$\dot{\mathbf{X}}(t) = \mathbf{f}(\mathbf{X}(t), \mathbf{Y}(t), t), \qquad t \in T; \qquad \mathbf{X}(t_0) = \mathbf{X}_0 \qquad (5.3)$$

where $\mathbf{X}(t)$ is an n-dimensional vector with components $X_i(t)$, $i = 1, 2, \ldots, n$. The n-dimensional vectors $\mathbf{f}$ and $\mathbf{X}_0$ and the m-dimensional vector $\mathbf{Y}(t)$ are defined in an analogous way.

As in the theory of differential equations of ordinary nonrandom functions, the development of random differential equations deals with, in some stochastic sense, the existence and uniqueness of the solution processes, and the study of certain properties associated with these solutions. In this chapter, we consider these questions in quite general terms. Techniques for solving random differential equations together with physical applications will be developed in several stages, commencing in Chapter 6.

5.1. Existence and Uniqueness

We are primarily concerned with the mean square theory of differential equations characterized by Eq. (5.3). In this development, all stochastic quantities involved in Eq. (5.3) take values in the L_2-space and all stochastic operations are operations in the mean square sense. Since we will be dealing extensively with vector stochastic quantities, let us first establish some definitions and notations for these vector quantities.

5.1.1. The L_2^n-Space

Let X_j, $j = 1, 2, \ldots, n$, be second-order r.v.'s. Following our remarks in Section 4.1, the n-dimensional random vector

$$\mathbf{X} = \begin{bmatrix} X_1 \\ X_2 \\ \cdot \\ \cdot \\ \cdot \\ X_n \end{bmatrix} \qquad (5.4)$$

constitutes a linear vector space if all equivalent r.v.'s are identified. This space with the norm [see Eq. (4.10) for definition]

$$\| \mathbf{X} \|_n = \max_{j=1,2,\ldots,n} \| X_j \| \qquad (5.5)$$

will be called the $L_2{}^n$-space. Since the L_2-space is complete, the $L_2{}^n$-space is also complete with the norm $\| \mathbf{X} \|_n$. It is thus a Banach space with respect to this norm.

The m.s. continuity, m.s. differentiation, and m.s. integration associated with a second-order vector stochastic process are defined with respect to the norm $\| \mathbf{X} \|_n$. Hence, a vector stochastic process $\mathbf{X}(t)$, $t \in T$, is m.s. continuous at t, for example, if

$$\lim_{\tau \to 0} \| \mathbf{X}(t + \tau) - \mathbf{X}(t) \|_n = 0, \qquad t + \tau \in T \qquad (5.6)$$

In view of this definition, it is clear that the vector s.p. $X(t)$ is m.s. continuous at $t \in T$ if, and only if, each of its component processes is m.s. continuous at $t \in T$.

Similar definitions and observations can be made in regard to m.s. differentiation and m.s. integration of the second-order vector s.p. $\mathbf{X}(t)$.

We conclude this section by introducing some new notations.

Let $\mathbf{X}(t)$, $t \in T$, be a second-order n-dimensional vector stochastic process. It is characterized by a mapping of the interval T into $L_2{}^n$. For the sake of convenience, we shall adopt the notation $\mathbf{X}(t): T \to L_2{}^n$ in what follows. Similarly, if the vector transformation $\mathbf{f}(\mathbf{X}(t), t)$ yields a second-order n-dimensional vector process for each $t \in T$, this property of the transformation will be denoted by $\mathbf{f}: L_2{}^n \times T \to L_2{}^n$. Clearly, the notation $\mathbf{X} \in L_2{}^n$ means that $\mathbf{X}$ is a second-order n-dimensional random vector.

5.1.2. Existence and Uniqueness of Mean Square Solutions

We now consider the solution of Eq. (5.3) in the mean square sense. The following definitions and theorems are natural generalizations of the corresponding deterministic theory of differential equations.

Let us first write Eq. (5.3) in the form

$$\dot{\mathbf{X}}(t) = \mathbf{f}(\mathbf{X}(t), t), \qquad t \in T; \qquad \mathbf{X}(t_0) = \mathbf{X}_0 \qquad (5.7)$$

Definition. Consider Eq. (5.7) where $\mathbf{f}: L_2^n \times T \to L_2^n$ continuously and $\mathbf{X}_0 \in L_2^n$. The s.p. $\mathbf{X}(t): T \to L_2^n$ is called a *mean square (m.s.) solution* of Eq. (5.7) on T if

1. $\mathbf{X}(t)$ is m.s. continuous on T;
2. $\mathbf{X}(t_0) = \mathbf{X}_0$;
3. $\mathbf{f}(\mathbf{X}(t), t)$ is the m.s. derivative of $\mathbf{X}(t)$ on T.

Whenever we speak of m.s. solutions, we shall assume that $\mathbf{f}$ and $\mathbf{X}_0$ satisfy the imposed conditions as stated above.

Theorem 5.1.1. $\mathbf{X}(t): T \to L_2^n$ is a m.s. solution of Eq. (5.7) if, and only if, for all $t \in T$,

$$\mathbf{X}(t) = \mathbf{X}_0 + \int_{t_0}^{t} \mathbf{f}(\mathbf{X}(s), s) \, ds \qquad (5.8)$$

where the integral is understood to be a m.s. integral.

Proof. The m.s. continuity of $\mathbf{f}$ as a function of m.s. continuous $\mathbf{X}(t)$ is sufficient to ensure the existence of the m.s. integral in Eq. (5.8). The only if part of the proof follows from the fundamental theorem of m.s. calculus (Section 4.5.1). The if part follows from Property (5) of m.s. integration, also given in Section 4.5.1.

Equations (5.7) and (5.8) are thus equivalent. We can now proceed to state a basic existence and uniqueness theorem for m.s. solutions to Eq. (5.7). The result given below is due to Strand [1, 2], which is a natural generalization of the classical Picard theorem based upon convergence of successive approximations (see, for example, Coddington and Levinson [3, p. 12]).

Theorem 5.1.2. Consider Eq. (5.7). If $\mathbf{f}: L_2^n \times T \to L_2^n$ satisfies the m.s. Lipschitz condition

$$\| \mathbf{f}(\mathbf{X}, t) - f(\mathbf{Y}, t) \|_n \leq k(t) \| \mathbf{X} - \mathbf{Y} \|_n \qquad (5.9)$$

where

$$\int_{t_0}^{a} k(t) \, dt < \infty,$$

then there exists a unique m.s. solution for any initial condition $\mathbf{X}_0 \in L_2^n$.

Proof. The proof is given by Strand [1, p. 28]. We omit it here because it closely parallels that of the classical Picard proof, the only significant difference being the replacement of ordinary R^n norms by the m.s. norms $\| \cdots \|_n$. Another proof, in Banach spaces but under more strict conditions on **f**, can be found in the work of Hille and Phillips [4, p. 67].

Unfortunately, the existence and uniqueness theorem in its general form has very limited applicability in the mean square theory of random differential equations. The first serious drawback is the difficulty with which to show that a function of second-order stochastic processes is second order itself. Secondly, the m.s. Lipschitz condition required in the theorem is too restrictive. Consider, for example, a simple first-order linear equation

$$\dot{X}(t) = AX(t), \qquad t \geq 0; \qquad X(0) = 1 \tag{5.10}$$

where A is a second-order r.v. Strand [1, 2] has shown that the m.s. Lipschitz condition (5.9) is satisfied if, and only if, A is bounded almost surely. Thus, Theorem 5.1.2 is not applicable to this simple case when A assumes a Gaussian or a Poisson distribution. This is certainly undesirable from a practical viewpoint.

In view of these limitations, it is more fruitful to consider existence and uniqueness questions associated with more restrictive classes of random differential equations. This will be done in Chapters 6, 7, and 8 where the mean square theory of three basic types of random differential equations will be developed. These types are classified according to the manner in which the s.p. $\mathbf{Y}(t)$ in Eq. (5.3) enters the equation.

5.2. The Ito Equation

A special class of random differential equations which has found important applications in control, filtering, and communication theory is one where the vector s.p. $\mathbf{Y}(t)$ in Eq. (5.3) has only white noise components. More specifically, we mean equations of the form

$$\dot{\mathbf{X}}(t) = \mathbf{f}(\mathbf{X}(t), t) + G(\mathbf{X}(t), t)\,\mathbf{W}(t), \qquad t \in T; \qquad \mathbf{X}(t_0) = \mathbf{X}_0 \tag{5.11}$$

where $\mathbf{W}(t)$ is an m-dimensional vector s.p. whose components are Gaussian white noise, $G(\mathbf{X}(t), t)$ is an $n \times m$ matrix function, and $\mathbf{X}_0$ is independent of $\mathbf{W}(t)$, $t \in T$.

The popularity of this model in control and filtering applications is due to two principal reasons. The first is the mathematical simplicity; it is a natural stochastic extension of the powerful state space approach in classical optimal control theory. Moreover, as we shall see, the solution process generated by Eq. (5.11) is Markovian, for which powerful techniques for obtaining its solution exist. The second reason is that, although white noise is a mathematical artifice, it approximates closely the behavior of a number of important noise processes in electrical and electronic systems [5]. The resistance noise in linear resistors, for example, yields a power spectral density for the equivalent voltage source which is "flat" up to very high frequencies. The use of white noise models leads to good results if we only deal with a small portion of the frequency spectrum in the analysis.

On account of these advantages, Eq. (5.11) has played a prominent and successful role in numerous engineering, biomedical, and other practical applications. Recent books which deal with these applications include Aoki [6], Kushner [7], Astrom [8], and Stratonovich [9] in control theory; and Middleton [10], Bucy and Joseph [11], and Jazwinski [12] in communication and filtering.

In view of the practical importance of this class of random differential equations, the mean square theory associated with it merits attention. However, we immediately see that, since white noise does not exist in the mean square sense (nor in the sample sense), discussions in Section 5.1 are not applicable. In what follows, we study Eq. (5.11) in the mean square sense following an interpretation due to Ito [13] (also see Doob [14] and Skorokhod [15]).

5.2.1. A Representation of the White Noise

Let us first give a useful representation for a white Gaussian process $W(t)$, $t \geq 0$.

Let $B(t)$, $t \geq 0$, be the Brownian motion process or the Wiener process (Section 3.3.4). It is Gaussian with mean zero and covariance (Problem 3.19)

$$\mu_B(t_1, t_2) = 2D \min(t_1, t_2), \qquad t_1, t_2 \geq 0 \tag{5.12}$$

We have noted in Section 4.4 that the s.p. $B(t)$, $t \geq 0$, is not m.s. differentiable. Formally, however, we can consider its m.s. derivative and determine its properties.

The formal derivative of $B(t)$, $\dot{B}(t)$, is clearly Gaussian with mean zero. Its covariance is given by

$$\mu_{\dot{B}}(t_1, t_2) = \partial^2 \mu_B(t_1, t_2)/\partial t_1 \, \partial t_2 \tag{5.13}$$

Thus,

$$\mu_{\dot{B}}(t_1, t_2) = 2D \, \partial^2 \min(t_1, t_2)/\partial t_1 \, \partial t_2$$
$$= 2D \, \partial U(t_1 - t_2)/\partial t_1 = 2D \, \delta(t_1 - t_2) \qquad (5.14)$$

In the above, the function $U(t_1 - t_2)$ is the Heaviside unit step function in t_1, that is, for a given t_2,

$$U(t_1 - t_2) = \begin{cases} 0, & t_1 < t_2 \\ 1, & t_1 > t_2 \end{cases} \qquad (5.15)$$

It is clear that the formal derivative $\dot{B}(t)$, $t \geq 0$, has the properties of a white Gaussian noise.

Thus, we can formally write

$$dB(t)/dt = W(t), \qquad t \geq 0 \qquad (5.16)$$

5.2.2. The Ito Equation and the Ito Integral

With the formal representation (5.16) for the white Gaussian noise, we can interpret Eq. (5.11) as formally equivalent to

$$d\mathbf{X}(t) = \mathbf{f}(\mathbf{X}(t), t) \, dt + \mathbf{G}(\mathbf{X}(t), t) \, d\mathbf{B}(t), \qquad t \in T; \qquad \mathbf{X}(t_0) = \mathbf{X}_0 \qquad (5.17)$$

and, in integral equation representation,

$$\mathbf{X}(t) - \mathbf{X}(t_0) = \int_{t_0}^{t} f(\mathbf{X}(s), s) \, ds + \int_{t_0}^{t} G(\mathbf{X}(s), s) \, d\mathbf{B}(s), \qquad t \in T$$
$$\mathbf{X}(t_0) = \mathbf{X}_0 \qquad (5.18)$$

where $\mathbf{X}_0$ is independent of the increment $d\mathbf{B}(t)$, $t \in T$.

Now, under suitable conditions, the first integral of Eq. (5.18) can be defined as a m.s. Riemann integral. The second integral, on the other hand, has no meaning in the mean square sense as a Riemann–Stieltjes integral. The reason is that, if we define the r.v.'s Y_n by

$$Y_n = \sum_{k=1}^{n} X(t_k')[B(t_k) - B(t_{k-1})], \qquad t_k' \in [t_{k-1}, t_k) \qquad (5.19)$$

this sequence of r.v.'s does not converge in the m.s. sense to a unique limit; the limit depends upon the specific choice of t_k'. Therefore, the integral

$$\int_{t_0}^{t} X(s) \, dB(s) \qquad (5.20)$$

does not exist as a m.s. integral in the usual sense. We will substantiate this statement in Example 5.2 following the introduction of the Ito integral.

In order to present a m.s. treatment of Eqs. (5.17) and (5.18), we turn to an interpretation of the second integral in Eq. (5.18) due to Ito. It is remarked that, with this interpretation, Eq. (5.17) or (5.18) is generally called the *Ito stochastic differential equation*. In fact, it is now generally accepted that the term "stochastic differential equations" means the Ito stochastic differential equations. For more general differential equations, the term "random differential equations" is preferred.

Consider now the integral (5.20) in the Ito sense. Let $B(t)$, $t \in T = [0, a]$, be a Wiener process with

$$E\{B(t)\} = 0, \qquad E\{[B(t) - B(s)]^2\} = |t - s|, \qquad t, s \in T \qquad (5.21)$$

and let $X(t): T \to L_2$ be m.s. continuous on T. At any $t \in T$, the s.p. $X(t)$ is independent of the increment $\{B(t_{k+1}) - B(t_k)\}$ for all t_k and t_{k+1} satisfying $0 \le t \le t_k \le t_{k+1} \le a$.

Let $\{p_n\}$ be a sequence of finite subdivisions of T and let

$$\Delta_n = \max_k (t_{k+1} - t_k)$$

We form the random variable

$$Y_n = \sum_{k=0}^{n-1} X(t_k)[B(t_{k+1}) - B(t_k)] \qquad (5.22)$$

It is clear that, since

$$\| X(t_k)[B(t_{k+1}) - B(t_k)] \| = \| X(t_k) \| \cdot \| B(t_{k+1}) - B(t_k) \| < \infty$$

each term of Eq. (5.22) belongs to L_2 and, hence, $Y_n \in L_2$.

Definition. If

$$\underset{\substack{n \to \infty \\ \Delta_n \to 0}}{\text{l.i.m.}} Y_n = Y \qquad (5.23)$$

exists, the r.v. Y is called the *Ito stochastic integral* or, in short, *Ito integral*, of $X(t)$ [with respect to $B(t)$] over the interval T. It is denoted by

$$(I) \int_0^a X(t)\, dB(t) = \underset{\substack{n \to \infty \\ \Delta_n \to 0}}{\text{l.i.m.}} Y_n \qquad (5.24)$$

An important observation to be made is that the values of $X(t)$ in Eq. (5.22) are not taken at arbitrary points in the intervals $[t_k, t_{k+1}]$ but at the points t_k. Thus, this definition is not one of a m.s. integral in the usual sense. As was mentioned above, we shall see that the m.s. limit Y depends upon this choice.

Theorem 5.2.1. Let $X(t)$ and $B(t)$ satisfy all the conditions given above. The Ito integral as defined by Eq. (5.24) exists and is unique.

Proof. Based upon the convergence in mean square criterion (Theorem 4.3.2), it is sufficient to show that the expectation $E\{Y_n Y_m\}$ converges to a finite number as $n, m \to \infty$ independently.

Consider

$$E\{Y_n Y_m\} = \sum_{i,j} E\{X(s_i) X(t_j)[B(s_{i+1}) - B(s_i)][B(t_{j+1}) - B(t_j)]\} \quad (5.25)$$

It is to be evaluated over the region $[0, a]^2$ as indicated in Fig. 5.1; each term in the sum is to be evaluated over a rectangle such as R_1 or R_2. In evaluating these terms, we distinguish two types of rectangles: those containing a segment of the diagonal line $t = s$, such as R_1, and those containing at most one point of the diagonal line, such as R_2.

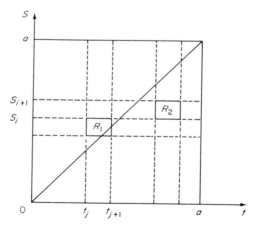

Fig. 5.1. The region $[0, a]^2$.

For the second type, a typical situation is one where $s_i < s_{i+1} \le t_j < t_{j+1}$. This implies that $[B(t_{j+1}) - B(t_j)]$ is independent of $X(s_i)$, $X(t_j)$, and $[B(s_{i+1}) - B(s_i)]$. Hence

$$E\{X(s_i) X(t_j)[B(s_{i+1}) - B(s_i)][B(t_{j+1}) - B(t_j)]\}$$
$$= E\{X(s_i) X(t_j)[B(s_{i+1}) - B(s_i)]\} E\{[B(t_{j+1}) - B(t_j)]\} = 0 \quad (5.26)$$

Hence, the contributions of all the terms in Eq. (5.25) over rectangles of the second type are zero.

Turning now to rectangles of the first type, consider a typical situation where $s_i < t_j < s_{i+1} < t_{j+1}$. We have

$$E\{X(s_i)\, X(t_j)[B(s_{i+1}) - B(s_i)][B(t_{j+1}) - B(t_j)]\}$$
$$= E\{X(s_i)\, X(t_j)[(B(s_{i+1}) - B(t_j)) + (B(t_j) - B(s_i))]$$
$$\times [(B(t_{j+1}) - B(s_{i+1})) + (B(s_{i+1}) - B(t_j))]\} \qquad (5.27)$$

It is clear that the only contributing term in the above is

$$E\{X(s_i)\, X(t_j)[B(s_{i+1}) - B(t_j)]^2\} = E\{X(s_i)\, X(t_j)\}(s_{i+1} - t_j) \qquad (5.28)$$

The s.p. $X(t)$, $t \in [0, a]$, is m.s. continuous. Hence, $E\{X(t)\, X(s)\}$ is continuous on the closed and bounded set $[0, a]^2$ and, therefore, is uniformly continuous on $[0, a]^2$. Then, to each $\varepsilon > 0$, there exists an $N(\varepsilon)$ such that its variation on the region $[s_i, s_{i+1}] \times [t_j, t_{j+1}]$ for all i and j is less than ε if $[s_i, s_{i+1}]$ and $[t_j, t_{j+1}]$ belong to, respectively, partitions p_m and p_n with $n, m > N(\varepsilon)$. Identifying s_i, t_j, and s_{i+1} by, respectively, u_{k-1}, u_k, and u_{k+1}, we see that

$$\left| \sum_k E\{X(u_{k-1})\, X(u_k)\}(u_{k+1} - u_k) - \sum_k E\{X^2(u_k)\}(u_{k+1} - u_k) \right|$$
$$\cdot \leq \varepsilon \sum_k (u_{k+1} - u_k) = \varepsilon a \qquad (5.29)$$

Since the second sum tends to the ordinary Riemann integral

$$\int_0^a E\{X^2(t)\}\, dt$$

Eq. (5.29) implies

$$E\{Y_n Y_m\} = \sum_{i,j} E\{X(s_i)\, X(t_j)[B(s_{i+1}) - B(s_i)][B(t_{j+1}) - B(t_j)]\}$$
$$= \sum_k E\{X(u_{k-1})\, X(u_k)\}(u_{k+1} - u_k) \to \int_0^a E\{X^2(t)\}\, dt \qquad (5.30)$$

as $n, m \to \infty$. Hence, it follows from the convergence in mean square criterion that the sequence $\{Y_n\}$ converges in mean square as $n \to \infty$. Clearly, the m.s. limit does not depend on the choice of the sequence $\{p_n\}$. The uniqueness proof follows directly from Theorem 4.2.2.

Example 5.1. Let us show that (see Doob [14, p. 442])

$$(I) \int_a^b B(t)\, dB(t) = \tfrac{1}{2}[B^2(b) - B^2(a)] - \tfrac{1}{2}(b - a) \qquad (5.31)$$

This Ito integral exists since $B(t)$, $t \in [a, b]$, satisfies all the conditions imposed on $X(t)$ considered above. Let $B_k = B(t_k)$, $a = t_0 < t_1 \cdots < t_n = b$, and $\Delta_n = \max_k(t_{k+1} - t_k)$. We have

$$\sum_{k=0}^{n-1} B_k(B_{k+1} - B_k) = -\sum_{k=0}^{n-1} B_k(B_k - B_{k+1})$$

$$= -[B_0{}^2 - B_0 B_1 + B_1{}^2 - B_1 B_2 + B_2{}^2 + \cdots + B_{n-1}^2 - B_{n-1} B_n]$$

$$= -\tfrac{1}{2}\Big[B_0{}^2 + \sum_{k=0}^{n-1}(B_{k+1} - B_k)^2 - B_n{}^2 \Big]$$

$$= \tfrac{1}{2}(B_b{}^2 - B_a{}^2) - \tfrac{1}{2}\sum_{k=0}^{n-1}(B_{k+1} - B_k)^2 \qquad (5.32)$$

Hence,

$$(I) \int_a^b B(t)\, dB(t) = \operatorname*{l.i.m.}_{\substack{n \to \infty \\ \Delta_n \to 0}} \sum_{k=0}^{n-1} B_k(B_{k+1} - B_k)$$

$$= \tfrac{1}{2}(B_b{}^2 - B_a{}^2) - \tfrac{1}{2}\operatorname*{l.i.m.}_{\substack{n \to \infty \\ \Delta_n \to 0}} \sum_{k=0}^{n-1}(B_{k+1} - B_k)^2 \qquad (5.33)$$

We now need to show that

$$\operatorname*{l.i.m.}_{\substack{n \to \infty \\ \Delta_n \to 0}} \sum_{k=0}^{n-1}(B_{k+1} - B_k)^2 = b - a \qquad (5.34)$$

Consider

$$\left\| \sum_k (\Delta B_k)^2 - (b - a) \right\|^2 = \left\| \sum_k (\Delta B_k)^2 - \Delta t_k \right\|^2 \qquad (5.35)$$

where $\Delta B_k = B_{k+1} - B_k$ and $\Delta t_k = t_{k+1} - t_k$. It is easily seen that, noting $E\{(\Delta B_k)^2\} = \Delta t_k$ and $E\{(\Delta B_k)^4\} = 3(\Delta t_k)^2$,

$$\left\| \sum_k (\Delta B_k)^2 - \Delta t_k \right\|^2 = \sum_k \| (\Delta B_k)^2 - \Delta t_k \|^2$$

$$+ 2\sum_{\substack{i,j \\ i \neq j}} E\{(\Delta B_i)^2 - \Delta t_i\}\, E\{(\Delta B_j)^2 - \Delta t_j\}$$

$$= \sum_k \| (\Delta B_k)^2 - \Delta t_k \|^2$$

$$= 2\sum_k (\Delta t_k)^2 \leq 2\Delta_n \sum_k \Delta t_k = 2\Delta_n(b - a) \to 0 \qquad (5.36)$$

as $n \to \infty$ and $\Delta_n \to 0$. Equation (5.36) establishes Eq. (5.34), which in turn gives Eq. (5.31).

We see from this example a surprising consequence of Ito integration. The formal integration rule does not apply as it would only yield the first term in Eq. (5.31). Intuitively, the presence of the "correction term," $-(b - a)/2$, is expected because it is needed to insure the obvious identity

$$E\left\{(I) \int_a^b B(t) \, dB(t)\right\} = 0 \tag{5.37}$$

Recently, Stratonovich [16] has defined another type of integral, the so-called *Stratonovich integral*, in which usual integration rules continue to apply. A number of conceptual differences exist between Ito's interpretation and Stratonovich's interpretation of stochastic integrals. The reader is referred to Mortensen [17] for a candid discussion of these differences.

Example 5.2. In this example we show that the m.s. limit of the sequence $\{Y_n\}$ defined by Eq. (5.22) generally depends upon the choice of t_k at which the values of $X(t)$ are taken. Thus, the Ito integral is not an ordinary m.s. integral.

Let us define Y_n and Z_n by

$$\begin{aligned} Y_n &= \sum_k B(t_k)[B(t_{k+1}) - B(t_k)] \\ Z_n &= \sum_k B(t_{k+1})[B(t_{k+1}) - B(t_k)] \end{aligned} \tag{5.38}$$

We see from Eq. (5.34) that, with $a = t_0 < t_1 \cdots < t_n = b$,

$$\underset{\substack{n \to \infty \\ \Delta_n \to 0}}{\text{l.i.m.}} (Z_n - Y_n) = \underset{\substack{n \to \infty \\ \Delta_n \to 0}}{\text{l.i.m.}} \sum_k [B(t_{k+1}) - B(t_k)]^2 = b - a \tag{5.39}$$

Hence, the sequences $\{Y_n\}$ and $\{Z_n\}$ converge in mean square to different limits.

5.2.3. Existence and Uniqueness of Solutions

The existence of the Ito integral permits us to discuss the existence and uniqueness of solutions to the Ito stochastic differential equation. We first need the following theorems.

Theorem 5.2.2. Let $\{X_n(t)\}$, $t \in T = [0, a]$, be a sequence of s.p.'s satisfying the properties:

(1) $X_n(t): T \to L_2$ is m.s. continuous on T;

(2) $X_n(t)$ is independent of the increment $\{B(t_{k+1}) - B(t_k)\}$ for all t_k and t_{k+1} satisfying $0 \leq t \leq t_k \leq t_{k+1} \leq a$.

If $X_n(t) \xrightarrow{\text{m.s.}} X(t)$ uniformly in T, then $X(t)$ also satisfies the properties given above and

$$Y_n(t) = (I) \int_0^t X_n(s) \, dB(s) \xrightarrow{\text{m.s.}} Y(t) = (I) \int_0^t X(s) \, dB(s) \qquad (5.40)$$

as $n \to \infty$, uniformly in $t \in T$.

Proof. As the m.s. limit of a uniformly convergent sequence of m.s. continuous second-order processes, the s.p. $X(t)$, $t \in T$, has all the properties stated above for $X_n(t)$. Hence, it follows from Theorem 5.2.1 that the Ito integral $Y(t)$ exists.

To see that $Y_n(t) \xrightarrow{\text{m.s.}} Y(t)$, we evaluate directly as follows.

$$\| Y_n(t) - Y(t) \|^2 = \left\| (I) \int_0^t [X_n(s) - X(s)] \, dB(s) \right\|^2$$

$$= \int_0^t \| X_n(s) - X(s) \|^2 \, ds \leq \int_0^a \| X_n(s) - X(s) \|^2 \, ds \to 0 \tag{5.41}$$

as $n \to \infty$. The second equality above follows from Eq. (5.30). The last quantity tends to zero because $X_n(t) \xrightarrow{\text{m.s.}} X(t)$ uniformly in $t \in T$.

Theorem 5.2.3. If $X(t)$ is m.s. continuous on $T = [0, a]$, then, at each $t \in T$,

$$\left\| \int_0^t X(s) \, ds \right\|^2 \leq t \int_0^t \| X(s) \|^2 \, ds \leq a \int_0^t \| X(s) \|^2 \, ds \tag{5.42}$$

Proof. By direct evaluation, we have

$$\left\| \int_0^t X(s) \, ds \right\|^2 = \int_0^t \int_0^t E\{X(s) \, X(r)\} \, ds \, dr$$

$$\leq \int_0^t \| X(s) \| \, ds \int_0^t \| X(r) \| \, dr$$

$$= \left[\int_0^t \| X(s) \| \, ds \right]^2 \leq \int_0^t dr \cdot \int_0^t \| X(s) \|^2 \, ds$$

$$\leq t \int_0^t \| X(s) \|^2 \, ds \leq a \int_0^t \| X(s) \|^2 \, ds \tag{5.43}$$

In the above, the first two inequalities are direct results of the Schwarz inequality [Eq. (4.1)].

We are now in the position to give an existence and uniqueness theorem for the solution of the Ito equation. For simplicity, the theorem is given for the scalar case. It can be generalized to the vector case following similar steps.

Consider the (scalar) Ito equation

$$X(t) = X_0 + \int_{t_0}^t f(X(s), s) \, ds + (I) \int_{t_0}^t g(X(s), s) \, dB(s)$$

$$X(t_0) = X_0, \qquad t \in T = [t_0, a] \tag{5.44}$$

where the initial condition $X_0 \in L_2$ is independent of $dB(t)$, $t \in T$.

Theorem 5.2.4. Let $f(x, t)$ and $g(x, t)$, $t \in T$, be two real functions satisfying the following conditions:

(a)   Both are continuous on $T \times (-\infty, \infty)$ and, moreover, uniformly continuous in t with respect to $x \in (-\infty, \infty)$.

(b)   The growth conditions

$$f^2(x, t) \leq K^2(1 + x^2), \qquad g^2(x, t) \leq K^2(1 + x^2) \tag{5.45}$$

(c)   The Lipschitz conditions

$$| f(x_2, t) - f(x_1, t) | \leq K | x_2 - x_1 |$$
$$| g(x_2, t) - g(x_1, t) | \leq K | x_2 - x_1 | \tag{5.46}$$

for a suitable $K > 0$. Then, Eq. (5.44) has a unique m.s. solution.

Proof. The existence will be shown by means of successive approximation. For all $t \in T$, we define the iterates

$$X_0(t) = X_0$$

$$X_{n+1}(t) = X_0 + \int_{t_0}^t f(X_n(s), s) \, ds + (I) \int_{t_0}^t g(X_n(s), s) \, dB(s), \qquad n \geq 0 \tag{5.47}$$

The existence proof follows from Theorem 5.2.2 if we prove that the sequence $\{X_n(t)\}$ satisfies all the conditions stated in Theorem 5.2.2 and that $X_n(t) \xrightarrow{\text{m.s.}} X(t)$ uniformly in T.

The zeroth iterate, $X_0(t)$, clearly has all the required properties. Consider now $X_1(t)$. Property (b) implies that $f(X_0(t), t)$ and $g(X_0(t), t)$ are second-order on T. Furthermore, they are m.s. continuous on T. This is seen by noting that

$$\| f(X_0(t + \Delta t), t + \Delta t) - f(X_0(t), t) \|$$
$$\leq \| f(X_0(t + \Delta t), t + \Delta t) - f(X_0(t + \Delta t), t) \|$$
$$+ \| f(X_0(t + \Delta t), t) - f(X_0(t), t) \| \tag{5.48}$$

As $\Delta t \to 0$, the first term tends to zero in view of Property (a). In view of Property (c), the second term can be written as

$$\| f(X_0(t + \Delta t), t) - f(X_0(t), t) \| \leq K \| X_0(t + \Delta t) - X_0(t) \| \tag{5.49}$$

which tends to zero as $\Delta t \to 0$ because $X_0(t)$ is m.s. continuous.

Same argument applies to $g(X_0(t), t)$. It is also clear that $g(X_0(t), t)$ is independent of $dB(t)$.

Hence, the integrals in Eq. (5.47) for $n = 0$ exist, the first as a m.s. Riemann integral and the second as the Ito integral. Moreover, the m.s. continuity of $f(X_0(t), t)$ and $g(X_0(t), t)$ ensures the m.s. continuity of $X_1(t)$.

Inductively, we see that all $X_n(t)$ exist and they all satisfy the conditions stated in Theorem 5.2.2.

Next, we show that $X_n(t) \xrightarrow{\text{m.s.}} X(t)$ uniformly on T by showing that the sequence $\{X_n(t)\}$ is m.s. fundamental, uniformly on T (Theorem 4.2.3).

Let

$$\Delta_n X(t) = X_{n+1}(t) - X_n(t)$$
$$\Delta_n f(t) = f(X_{n+1}(t), t) - f(X_n(t), t) \tag{5.50}$$
$$\Delta_n g(t) = g(X_{n+1}(t), t) - g(X_n(t), t)$$

Equations (5.47) can then be written as

$$\Delta_0 X(t) = \int_{t_0}^t f(X_0(s), s)\, ds + (\text{I}) \int_{t_0}^t g(X_0(s), s)\, dB(s),$$
$$\Delta_n X(t) = \int_{t_0}^t \Delta_{n-1} f(s)\, ds + (\text{I}) \int_{t_0}^t \Delta_{n-1} g(s)\, dB(s), \qquad n \geq 1 \tag{5.51}$$

With the aid of Theorem 5.2.3 and Property (c), we can show that

$$\left\| \int_{t_0}^t f(X_0(s), s)\, ds \right\| \leq A(t - t_0)^{1/2}$$
$$\left\| \int_{t_0}^t \Delta_{n-1} f(s)\, ds \right\| \leq K\sqrt{T} \left[\int_{t_0}^t \| \Delta_{n-1} X(s) \|^2\, ds \right]^{1/2} \tag{5.52}$$

Similarly, it follows from Eq. (5.30) and Property (c) that

$$\left\| (I) \int_{t_0}^{t} g(X_0(s), s) \, dB(s) \right\| \leq B(t - t_0)^{1/2}$$

$$\left\| (I) \int_{t_0}^{t} \Delta_{n-1} g(s) \, dB(s) \right\| \leq K \left[\int_{t_0}^{t} \| \Delta_{n-1} X(s) \|^2 \, ds \right]^{1/2} \qquad (5.53)$$

where A and B are positive constants and $T = a - t_0$. Thus,

$$\| \Delta_n X(t) \| \leq K(1 + \sqrt{T}) \left[\int_{t_0}^{t} \| \Delta_{n-1} X(s) \|^2 \, ds \right]^{1/2} \qquad (5.54)$$

with

$$\Delta_0 X(t) \leq A(t - t_0)^{1/2} + B(t - t_0)^{1/2} = C(t - t_0)^{1/2} \qquad (5.55)$$

By means of successful substitution, we obtain

$$\| \Delta_n X(t) \| \leq [K(1 + \sqrt{T})]^n [C^2(t - t_0)^{n+1}/(n + 1)!]^{1/2} \qquad (5.56)$$

It is now easy to see that

$$\| X_n(t) - X_m(t) \| = \left\| \sum_{k=m}^{n-1} \Delta_k X(t) \right\| \leq \sum_{k=m}^{n-1} \| \Delta_k X(t) \| \to 0 \qquad (5.57)$$

as $m, n \to \infty$ independent of $t \in T$. Theorem 4.2.3 thus implies that a limiting process $X(t)$ exists such that

$$\underset{n \to \infty}{\text{l.i.m.}} \, X_n(t) = X(t) \qquad (5.58)$$

uniformly on T. According to Theorem 5.2.2, this limiting process $X(t)$ is given by Eq. (5.44). The existence proof is thus complete.

In order to show the uniqueness of this solution, suppose both $X(t)$ and $Y(t)$, with initial conditions $X(t_0) = Y(t_0) = X_0$, satisfy Eq. (5.44). Then, if we set

$$\Delta_n(t) = X_n(t) - Y_n(t)$$
$$\Delta_n f(t) = f(X_n(t), t) - f(Y_n(t), t) \qquad (5.59)$$
$$\Delta_n g(t) = g(X_n(t), t) - g(Y_n(t), t)$$

the nth iterate gives

$$\Delta_n(t) = \int_{t_0}^{t} \Delta_n f(s) \, ds + (I) \int_{t_0}^{t} \Delta_n g(s) \, dB(s) \qquad (5.60)$$

Following the same steps we used in obtaining the inequality (5.56), we get

$$\| \Delta_n(t) \| \leq [K(1 + \sqrt{T})]^n [D^2(t - t_0)^{n+1}/(n + 1)!]^{1/2} \to 0 \qquad (5.61)$$

as $n \to \infty$, where D is a constant. Hence,

$$X(t) = Y(t) \qquad (5.62)$$

with probability one.

As we mentioned earlier, an important property associated with the solution of the Ito equation is the Markov property. This is verified in the next theorem for the scalar case.

Theorem 5.2.5. The solution process $X(t)$ generated by Eq. (5.44) is Markovian.

Proof. Let $t_0 < s < t < a$. We need to show that

$$P\{X(t) \leq x(t) \mid X(s) = x(s)\}$$
$$= P\{X(t) \leq x(t) \mid X(s) = x(s), X(u) = x(u); t_0 \leq u \leq s\} \qquad (5.63)$$

By the additive property of the integrals, we can write

$$X(t) = X(s) + \int_s^t f(X(v), v) \, dv + (\mathrm{I}) \int_s^t g(X(v), v) \, dB(v) \qquad (5.64)$$

As a result of our construction of the successive iterates in the foregoing existence proof, $X(t)$ depends only upon $X(s)$ and the increments $dB(v)$, $v \geq s$. These increments are independent of X_0 by hypothesis and are independent of $dB(u)$, $u \leq s$, as $B(t)$ has independent increments. Hence, they are independent of $X(u)$, $u \leq s$. Equation (5.63) is thus verified.

The same argument shows that the vector Ito equation (5.17) generates a vector Markov solution process. Because of this Markov property, considerable mathematical machinery exists for studying equations of the Ito type. In particular, we have the Kolmogorov and Fokker–Planck equations from which considerable information about the solution process can be obtained. These considerations will be taken up in detail in Chapters 7 and 8.

5.3. General Remarks on the Solution Process

Let us now return to our basic system of random differential equations given by Eqs. (5.1) or Eq. (5.3). For convenience, we reproduce Eq. (5.3) here

$$\dot{\mathbf{X}} = f(\mathbf{X}(t), \mathbf{Y}(t), t), \qquad t \in T; \qquad \mathbf{X}(t_0) = \mathbf{X}_0 \tag{5.65}$$

Some general remarks are in order on what is meant by a solution of the random equation above and what properties of the solution process does one seek in a given situation.

As in the theory of differential equations of nonrandom functions, both quantitative and qualitative theories of random differential equations are important from mathematical as well as physical point of view. In general, certain joint statistical behavior of the initial conditions $\mathbf{X}_0$ and the vector process $\mathbf{Y}(t)$ is specified, in terms of these known statistical properties, one is required to determine certain statistical properties of the solution on a quantitative or qualitative basis.

In the quantitative theory, since the solution process $\mathbf{X}(t)$ is defined by the joint probability distributions of the r.v.'s $\mathbf{X}(t_j)$ for all finite sets $\{t_j\} \in T$, the determination of this family of distribution functions constitutes the solution. It is the solution in the sense that the knowledge of this family of distribution functions permits us to calculate the probabilities of many useful events associated with the process.

This goal, however, is in general difficult, if not impossible, to achieve. We have seen in Section 4.6 the mathematical complexities encountered in computing the distribution functions associated with simple integrals and differentials. Much greater difficulties are certainly expected in dealing with solutions of differential equations. Aside from mathematical difficulties, this requirement for the solution process is also unrealistic from a practical point of view. The evaluation of the joint distribution functions for $\mathbf{X}(t)$ requires considerable information on the joint probabilistic behavior of the initial conditions $\mathbf{X}_0$ together with the stochastic process $\mathbf{Y}(t)$, and this information is generally unavailable in practical situations.

Hence, solutions of random differential equations usually mean much less than joint distribution functions. In most cases, one is able to determine only a very limited number of the statistical properties associated with the solution process. In solving a random differential equation, therefore, it is important for us to recognize first what types of statistical properties of $\mathbf{X}(t)$ are pertinent to the problem at hand, and then proceed to make the attempt to find only these properties.

Relatively speaking, the determination of the moments of a solution process is least troublesome. The moments, particularly the mean and the correlation function, are certainly among the most important features associated with a stochastic process. The mean of a solution process gives an average around which its realizations are grouped. The correlation function gives an indication of the degree to which the realizations at one point t_i are influenced by those at another. At any given point, the variance or the second moment provides a measure of dispersion for possible values the various realizations of the solution process can take.

In many problems, properties of a certain random variable associated with the solution process are of importance. For example, in the study of certain modes of failure in engineering and biological systems, it is important to consider the properties of the random variable which represents the number of crossings of the solution process at a given level in a given time interval. It can be shown that the knowledge of the joint probability distribution of $X(t)$ and $\dot{X}(t)$ over the interval is required for the determination of the properties of this random variable.

The problem referred to above is the so-called threshold crossing problem. Other examples of random variables having considerable practical significance include those characterizing extreme values, lengths of intervals between zeroes, and first passage time. Motivated by their usefulness in applications, these problems have received considerable attention in the applied literature. They will be briefly reviewed in Appendix *B*.

On the qualitative side, the development of stability concepts associated with random differential equations is equally relevant and important but, at this time, is far from being complete. We defer comments on this topic to Chapter 9, in which some aspects of a stochastic stability theory will be explored.

Finally, let us remark that another important but difficult consideration of random differential equations is the sample function approach. As we have pointed out at the beginning of Chapters 3 and 4, stochastic processes describing natural phenomena represent families of ordinary functions, and random differential equations in nature represent families of deterministic sample differential equations. It is thus meaningful to consider sample functions of stochastic processes contained in Eq. (5.65) and its sample solution properties. This line of approach will be considered in Appendix A.

5.4. Classification of Random Differential Equations

The mean square treatment of random differential equations and its applications will be developed in the following chapters. For convenience, this is done in a sequential fashion by starting with the simplest type and proceeding to the most general case. Accordingly, we follow Syski [18] and classify Eq. (5.65) into three basic types:

(a)   Random initial conditions,
(b)   Random inhomogeneous parts,
(c)   Random coefficients.

Chapter 6 is primarily devoted to the treatment of Eq. (5.65) when the initial condition X_0 is the only random element of the equation. This is the simplest case in the sense that it requires a minimum of probabilistic concepts, and the treatment follows very closely that for deterministic differential equations.

Random differential equations with random inhomogeneous parts are considered in Chapter 7. Referring to Eq. (5.65), the vector process $Y(t)$ is present but it enters the equation only as inhomogeneous terms. Thus, the s.p. $Y(t)$ is not coupled with the (unknown) solution process $X(t)$.

The mean square treatment of differential equations with random coefficients is the most difficult and is not as well developed as the first two cases. Chapter 8 is concerned with problems in this area together with some problems of the mixed type where random initial conditions, random inhomogeneous parts, and random coefficients are all present.

When the coefficients of a differential equation are represented by stochastic processes, the passage of stability theory of differential equations from the deterministic case to the random case is interesting and relevant, and merits detailed consideration. This is taken up in Chapter 9. As we shall see, the development in this area is also far from being complete and we will be able to consider only a limited number of particular cases.

References

1. J. L. Strand, *Stochastic Ordinary Differential Equations*. Ph. D. Thesis, Univ. of California, Berkeley, California, 1968.
2. J. L. Strand, Random ordinary differential equations. *J. Differential Equations* **7**, 538–553 (1970).
3. E. A. Coddington and N. Levinson, *Theory of Ordinary Differential Equations*. McGraw-Hill, New York, 1955.

4. E. Hille and R. Phillips, *Functional Analysis and Semi-Groups* (Amer. Math. Soc. Colloq. Publ., No. 31) Amer. Math. Soc., Providence, Rhode Island, 1957.

5. D. A. Bell, *Electrical Noise*. Van Nostrand-Reinhold, Princeton, New Jersey, 1960.

6. M. Aoki, *Optimization of Stochastic Systems*. Academic Press, New York, 1967.

7. H. J. Kushner, *Stochastic Stability and Control*. Academic Press, New York, 1967.

8. K. J. Astrom, *Introduction to Stochastic Control Theory*. Academic Press, New York, 1970.

9. R. L. Stratonovich, *Conditional Markov Processes and Their Applications to the Theory of Optimal Control*. Amer. Elsevier, New York, 1968.

10. D. Middleton, *An Introduction to Statistical Communication Theory*. McGraw-Hill, New York, 1960.

11. R. S. Bucy and P. D. Joseph, *Filtering for Stochastic Processes with Application to Guidance*. Wiley (Interscience), New York, 1968.

12. A. H. Jazwinski, *Stochastic Processes and Filtering Theory*. Academic Press, New York, 1970.

13. K. Ito, *Lectures on Stochastic Processes* (Lecture Notes). Tata Inst. of Fundamental Res., Bombay, India, 1961.

14. J. L. Doob, *Stochastic Processes*. Wiley, New York, 1953.

15. Y. Skorokhod, *Studies in the Theory of Random Processes*. Addison-Wesley, Reading, Massachusetts, 1965.

16. R. L. Stratonovich, A new representation for stochastic integrals and equations. *SIAM J. Control* **4**, 362–371 (1966).

17. R. E. Mortensen, *Mathematical problems of modeling stochastic nonlinear dynamic systems*. NASA CR-1168. NASA, Washington, D. C., 1968.

18. R. Syski, Stochastic differential equation. In *Modern Nonlinear Equations* (T. L. Saaty, ed.), Chapter 8. McGraw-Hill, New York, 1967.

PROBLEMS

5.1. Following the classical Picard existence proof in the deterministic theory of differential equations, supply a proof of Theorem 5.1.2 for the scalar case.

5.2. Consider the linear random vector differential equation

$$\dot{\mathbf{X}}(t) = A(t)\,\mathbf{X}(t) + \mathbf{Y}(t), \qquad t \geq 0; \qquad \mathbf{X}(0) = \mathbf{X}_0$$

where $A(t)$ is an $n \times n$ deterministic and continuous coefficient matrix, $\mathbf{Y}(t)$ is a vector second-order s.p. and $\mathbf{X}_0 \in L_2{}^n$. Show that, by means of Theorem 5.1.2, the mean-square solution of the equation above exists and is unique.

5.3. The second-order Ito integral is defined by

$$(I) \int_0^a X(t)\, d^2 B(t) = \underset{\substack{n \to \infty \\ \Delta_n \to 0}}{\text{l.i.m.}} \sum_{k=0}^{n-1} X(t_k)[B(t_{k+1}) - B(t_k)]^2$$

where $X(t)$ and $B(t)$ satisfy all the conditions given in Section 5.2.2. Show that

$$\text{(I)} \int_0^a X(t) \, d^2B(t) = \int_0^a X(t) \, dt$$

5.4. Consider the Ito integral Eq. (5.24). Show that, if the integrand $X(t)$ is a deterministic continuous function of $t \in T$, the Ito integral reduces to an ordinary m.s. integral.

5.5. Evaluate the expectations

$$E\left\{ \text{(I)} \int_0^a X(t) \, dB(t) \right\}$$

and

$$E\left\{ \left[\text{(I)} \int_0^a X(t) \, dB(t) \right] \left[\text{(I)} \int_0^a Y(t) \, dB(t) \right] \right\}$$

where the s.p.'s $X(t)$ and $Y(t)$ are m.s. continuous on $T = [0, a]$ and they are independent of the increment $\{B(t_{k+1}) - B(t_k)\}$ for all t_k and t_{k+1} satisfying $0 \le t \le t_k \le t_{k+1} \le a$.

5.6. Evaluate the Ito integral

$$\text{(I)} \int_a^b [B(t) - B(a)] \, dB(t)$$

5.7. Show that the existence and uniqueness theorem of the solution to the Ito equation (Theorem 5.2.4) can be generalized to cover the equation

$$dX(t) = f(X(t), t) \, dt + \sum_{k=1}^n g_k(X(t), t) \, dB_k(t)$$

where $f(X(t), t)$ and $g_k(X(t), t)$ satisfy Conditions (a), (b), and (c) in Theorem 5.2.4 and $B_k(t)$ represents a system of Wiener processes.

5.8. Consider the linear Ito equation

$$dX(t) = [f(t) \, X(t) + g(t)] \, dt + [p(t) \, X(t) + q(t)] \, dB(t), \qquad t \in T$$

where $f(t)$, $g(t)$, $p(t)$, and $q(t)$ are continuous on T. Give an existence and uniqueness proof of its m.s. solution.

Chapter 6

Differential Equations with Random Initial Conditions

We consider in this chapter random ordinary differential equations where the randomness enters into the equations only through their initial conditions. This class is treated first because of its mathematical simplicity. The passage from the deterministic situation to the random one for these equations is also most transparent. Furthermore, as we shall see in the next chapter, the methods of solution developed here are useful in more complex situations as well.

It is well known that equations of this type play a prominent role in classical statistical mechanics, statistical thermodynamics, and kinetic theory. In statistical mechanics, for example, the investigation of Lagrange's equations of motion of a system of particles, with a probabilistic description of their initial states in the phase space, forms the basis of the Gibbs–Liouville theory [1]. More recently, the study of differential equations with random initial conditions has been motivated by their usefulness in many other applications. Let us give two examples.

It was briefly mentioned in Chapter 1 that an application of this class of differential equations is found in the analysis of space missions. To elaborate, let us consider in some detail preflight guidance analysis of space trajectories. Due to error sources in the injection guidance system, a spacecraft designed to reach a desired terminal point will not in general achieve it successfully unless corrective maneuvers are made. The preflight guidance analysis of an interplanetary trajectory deals primarily with the problems of examining target dispersions arising from errors in the injection guidance system, the orbit determination process, and other error sources, and of determining the amount of propellent to carry aboard the space vehicle for corrective maneuvers. Hence, the guidance analysis is based on the study of the motion of the spacecraft as described by the equations of motion

$$\dot{\mathbf{X}}(t) = \mathbf{f}(\mathbf{X}(t), t), \qquad \mathbf{X}(t_0) = \mathbf{X}_0 \tag{6.1}$$

where $\mathbf{X}(t)$ is the six-dimensional (position and velocity) state vector of the spacecraft. The initial condition $\mathbf{X}_0$ represents its injection or burn-off conditions. Since most of the error sources mentioned above can only be described statistically, they lead to a statistical description of the initial conditions. Our problem is then one of determining the statistical properties of the state vector $\mathbf{X}(t)$ at some terminal time in terms of the statistics of $\mathbf{X}_0$. This information is then used to estimate the amount of fuel needed for making in-flight trajectory corrections. Problems of this type have been considered by Braham and Skidmore [2], also by Soong and Pfeiffer [3].

Another area in which problems of this genre arise naturally is biosciences. Consider, for example, the mathematical study of chemotherapy or drug administration. Based upon compartmental analysis [4], a basic n-compartment model in chemotherapy is represented by a system of differential equations of the form

$$\dot{X}_i = -a_{ii}X_i + \sum_{\substack{j=1 \\ i \neq j}}^{n} a_{ij}X_j, \qquad i = 1, 2, \ldots, n; \qquad X_i(0) = X_{i0} \tag{6.2}$$

where $X_i(t)$ is the drug or chemical concentration in the ith compartment of a human system, the coefficients a_{ij} are the rate constants, and X_{i0} is the drug concentration in the ith compartment at some initial time instant. In the analysis, the values of X_{i0} at some convenient initial time are often not known with certainty. Furthermore, in using Eq. (6.2) for a large population of human patients, differences in the behavior of individual patients also necessitate the use of probabilistic models for the initial conditions.

The reader can no doubt draw analogies between these examples and many other practical problem areas.

6.1. The Mean Square Theory

In order to show a clear parallel between the mean square theory of this class of random differential equations and the corresponding deterministic theory, it is instructive to consider the existence and uniqueness questions of their mean square solutions. For the sake of explictness, we restrict ourselves to the linear vector–matrix equation

$$\dot{\mathbf{X}}(t) = F(t)\,\mathbf{X}(t), \qquad t \in T = [t_0, a] \tag{6.3}$$

with the random initial condition

$$\mathbf{X}(t_0) = \mathbf{X}_0, \qquad \mathbf{X}_0 \in L_2{}^n \tag{6.4}$$

In Eq. (6.3), we assume that the elements of the $n \times n$ real matrix $F(t)$ are continuous functions of $t \in T$.

Let us first recall some results in the deterministic situation. Suppose for the moment that the initial condition (6.4) takes a deterministic value $\mathbf{x}_0$. With the continuity assumption for $F(t)$, it is well known that Eq. (6.3) has a unique deterministic solution $\mathbf{x}(t)$ which can be put in the form

$$\mathbf{x}(t) = \Phi(t, t_0)\mathbf{x}_0, \qquad t \in T \tag{6.5}$$

The $n \times n$ matrix $\Phi(t, t_0)$ is called the principal matrix associated with $F(t)$. It has the following properties on T:

(a) It is continuous and has a continuous derivative; it is never singular.

(b) For all t_0, we have

$$\Phi(t_0, t_0) = I \tag{6.6}$$

where I is the identity matrix.

(c) For all t_0, t_1, and t_2

$$\Phi(t_2, t_0) = \Phi(t_2, t_1)\,\Phi(t_1, t_0) \tag{6.7}$$

(d) It satisfies its own differential equation

$$\dot{\Phi}(t, t_0) = F(t)\,\Phi(t, t_0) \tag{6.8}$$

We can show that the properties listed above uniquely determine the matrix $\Phi(t, t_0)$ [5]. In the case where $F(t)$ is a constant matrix, it has the explicit form

$$\Phi(t, t_0) = \exp[(t - t_0)F] = \sum_{j=0}^{\infty} [(t - t_0)F]^j/j! \tag{6.9}$$

We now return to Eq. (6.3) with the initial condition (6.4). The existence and uniqueness of its m.s. solution are considered in the next two theorems. As expected, these results follow directly from classical fixed point theorems in Banach spaces [6, 7].

Theorem 6.1.1. Consider Eq. (6.3) whose coefficient matrix $F(t)$ is continuous on T. Let the initial condition be

$$\mathbf{X}(t_0) = \mathbf{0} \tag{6.10}$$

with probability one. Then,

$$\mathbf{X}(t) = \mathbf{0}, \qquad t \in T, \tag{6.11}$$

is its unique m.s. solution. The derivative in Eq. (6.3) is understood to be a m.s. derivative.

Proof. Let us first remark that, based upon Theorem 5.1.1, the random system described by Eqs. (6.3) and (6.10) is equivalent to the system of integral equations

$$\mathbf{X}(t) = \int_{t_0}^{t} F(s)\, \mathbf{X}(s)\, ds, \qquad t \in T \tag{6.12}$$

where the integral is a m.s. integral.

It is easily verified that $\mathbf{X}(t) = 0$ is a m.s. solution. To show it is unique, let us assume that $\mathbf{X}(t)$ is also a m.s. solution. If $L(t)$ is a m.s. integral operator defined by

$$L(t)\, \mathbf{X}(t) \equiv \int_{t_0}^{t} F(s)\, \mathbf{X}(s)\, ds \tag{6.13}$$

we see that

$$\| L(t)\, \mathbf{X}(t) \|_n \leq \int_{t_0}^{t} | F(s) | \cdot \| \mathbf{X}(s) \|_n\, ds \leq ab(t - t_0) \tag{6.14}$$

where

$$a = \max_{s \in T} | F(s) |, \qquad b = \max_{s \in T} \| \mathbf{X}(s) \|_n \tag{6.15}$$

Similarly,

$$\| L^2(t)\, \mathbf{X}(t) \|_n = \left\| \int_{t_0}^{t} F(s)\, L(s)\, \mathbf{X}(s)\, ds \right\|_n$$

$$\leq \int_{t_0}^{t} | F(s) | \cdot \| L(s)\, \mathbf{X}(s) \|_n\, ds$$

$$\leq \int_{t_0}^{t} a^2 b(s - t_0)\, ds = a^2 b(t - t_0)^2/2 \tag{6.16}$$

Inductively, we thus have

$$\| L^m(t) \, \mathbf{X}(t) \|_n \leq a^m b(t - t_0)^m / m! \tag{6.17}$$

Being a m.s. solution, it follows from Eq. (6.12) that

$$\mathbf{X}(t) = L^m(t) \, \mathbf{X}(t), \qquad t \in T \tag{6.18}$$

for all m and

$$\| \mathbf{X}(t) \|_n = \| L^m(t) \, \mathbf{X}(t) \|_n \leq a^m b(t - t_0)^m / m! \tag{6.19}$$

Hence, it is necessary that $\mathbf{X}(t) = 0$, $t \in T$.

Theorem 6.1.2. Consider the random system described by Eqs. (6.3) and (6.4). Let $F(t)$ be continuous on T. Then, there exists a unique m.s. solution which can be represented by

$$\mathbf{X}(t) = \Phi(t, t_0) \, \mathbf{X}_0, \qquad t \in T \tag{6.20}$$

where $\Phi(t, t_0)$ is the (unique) principal matrix associated with the coefficient matrix $F(t)$.

Proof. The equivalent integral equation representation in this case is

$$\mathbf{X}(t) = \mathbf{X}_0 + \int_{t_0}^{t} F(s) \, \mathbf{X}(s) \, ds \tag{6.21}$$

To show that the expression (6.20) is a m.s. solution, we first note that it is second-order and is m.s. continuous on T. Secondly, the substitution of Eq. (6.20) into Eq. (6.21) results

$$\Phi(t, t_0) \mathbf{X}_0 = \mathbf{X}_0 + \int_{t_0}^{t} F(s) \, \Phi(s, t_0) \mathbf{X}_0 \, ds \tag{6.22}$$

This is clearly an identity for any $\mathbf{X}_0 \in L_2^n$ because, by virtue of Eqs. (6.6) and (6.8), the principal matrix $\Phi(t, t_0)$ admits the integral equation representation

$$\Phi(t, t_0) = I + \int_{t_0}^{t} F(s) \, \Phi(s, t_0) \, ds \tag{6.23}$$

Hence, $\mathbf{X}(t) = \Phi(t, t_0) \mathbf{X}_0$ is a m.s. solution.

The uniqueness of this solution follows from Theorem 6.1.1 since, if $\mathbf{Y}(t)$ is also a m.s. solution, the s.p. $\mathbf{X}(t) - \mathbf{Y}(t)$ satisfies the random system considered in Theorem 6.1.1. Hence, $\mathbf{X}(t) - \mathbf{Y}(t) = \mathbf{0}$ identically.

Two important observations can be made at this point regarding the m.s. solutions of differential equations with random initial conditions. Comparing the m.s. solution (6.20) with the deterministic solution (6.5), the m.s. solution is obtainable from its corresponding deterministic solution by simply replacing $\mathbf{x}_0$ by $\mathbf{X}_0$. Thus, clear analogy exists between the mean square theory and the corresponding deterministic theory for this class of random differential equations.

This observations can be extended to the nonlinear case as well. Consider the general system defined by Eq. (6.1). If its corresponding deterministic solution has the form

$$\mathbf{x}(t) = \mathbf{h}(\mathbf{x}_0, t) \tag{6.24}$$

then, the m.s. solution of Eq. (6.1) is

$$\mathbf{X}(t) = \mathbf{h}(\mathbf{X}_0, t) \tag{6.25}$$

provided, of course, that both the deterministic and the m.s. solutions exist.

The second point is that, as seen from Eq. (6.25), the solution process for this class of differential equations has a simple structure; its components are s.p.'s with at most n degrees of randomness. Equation (6.25) basically defines an algebraic transformation of a random vector $\mathbf{X}_0$ into another random vector $\mathbf{X}(t)$, with t as a parameter. Thus, elementary mathematical tools exist for determining the statistical properties of $\mathbf{X}(t)$ in terms of those associated with $\mathbf{X}_0$.

Because of this simple structure, differential equations with only random initial conditions are sometimes called "crypto-deterministic." Whittaker [8] used this name to describe motion of particles which is subject to uncertainties at the initial position, but its evolution with time follows a deterministic law.

6.2. Statistical Properties of the Solution Process

Consider the random system described by

$$\dot{\mathbf{X}}(t) = \mathbf{f}(\mathbf{X}(t), t), \qquad t \in T; \qquad \mathbf{X}(t_0) = \mathbf{X}_0 \tag{6.26}$$

where the only random element involved is the initial condition $\mathbf{X}_0$. Assuming that the solution process $\mathbf{X}(t)$, $t \in T$, exists, we are now interested in the determination of some of the probability distributions associated with the solution process given the joint probability distributions of $\mathbf{X}_0$. In what follows, we consider two basic tools for doing this.

For additional references, the reader is referred to Syski [9]. Some of the examples given below are due to him.

6.2.1. Transformation of Random Variables

Let us assume that an explicit m.s. solution of Eq. (6.26) can be found. Based upon our discussion in the preceding section, the solution process takes the general form

$$\mathbf{X}(t) = \mathbf{h}(\mathbf{X}_0, t) \tag{6.27}$$

which essentially represents an algebraic transformation of a set of random variables into another. We recall that a technique for determining the relationship between the joint density functions of two sets of random variables has been discussed in Section 2.4.2. This can obviously be used to advantage here.

The main results contained in Section 2.4.2 immediately lead to a useful procedure for determining the joint density function $f(\mathbf{x}, t)$ of solution process $\mathbf{X}(t)$ at a given t in terms of the joint density function $f_0(\mathbf{x}_0)$ of the initial condition $\mathbf{X}_0$. This procedure is stated below as a theorem.

Theorem 6.2.1. Suppose that the transformation defined by Eq. (6.27) is continuous in $\mathbf{X}_0$, has continuous partial derivatives with respect to $\mathbf{X}_0$, and defines a one-to-one mapping. Then, if the inverse transform is written as

$$\mathbf{X}_0 = \mathbf{h}^{-1}(\mathbf{X}, t), \tag{6.28}$$

the joint density function $f(\mathbf{x}, t)$ of $\mathbf{X}(t)$ is given by

$$f(\mathbf{x}, t) = f_0[\mathbf{x}_0 = \mathbf{h}^{-1}(\mathbf{x}, t)] \, |J| \tag{6.29}$$

where J is the Jacobian

$$J = \left| \frac{\partial \mathbf{x}_0^{\mathrm{T}}}{\partial \mathbf{x}} \right| \tag{6.30}$$

Let us now give some examples.

Example 6.1. Consider a linear oscillator whose equation of motion is

$$\ddot{X}(t) + \omega^2 X(t) = 0, \qquad 0 \le t < \infty \tag{6.31}$$

with random initial conditions

$$X(0) = X_0 \qquad \text{and} \qquad \dot{X}(0) = \dot{X}_0 \qquad (6.32)$$

The circular frequency ω is assumed to be deterministic and the joint density function $f_0(x_0, \dot{x}_0)$ of X_0 and $\dot{X}_0$ is assumed known.

Let

$$\mathbf{X}(t) = \begin{bmatrix} X(t) \\ \dot{X}(t) \end{bmatrix} \qquad (6.33)$$

Equation (6.31) can be written in the vector form

$$\dot{\mathbf{X}}(t) = \begin{bmatrix} 0 & 1 \\ -\omega^2 & 0 \end{bmatrix} \mathbf{X}(t) \qquad (6.34)$$

The well-known solution of Eq. (6.34) is

$$\mathbf{X}(t) = \Phi(t)\, \mathbf{X}_0 = \begin{bmatrix} \cos \omega t & (1/\omega) \sin \omega t \\ -\omega \sin \omega t & \cos \omega t \end{bmatrix} \mathbf{X}_0 \qquad (6.35)$$

with

$$\Phi^{-1}(t) = \begin{bmatrix} \cos \omega t & -(1/\omega) \sin \omega t \\ \omega \sin \omega t & \cos \omega t \end{bmatrix} \qquad (6.36)$$

According to Eqs. (6.29) and (6.30), the joint density function $f(\mathbf{x}, t) = f(x, t; \dot{x}, t)$ of $X(t)$ and $\dot{X}(t)$ is then given by

$$f(\mathbf{x}, t) = f_0[\mathbf{x}_0 = \Phi^{-1}(t)\, \mathbf{x}]\, |J| \qquad (6.37)$$

with

$$J = \left| \frac{\partial \mathbf{x}_0^{\mathrm{T}}}{\partial \mathbf{x}} \right| = |\Phi^{-1}(t)| = 1 \qquad (6.38)$$

Hence, the general form of $f(\mathbf{x}, t)$ is

$$\begin{aligned} f(\mathbf{x}, t) &= f(x, t; \dot{x}, t) \\ &= f_0(x \cos \omega t - (\dot{x}/\omega) \sin \omega t,\ \omega x \sin \omega t + \dot{x} \cos \omega t) \end{aligned} \qquad (6.39)$$

As an illustration, let $X(0)$ and $\dot{X}(0)$ be two independent random variables, both are Gaussian with means zero and variances σ_1^2 and σ_2^2, respectively. The joint density function $f_0(\mathbf{x}_0)$ is

$$f_0(\mathbf{x}_0) = f_0(x_0, \dot{x}_0) = \frac{1}{2\pi\sigma_1\sigma_2} \exp\left[-\tfrac{1}{2}\left(\frac{x_0^2}{\sigma_1^2} + \frac{\dot{x}_0^2}{\sigma_2^2} \right) \right] \qquad (6.40)$$

Equation (6.39) then gives

$$f(x, t; \dot{x}, t) = \frac{1}{2\pi\sigma_1\sigma_2} \exp\left[-\tfrac{1}{2}\left\{\frac{(x\cos\omega t - (\dot{x}/\omega)\sin\omega t)^2}{\sigma_1{}^2}\right.\right.$$

$$\left.\left. + \frac{(\omega x\sin\omega t + \dot{x}\cos\omega t)^2}{\sigma_2{}^2}\right\}\right] \tag{6.41}$$

and, upon simplifying,

$$f(x, t; \dot{x}, t) = \frac{1}{2\pi\sigma_X\sigma_{\dot{X}}(1 - \varrho^2)^{1/2}} \exp\left[-\frac{1}{2\sigma_X{}^2\sigma_{\dot{X}}{}^2(1 - \varrho^2)}(x^2\sigma_{\dot{X}}{}^2\right.$$

$$\left. - 2\varrho\sigma_X\sigma_{\dot{X}}x\dot{x} + \dot{x}^2\sigma_X{}^2)\right] \tag{6.42}$$

where

$$\sigma_X{}^2 = \sigma_1{}^2\cos^2\omega t + \sigma_2{}^2\sin^2\omega t/\omega^2$$

$$\sigma_{\dot{X}}{}^2 = \sigma_2{}^2\cos^2\omega t + \omega^2\sigma_1{}^2\sin^2\omega t \tag{6.43}$$

$$\varrho = (\sigma_2{}^2 - \omega^2\sigma_1{}^2)\sin\omega t\cos\omega t/\sigma_1\sigma_2\omega$$

The joint density function of $X(t)$ and $\dot{X}(t)$ is seen to be bivariate Gaussian. They are statistically dependent with a time-dependent correlation coefficient.

We note in passing that the results given above for the Gaussian case can be obtained directly from Eq. (6.35). Since $X(t)$ and $\dot{X}(t)$ are linear in X_0 and $\dot{X}_0$, their joint density function must be of the form given by Eq. (6.42). Their means, variances, and covariance can be easily computed from Eq. (6.35).

It is clear that the marginal density functions of $X(t)$ and $\dot{X}(t)$ can be obtained easily from the general result Eq. (6.39) by integration.

We may also obtain other joint density functions such as the one of $X(t_1)$ and $X(t_2)$. For this case, the only change in the procedure is that, if we let

$$\mathbf{X}(t_{12}) = \begin{bmatrix} X(t_1) \\ X(t_2) \end{bmatrix} = \Phi(t_{12})\,\mathbf{X}_0 \tag{6.44}$$

the matrix $\Phi(t_{12})$ replaces $\Phi(t)$ in Eq. (6.35) with

$$\Phi(t_{12}) = \begin{bmatrix} \cos\omega t_1 & (1/\omega)\sin\omega t_1 \\ \cos\omega t_2 & (1/\omega)\sin\omega t_2 \end{bmatrix} \tag{6.45}$$

followed by corresponding changes from that point on.

Example 6.2. Consider a scalar Riccati equation

$$\dot{X}(t) = -aX^2(t), \qquad a > 0, \qquad 0 \le t < \infty \tag{6.46}$$

with a random initial condition

$$X(0) = X_0 \tag{6.47}$$

The explicit form of the solution process $X(t)$ is

$$X(t) = X_0/(1 + aX_0 t) \tag{6.48}$$

We thus have

$$X_0 = X/(1 - aXt) \tag{6.49}$$

The Jacobian J defined by Eq. (6.30) is now

$$J = |\,\partial x_0/\partial x\,| = 1/(1 - axt)^2 \tag{6.50}$$

and, as seen from Eq. (6.29), the density function $f(x, t)$ of $X(t)$ has the form

$$f(x, t) = f_0(x/(1 - axt))/(1 - axt)^2 \tag{6.51}$$

where $f_0(x_0)$ is the probability density function of X_0.

6.2.2. Applications of the Liouville Equation

As we have seen in the foregoing, the use of Theorem 6.2.1 in the determination of $f(\mathbf{x}, t)$ requires the explicit form of the solution process. An alternate method of solution for $f(\mathbf{x}, t)$ is to make use of the fundamental Liouville's theorem in the theory of dynamic systems. This approach converts the problem to one of solving an initial value problem involving a first-order partial differential equation.

Theorem 6.2.2 (Liouville's Theorem). Assume that the m.s. solution process $\mathbf{X}(t)$ of Eq. (6.26) exists. Then, the joint density function $f(\mathbf{x}, t)$ of $\mathbf{X}(t)$ satisfies the Liouville equation

$$\frac{\partial f}{\partial t} + \sum_{j=1}^{n} \frac{\partial (ff_j)}{\partial x_j} = 0 \tag{6.52}$$

where $\mathbf{x}^{\mathrm{T}} = [x_1\, x_2 \cdots x_n]$ and $[\mathbf{f}(\mathbf{x}, t)]^{\mathrm{T}} = [f_1\, f_2 \cdots f_n]$. The superscript T denotes transpose.

Proof. This theorem is a form of the Liouville's Theorem for classical dynamic systems where the density function $f(\mathbf{x}, t)$ is an integral invariant of the system defined by Eq. (6.26); that is, for any region $S(t)$ at $t \in T$ the integral

$$\int \cdots \int_{S(t)} f(\mathbf{x}, t) \, dx_1 \, dx_2 \cdots dx_n = \int_{S(t)} f(\mathbf{x}, t) \, d\mathbf{x}$$

is independent of t. A proof based upon this property is given by Syski [9].

We shall give an alternate proof of this theorem from a probabilistic point of view using the concept of characteristic functions. This approach is due to Kozin [10].

Let us denote the joint characteristic function of $X_1(t)$, $X_2(t)$, $\ldots$, $X_n(t)$ by $\phi(u_1, t; u_2, t; \ldots; u_n, t) = \phi(\mathbf{u}, t)$. Then

$$\phi(\mathbf{u}, t) = E\left\{\exp\left[i \sum_{j=1}^{n} u_j X_j(t)\right]\right\} = \int_{-\infty}^{\infty} \exp(i\mathbf{u}^T \mathbf{x}) f(\mathbf{x}, t) \, d\mathbf{x} \qquad (6.53)$$

Differentiating Eq. (6.53) with respect to t gives

$$\dot{\phi}(\mathbf{u}, t) = E\left\{i \sum_{m=1}^{n} u_m \dot{X}_m(t) \exp\left[i \sum_{j=1}^{n} u_j X_j(t)\right]\right\} \qquad (6.54)$$

and, using Eq. (6.26),

$$\dot{\phi}(\mathbf{u}, t) = E\left\{i \sum_{m=1}^{n} u_m f_m(\mathbf{X}(t), t) \exp\left[i \sum_{j=1}^{n} u_j X_j(t)\right]\right\}$$

$$= i \sum_{m=1}^{n} u_m E\left\{f_m \exp\left[i \sum_{j=1}^{n} u_j X_j(t)\right]\right\} \qquad (6.55)$$

Now, the functions $\phi(\mathbf{u}, t)$ and $f(\mathbf{x}, t)$ form an n-dimensional Fourier transform pair. The inverse Fourier transform of Eq. (6.55) hence gives the partial derivative of $f(\mathbf{x}, t)$ with respect to t. It is left as an exercise for the reader to show that this inverse Fourier transform is

$$\frac{\partial f}{\partial t} = -\sum_{j=1}^{n} \frac{\partial(ff_j)}{\partial x_j} \qquad (6.56)$$

and the theorem is proved.

We thus see that the problem of determining the joint probability density function $f(\mathbf{x}, t)$ using the Liouville's theorem is an initial value problem for first-order partial differential equations, the initial value being the joint probability density function of the initial conditions.

A general explicit solution of the Liouville equation (6.52) can be found by examining its associated Lagrange system. Let us write Eq. (6.52) in the form

$$\frac{\partial f}{\partial t} + f \sum_{j=1}^{n} \frac{\partial f_j}{\partial x_j} + \sum_{j=1}^{n} f_j \frac{\partial f}{\partial x_j} = 0 \tag{6.57}$$

or

$$\frac{\partial f}{\partial t} + f \boldsymbol{V} \cdot \mathbf{f}(\mathbf{x}, t) + \sum_{j=1}^{n} f_j \frac{\partial f}{\partial x_j} = 0 \tag{6.58}$$

where the sum $\sum_{j=1}^{n} \partial f_j / \partial x_j$ is identified with the divergence of the vector $\mathbf{f}(\mathbf{x}, t)$.

The Lagrange system of Eq. (6.58) is

$$\frac{dt}{1} = - \frac{df}{f \boldsymbol{V} \cdot \mathbf{f}(\mathbf{x}, t)} = \frac{dx_1}{f_1} = \frac{dx_2}{f_2} = \cdots = \frac{dx_n}{f_n} \tag{6.59}$$

The first equality of the above yields the solution

$$f(\mathbf{x}, t) = f_0(\mathbf{x}_0) \exp\left\{- \int_{t_0}^{t} \boldsymbol{V} \cdot \mathbf{f}[\mathbf{x} = \mathbf{h}(\mathbf{x}_0, \tau), \tau] \, d\tau\right\}\Bigg|_{\mathbf{x}_0 = \mathbf{h}^{-1}(\mathbf{x}, t)} \tag{6.60}$$

where $f_0(\mathbf{x}_0)$ is the joint density function of the initial condition $\mathbf{X}_0$.

Example 6.3. Let us consider again the linear oscillator studied in Example 6.1. As seen from Eq. (6.34),

$$f_1(\mathbf{x}, t) = x_2, \qquad f_2(\mathbf{x}, t) = -\omega^2 x_1 \tag{6.61}$$

Hence,

$$\boldsymbol{V} \cdot \mathbf{f}(\mathbf{x}, t) = \frac{\partial f_1}{\partial x_1} + \frac{\partial f_2}{\partial x_2} = 0 \tag{6.62}$$

and Eq. (6.60) gives the solution

$$f(\mathbf{x}, t) = f_0[\mathbf{x}_0 = \mathbf{h}^{-1}(\mathbf{x}, t)] \tag{6.63}$$

which is in agreement with the previous result.

It is of interest to note that Theorem 6.2.1 leads to the general formula (6.29) for the determination of $f(\mathbf{x}, t)$, and Theorem 6.2.2 leads to the general formula (6.60). Comparing these two formulas, we find that the absolute value of the Jacobian in Eq. (6.29) can be evaluated by

$$|J| = \exp\left\{- \int_{t_0}^{t} \boldsymbol{V} \cdot \mathbf{f}[\mathbf{x} = \mathbf{h}(\mathbf{x}_0, \tau), \tau] \, d\tau\right\}\Bigg|_{\mathbf{x}_0 = \mathbf{h}^{-1}(\mathbf{x}, t)} \tag{6.64}$$

This formula is useful particularly when the dimension of the Jacobian is large.

We have presented two methods for determining the joint density function of the m.s. solution $\mathbf{X}(t)$ when the initial conditions of a differential equations are random. In general, this determination presents no conceptual difficulty if the solution of the corresponding deterministic differential equation can be found. Thus, for this class of random differential equations, a great deal of information about the statistical behavior of its m.s. solution can often be obtained.

The statistical moments of the solution process can, of course, be found in a much simpler way. They can be determined directly from the explicit form of the solution process in terms of the initial conditions. Hence, if Eq. (6.27) represents the solution process for the general problem, the nth moment of $X_i(t)$, the ith component of $\mathbf{X}(t)$, is immediately given by

$$E\{X_i{}^n(t)\} = \int_{-\infty}^{\infty} h_i{}^n(\mathbf{x}_0, t) f_0(\mathbf{x}_0) \, d\mathbf{x}_0 \qquad (6.65)$$

For a system of linear equations, the solution process is a linear function of the initial conditions. It is easily seen that the mean of the solution process has the following property:

Consider the linear case

$$\dot{\mathbf{X}}(t) = F(t) \, \mathbf{X}(t), \qquad t \in T \qquad (6.66)$$

The initial condition $\mathbf{X}(t_0) = \mathbf{X}_0$ is random with a joint density function $f_0(\mathbf{x}_0)$.

Lemma. Assume that the m.s. solution of Eq. (6.66) exists. Then, the mean of the solution process $\mathbf{X}(t)$, $E\{\mathbf{X}(t)\}$, satisfies the same differential equation

$$dE\{\mathbf{X}(t)\}/dt = F(t) \, E\{\mathbf{X}(t)\} \qquad (6.67)$$

with the initial condition

$$E\{\mathbf{X}(t_0)\} = E\{\mathbf{X}_0\} \qquad (6.68)$$

The proof is immediate in view of the commutability of the m.s. derivative and expectation.

References

1. J. W. Gibbs, *The Collected Works of J. Willard Gibbs*, Vol. II. Yale Univ. Press, New Haven, Connecticut, 1948.
2. H. S. Braham and L. J. Skidmore, Guidance-error analysis of satellite trajectories. *J. Aerosp. Sci.* **29**, 1091–1097 (1962).
3. T. T. Soong and C. G. Pfeiffer, Unified guidance analysis in design of space trajectories. *J. Spacecr. Rockets* **3**, 98–103 (1966).
4. T. Teorell, Kinetics of distribution of substances administered in the body. *Arch. Internat. Pharmacodyn. Ther.* **57**, 205–240 (1937).
5. E. A. Coddington and N. Levinson, *Theory of Ordinary Differential Equations.* McGraw-Hill, New York, 1955.
6. E. Hille and R. Phillips, *Functional Analysis and Semi-Groups* (Amer. Math. Soc. Colloq. Publ., No. 31). Amer. Math. Soc., Providence, Rhode Island, 1957.
7. P. A. Ruymgaart, *The Kalman-Bucy Filter and Its Behavior with Respect to Smooth Perturbations of the Involved Wiener-Levy Processes.* Ph. D. Thesis, Dept. of Math., Technol. Univ. Delft, Delft, The Netherlands, 1971.
8. E. T. Whittaker, *Analytical Dynamics*, 4th ed. Cambridge Univ. Press, London and New York, 1937.
9. R. Syski, Stochastic differential equations. In *Modern Nonlinear Equations* (T. L. Saaty, ed.), Chapter 8. McGraw-Hill, New York, 1967.
10. F. Kozin, On the probability densities of the output of some random systems. *J. Appl. Mech.* **28**, 161–165 (1961).

PROBLEMS

6.1. Consider the first-order linear differential equation

$$\dot{X}(t) + aX(t) = 0, \qquad t \geq 0$$

where $X(0) = X_0$ is a Gaussian r.v. with mean m and variance σ^2. Determine the joint density function $f(x, t; \dot{x}, t)$ of $X(t)$ and $\dot{X}(t)$.

6.2. A two-compartment model in chemical kinetics or in chemotherapy is described by

$$\dot{\mathbf{X}}(t) = \begin{bmatrix} -a_1 & a_2 \\ a_1 & -a_2 \end{bmatrix} \mathbf{X}(t), \qquad a_1, a_2 > 0, t \geq 0$$

with random initial condition $\mathbf{X}(0) = \mathbf{X}_0$.

(a)   Express the joint density function $f(\mathbf{x}, t)$ of $\mathbf{X}(t)$ in terms of $f_0(\mathbf{x}_0)$, the joint density function of $\mathbf{X}_0$.

(b)   Express the joint density function $f(x_1, t_1; x_2, t_2)$ of $X_1(t_1)$ and $X_2(t_2)$ in terms of $f_0(\mathbf{x}_0)$.

(c) Let $a_1 = 1$, $a_2 = 0.5$, $X_{02} = 1 - X_{01}$, and X_{01} is uniformly distributed between 0.7 and 1.0. Determine $f(\mathbf{x}, t)$ and plot the marginal density function $f(x_1, t)$ at $t = 10$.

6.3. Consider Example 6.2. Let $f_0(x_0)$ be uniformly distributed between a and b, a, $b > 0$. Determine $f(x, t)$, $E\{X(t)\}$, and $E\{X^2(t)\}$.

6.4. Show in detail that the inverse Fourier transform of Eq. (6.55) is Eq. (6.56).

6.5. Equations (6.67) and (6.68) define the mean equation for the solution of the random differential equation (6.66). Derive the corresponding correlation function equation, that is, a deterministic differential equation satisfied by the correlation function matrix of $\mathbf{X}(t)$.

6.6. Verify the result of Example 6.2 using Theorem 6.2.2.

6.7. Let $X(t)$ be the m.s. solution of

$$\sum_{j=0}^{n} a_j(t) \, X^{(j)}(t) = 0, \qquad a_n(t) \neq 0, t \geq 0$$

where $X^{(j)}(t) \equiv d^j X(t)/dt^j$. Show that the joint density function of $X(t)$ together with its $(n - 1)$ m.s. derivatives at t is given by

$$f(\mathbf{x}, t) \equiv f(x, t; \dot{x}, t; \ldots; x^{(n-1)}, t)$$

$$= f_0(\mathbf{x}_0 = \mathbf{h}^{-1}(\mathbf{x}, t)) \exp\left[\int_0^t \frac{a_{n-1}(s)}{a_n^{(s)}} \, ds\right]$$

Chapter 7

Differential Equations with Random Inhomogeneous Parts

This chapter is devoted to the study of differential equations of the type

$$\dot{\mathbf{X}}(t) = \mathbf{f}(\mathbf{X}(t), t) + \mathbf{Y}(t), \, t \in T = [t_0, a] \tag{7.1}$$

with the initial condition

$$\mathbf{X}(t_0) = \mathbf{X}_0 \tag{7.2}$$

where $\mathbf{Y}(t)$ is a vector stochastic process which enters the equation as an inhomogeneous term. The initial condition $\mathbf{X}_0$ can be taken to be random or deterministic.

As is well known, the solution process $\mathbf{X}(t)$ can be considered as the output of a dynamical system governed by a set of ordinary differential equations with an input $\mathbf{Y}(t)$. Schematically, Eq. (7.1) describes the situation shown in Fig. 7.1, where the dynamics of the system S is assumed known precisely.

Our main concern is the determination of the statistical properties of the output process $\mathbf{X}(t)$ in terms of the joint statistical properties of $\mathbf{Y}(t)$ and $\mathbf{X}_0$. We note in passing that, in view of the development in Chapter 6, we are not particularly concerned with the part which $\mathbf{X}_0$ plays in this general problem.

152

Fig. 7.1. The system description of Eq. (7.1).

The study of random differential equations of this type began with Langevin's investigation in 1908 concerning the motion of a particle executing Brownian motion. The differential equation investigated by Langevin is

$$m\dot{X}(t) = -fX(t) + Y(t) \tag{7.3}$$

where $X(t)$ is the velocity of a free particle in one-dimensional motion. The influence of the surrounding medium on the particle consists of a friction term, $-fX(t)$, and a random forcing term, $Y(t)$, representing the molecular impact.

The development of the theory of this class of differential equations since that time has been progressing at an amazing pace. In recent years, this subject has become an important and a natural mathematical discipline in the study of a wide variety of applied problems. Indeed, as shown in Fig. 7.1, equations of this type now describe systems of all descriptions subject to random disturbances or inputs. Some examples of these inputs are environmental effects in vibrational studies of mechanical and electrical systems, earthquake disturbance and wind load in structural analysis, noise-corrupted signals in communication theory, the motion of the sea or ground roughness in vehicle dynamics considerations, economic and political stimuli in the analysis of economical systems, biochemical inputs to biological systems, and human behavior in the study of man–machine systems. In these examples, inability to determine the future values of an input from a knowledge of the past has forced the introduction of random inputs and random forcing functions.

7.1. The Mean Square Theory

Following our development in Section 6.1, we shall discuss first in some detail the mean square solutions associated with linear random differential equations of this type. Our aim is to give an explicit solution representation for the linear case under certain conditions and to demonstrate again the close tie between the mean square theory and the corresponding deterministic theory.

Let us consider the vector linear differential equation

$$\dot{\mathbf{X}}(t) = F(t)\,\mathbf{X}(t) + \mathbf{Y}(t), \qquad t \in T; \qquad \mathbf{X}(t_0) = \mathbf{X}_0 \qquad (7.4)$$

where $F(t)$ is an $n \times n$ real matrix whose elements are continuous functions of $t \in T$. The s.p. $\mathbf{Y}(t): T \to L_2^n$ continuously and $\mathbf{X}_0 \in L_2^n$. We have no need to assume the independence of $\mathbf{Y}(t)$ and $\mathbf{X}_0$; but the initial condition is generally independent of the forcing term or the input in practice.

To review the corresponding deterministic theory, consider the deterministic version of Eq. (7.4) which has the form

$$\dot{\mathbf{x}}(t) = F(t)\,\mathbf{x}(t) + \mathbf{y}(t), \qquad t \in T; \qquad \mathbf{x}(t_0) = \mathbf{x}_0 \qquad (7.5)$$

where $\mathbf{y}(t)$ is continuous in $t \in T$. It is well known [1] that a unique solution of Eq. (7.5) exists. It may be written in the form

$$\mathbf{x}(t) = \Phi(t, t_0)\mathbf{x}_0 + \int_{t_0}^{t} \Phi(t, s)\,\mathbf{y}(s)\,ds, \qquad t \in T \qquad (7.6)$$

where, as defined in Section 6.1, $\Phi(t, t_0)$ is the principal matrix associated with $F(t)$.

Our mean square theory for Eq. (7.4) is now developed along the same lines as in Section 6.1.

Theorem 7.1.1. Consider Eq. (7.4) where the coefficient matrix $F(t)$, the s.p. $\mathbf{Y}(t)$, and the initial condition $\mathbf{X}_0$ satisfy all the stated conditions. Then, there exists a unique m.s. solution which can be represented by

$$\mathbf{X}(t) = \Phi(t, t_0)\mathbf{X}_0 + \int_{t_0}^{t} \Phi(t, s)\,\mathbf{Y}(s)\,ds, \qquad t \in T \qquad (7.7)$$

The integral above is a m.s. integral.

Proof. The uniqueness proof follows directly from Theorem 6.1.1, and it is exactly the same as the one we gave in the proof of Theorem 6.1.2.

In order to show that Eq. (7.7) is the m.s. solution, we first note from Theorem 6.1.2 that the first term in Eq. (7.7) is the m.s. complementary solution of Eq. (7.4). Next, consider the integral

$$\mathbf{Z}(t) = \int_{t_0}^{t} \Phi(t, s)\,\mathbf{Y}(s)\,ds \qquad (7.8)$$

It exists as a m.s. integral by virtue of the properties possessed by $\Phi(t, s)$ and $\mathbf{Y}(s)$. Furthermore, the s.p. $\mathbf{Z}(t)$ is m.s. continuous on T (see Section 4.5.1). According to the m.s. Leibniz rule [Eq. (4.160)], the m.s. derivative of $\mathbf{Z}(t)$ exists and

$$\dot{\mathbf{Z}}(t) = \int_{t_0}^{t} \frac{\partial \Phi(t, s)}{\partial t} \mathbf{Y}(s) \, ds + \Phi(t, t) \mathbf{Y}(t) \tag{7.9}$$

Now, we have indicated in Section 6.1 that

$$\Phi(t, t) = I \qquad \text{and} \qquad \dot{\Phi}(t, s) = \frac{\partial \Phi(t, s)}{\partial t} = F(t) \Phi(t, s) \tag{7.10}$$

The substitution of Eqs. (7.10) into Eq. (7.9) thus yields

$$\dot{\mathbf{Z}}(t) = F(t) \int_{t_0}^{t} \Phi(t, s) \mathbf{Y}(s) \, ds + \mathbf{Y}(t) \tag{7.11}$$

Hence, the m.s. derivative of $\mathbf{X}(t)$, given by Eq. (7.7), exists and we get

$$\begin{aligned}
\dot{\mathbf{X}}(t) &= \dot{\Phi}(t, t_0) \mathbf{X}_0 + \dot{\mathbf{Z}}(t) \\
&= F(t) \left[\Phi(t, t_0) \mathbf{X}_0 + \int_{t_0}^{t} \Phi(t, s) \mathbf{Y}(s) \, ds \right] + \mathbf{Y}(t) \\
&= F(t) \mathbf{X}(t) + \mathbf{Y}(t) \tag{7.12}
\end{aligned}$$

We thus see that the differential equation (7.4) is satisfied in the m.s. sense by the s.p. $\mathbf{X}(t)$ given by Eq. (7.7). It is obvious that the initial condition is also satisfied. Our proof is now complete.

This theorem again points out that, at least in the linear case, the deterministic solution and the m.s. solution admit the same solution representation when they exist. Again, this observation is important because it allows us to construct m.s. solutions for this class of random differential equations along traditional deterministic lines.

The development of a practical mean square theory for the general nonlinear case give rise to a number of difficulties. These difficulties have been commented upon in Chapter 5.

7.1.1. Equations of the Ito Type

It has been stressed repeatedly that random differential equations involving white noise are of great practical importance. Considerable attention will be paid to these equations in this chapter. In order to do this, it is pertinent to advance several remarks concerning the m.s. solution of Eq. (7.4) when $\mathbf{Y}(t)$ contains Gaussian white noise components. We are particularly interested in arriving at a convenient solution representation from which properties of the solution process can be determined.

Let us consider Eq. (7.4) where the input process $\mathbf{Y}(t)$ has Gaussian white noise components, that is,

$$\dot{\mathbf{X}}(t) = F(t)\,\mathbf{X}(t) + \mathbf{W}(t), \qquad t \geq 0; \qquad \mathbf{X}(0) = \mathbf{0} \qquad (7.13)$$

where $\mathbf{W}(t)$, $t \geq 0$, is a vector Gaussian white noise, and we have taken $\mathbf{X}(0) = \mathbf{0}$ for convenience. Theorem 7.1.1 does not apply in this case since Eq. (7.13) does not exist in the mean square sense. Following our discussion of the Ito formulation given in Section 5.2, a formal interpretation of Eq. (7.13) is

$$d\mathbf{X}(t) = F(t)\,\mathbf{X}(t)\,dt + d\mathbf{B}(t), \qquad t \geq 0; \qquad \mathbf{X}(0) = \mathbf{0} \qquad (7.14)$$

and, in integral equation form,

$$\mathbf{X}(t) = \int_0^t F(s)\,\mathbf{X}(s)\,ds + \mathbf{B}(t) \qquad (7.15)$$

We note that Ito integrals do not appear in the above.

We now can consider a formal m.s. solution of Eq. (7.13) in the sense that it satisfies the integral equation (7.15). The uniqueness and existence of this solution has already been established in Theorem 5.2.4. Moreover, it is easy to show that it has an explicit solution representation

$$\mathbf{X}(t) = \int_0^t \Phi(t, s)\,d\mathbf{B}(s) \qquad (7.16)$$

The integral above exists as a m.s. Riemann–Stieltjes integral because the elements of the covariance matrix associated with $\mathbf{B}(t)$ are of bounded variation on $T \times T$ (Theorem 4.5.2). According to Theorem 4.5.3, the integration by parts of the integral above yields

$$\mathbf{X}(t) = \mathbf{B}(t) - \int_0^t [\partial\Phi(t, s)/\partial s]\,\mathbf{B}(s)\,ds \qquad (7.17)$$

The solution process $\mathbf{X}(t)$ as expressed by Eq. (7.16) or Eq. (7.17) exists and is m.s. continuous. We point out again that it is only a formal solution of our original differential equation (7.13) since it does not have a m.s. derivative.

It is of interest to compare the results obtained above with another "formal" solution of Eq. (7.13) in the form

$$\mathbf{X}(t) = \int_0^t \Phi(t, s)\,\mathbf{W}(s)\,ds \qquad (7.18)$$

which we would have obtained if we had proceeded formally according to Theorem 7.1.1 without worrying about its mean square properties. It turns out that, and we will demonstrate it in Example 7.1, the statistical properties of $\mathbf{X}(t)$ as obtained from Eq. (7.16) or Eq. (7.17) are the same as those calculated from Eq. (7.18). This observation is a useful one since it implies that, although Eq. (7.13) is meaningless as it stands, we still get the "correct" solution by treating the white noise as one satisfying all the conditions set forth in Theorem 7.1.1. This situation is somewhat analogous to the treatment of the Dirac delta function in the real analysis. The Dirac delta function itself is meaningless as an ordinary function, but its integrals make mathematical sense.

7.2. Solution Processes for Linear Equations

The linear case is important and has been considered in connection with numerous applied problems. We take up this case first and begin by determining the moments of its solution process.

7.2.1. Means, Correlation Functions, and Spectral Densities

Let us return to the linear vector differential equation given by Eq. (7.4) and consider its solution process in the form

$$\mathbf{X}(t) = \Phi(t, t_0)\mathbf{X}_0 + \int_{t_0}^{t} \Phi(t, s)\, \mathbf{Y}(s)\, ds, \qquad t \in T \tag{7.19}$$

Clearly, a fruitful procedure for determining the moments of the solution process $\mathbf{X}(t)$ is to make use of this explicit solution representation. The first term introduces nothing new. For convenience, we will consider only the particular solution

$$\mathbf{X}(t) = \int_{t_0}^{t} \Phi(t, s)\, \mathbf{Y}(s)\, ds, \qquad t \in T \tag{7.20}$$

Noting that the integral above is a m.s. integral, the relationships between the means and correlation function matrices of $\mathbf{X}(t)$ and $\mathbf{Y}(t)$ have already been outlined in Section 4.5.2 for the scalar case. For the vector case, we easily get

$$E\{\mathbf{X}(t)\} = \int_{t_0}^{t} \Phi(t, s)\, E\{\mathbf{Y}(s)\}\, ds, \qquad t \in T \tag{7.21}$$

for the mean. The determination of the correlation function matrix $\Gamma_{XX}(t, s)$ $= E\{\mathbf{X}(t)\,\mathbf{X}^{\mathrm{T}}(s)\}$ is also straightforward. We have

$$\Gamma_{XX}(t, s) = E\left\{\left[\int_{t_0}^{t} \Phi(t, u)\,\mathbf{Y}(u)\,du\right]\left[\int_{t_0}^{s} \Phi(s, v)\,\mathbf{Y}(v)\,dv\right]^{\mathrm{T}}\right\}$$

$$= \int_{t_0}^{s}\int_{t_0}^{t} \Phi(t, u)\,\Gamma_{YY}(u, v)\,\Phi^{\mathrm{T}}(s, v)\,du\,dv \qquad (7.22)$$

Higher-order moments associated with the solution process $\mathbf{X}(t)$ can be obtained following the same procedure; the mean and the correlation function matrix, however, are by far more important.

The mean of $\mathbf{X}(t)$ can, of course, also be determined directly from the original differential equation by writing

$$\frac{d}{dt} E\{\mathbf{X}(t)\} = F(t)\,E\{\mathbf{X}(t)\} + E\{\mathbf{Y}(t)\}, \quad t \in T; \quad E\{\mathbf{X}(t_0)\} = E\{\mathbf{X}_0\} \quad (7.23)$$

whose solution is the mean of $\mathbf{X}(t)$ as a function of $E\{\mathbf{Y}(t)\}$ and $E\{\mathbf{X}_0\}$.

It is noted that, while a relation between $\Gamma_{XX}(t, s)$ and $\Gamma_{YY}(t, s)$ can be easily established, the task of evaluating the double integral in Eq. (7.22) is in general a tedious one.

Example 7.1. Consider the response of a mass–spring linear oscillator to a white noise random excitation. The equation of motion is

$$\ddot{X}(t) + \omega_0^2\,X(t) = W(t), \qquad t \geq 0 \qquad (7.24)$$

where $W(t)$ is a Gaussian white noise process with mean zero and correlation function $\Gamma_{WW}(t, s) = 2D\,\delta(t - s)$. For convenience, we set $X(0) = 0$.

We wish to demonstrate by means of this example that Eqs. (7.16) and (7.18) give the same properties for the solution process $X(t)$. Let $X(t) = X_1(t)$, $\dot{X}(t) = X_2(t)$, and

$$\mathbf{X}(t) = \begin{bmatrix} X_1(t) \\ X_2(t) \end{bmatrix} \qquad (7.25)$$

The vector–matrix form of Eq. (7.24) is

$$\dot{\mathbf{X}}(t) = F(t)\,\mathbf{X}(t) + \mathbf{W}(t) \qquad (7.26)$$

where

$$F(t) = \begin{bmatrix} 0 & 1 \\ -\omega_0^2 & 0 \end{bmatrix}, \qquad \mathbf{W}(t) = \begin{bmatrix} 0 \\ W(t) \end{bmatrix} \qquad (7.27)$$

Using Eq. (7.16), the solution $\mathbf{X}(t)$ is given by

$$\mathbf{X}(t) = \int_0^t \Phi(t, s) \, d\mathbf{B}(s) \tag{7.28}$$

where

$$d\mathbf{B}(s) = \begin{bmatrix} 0 \\ dB(s) \end{bmatrix} \tag{7.29}$$

with $E\{B(t)\} = 0$ and $E\{B(t) B(s)\} = 2D \min(t, s)$. It is easy to show that

$$\Phi(t, s) = \Phi(t - s) = \begin{bmatrix} \cos \omega_0(t - s) & \sin \omega_0(t - s)/\omega_0 \\ -\omega_0 \sin \omega_0(t - s) & \cos \omega_0(t - s) \end{bmatrix} \tag{7.30}$$

Hence, the solution for $X(t)$ as obtained from Eq. (7.28) is

$$X_1(t) = X(t) = (1/\omega_0) \int_0^t \sin \omega_0(t - s) \, dB(s) \tag{7.31}$$

Since the s.p. $B(t)$, $t \geq 0$, is a Gaussian process, it follows from Theorem 4.6.4 that $X(t)$ is also Gaussian. Its joint distributions are then determined by its mean and the correlation function.

Clearly, we have

$$E\{X(t)\} = (1/\omega_0) \int_0^t \sin \omega_0(t - s) \, dE\{B(s)\} = 0 \tag{7.32}$$

As for the correlation function, we need to evaluate

$$\Gamma_{XX}(t, s) = (1/\omega_0^2) \int_0^s \int_0^t \sin \omega_0(t - u) \sin \omega_0(s - v) \, dd E\{B(u) B(v)\}$$

$$= (1/\omega_0^2) \int_0^s \int_0^t \sin \omega_0(t-u) \sin \omega_0(s-v) \, dd[2D \min(u, v)] \tag{7.33}$$

Suppose that $s \geq t$, Eq. (7.33) becomes

$$\Gamma_{XX}(t, s) = (2D/\omega_0^2) \int_0^s \int_0^t \sin \omega_0(t - u) \sin \omega_0(s - v) \, d_u d_v[\min(u, v)]$$

$$= (2D/\omega_0^2) \int_0^t \sin \omega_0(t - u) \, d_u \left[\int_0^u \sin \omega_0(s - v) \, d_v(v) \right]$$

$$= (2D/\omega_0^2) \int_0^t \sin \omega_0(t - u) \sin \omega_0(s - u) \, du$$

$$= \frac{D}{\omega_0^2} \left[t \cos \omega_0(t - s) - \frac{1}{2\omega_0} \sin \omega_0(t - s) - \frac{1}{2\omega_0} \sin \omega_0(t + s) \right] \tag{7.34}$$

The result corresponding to the case $s < t$ is obtained from above by simply interchanging t and s.

Let us now use the formal representation of $X(t)$ as given by Eq. (7.18), that is,

$$X(t) = (1/\omega_0) \int_0^t \sin \omega_0(t - s) \, W(s) \, ds \qquad (7.35)$$

Again, the Gaussianity of $W(t)$ implies that $X(t)$, $t \geq 0$, is a Gaussian process. Its mean is

$$E\{X(t)\} = (1/\omega_0) \int_0^t \sin \omega_0(t - s) \, E\{W(s)\} \, ds = 0 \qquad (7.36)$$

The correlation function is given by

$$\Gamma_{XX}(t, s) = (1/\omega_0{}^2) \int_0^s \int_0^t \sin \omega_0(t - u) \sin \omega_0(s - v) \, E\{W(u) \, W(v)\} \, du \, dv$$

$$= (2D/\omega_0{}^2) \int_0^s \int_0^t \sin \omega_0(t - u) \sin \omega_0(s - v) \, \delta(u - v) \, du \, dv \qquad (7.37)$$

For the case $s \geq t$, we get

$$\Gamma_{XX}(t, s) = (2D/\omega_0{}^2) \int_0^t \sin \omega_0(t - u) \sin \omega_0(s - u) \, du \qquad (7.38)$$

It is seen that the results given by Eqs. (7.36) and (7.38) are in agreement with, respectively, those given by Eqs. (7.32) and (7.34). This serves to demonstrate that Eq. (7.18) indeed leads to the correct result.

Let us remark in passing that the solution of Eq. (7.24) can be written down directly without the necessity of converting it to a system of first-order differential equations. It is well known that Eq. (7.31) can be obtained from

$$X(t) = \int_0^t h(t - s) \, dB(s) \qquad (7.39)$$

where $h(t)$ is the *weighting function* or the *impulse response function* of the system. The function $h(t)$ admits either one of the following equivalent interpretations: (a) $h(t)$ is the solution of the differential equation

$$\ddot{h}(t) + \omega_0{}^2 \, h(t) = \delta(t), \qquad t \geq 0 \qquad (7.40)$$

with $\dot{h}(0) = h(0) = 0$; (b) it is the solution of the homogeneous equation

$$\ddot{h}(t) + \omega_0^2 \, h(t) = 0, \qquad t \geq 0 \tag{7.41}$$

with $\dot{h}(0) = 1$ and $h(0) = 0$. In both cases, $h(t) = 0$, $t < 0$.

Example 7.2. Let us consider a more general case where the input process is nonwhite. The equation of motion for a damped linear oscillator is

$$\ddot{X}(t) + 2\omega_0 \zeta \, \dot{X}(t) + \omega_0^2 \, X(t) = Y(t), \qquad t \geq 0; \qquad \dot{X}(0) = X(0) = 0 \tag{7.42}$$

where the s.p. $Y(t)$ is assumed to be m.s. continuous and wide-sense stationary with mean zero and correlation function $\Gamma_{YY}(\tau)$. Only the case where $\zeta < 1$, the underdamped case, is considered.

According to Theorem 7.1.1, the solution of Eq. (7.42) is

$$X(t) = \int_0^t h(t - s) \, Y(s) \, ds \tag{7.43}$$

where the impulse response $h(t)$ has the form

$$h(t) = (1/\omega_1) \exp(-\zeta\omega_0 t) \sin \omega_1 t, \qquad t \geq 0 \tag{7.44}$$

with $\omega_1 = \omega_0(1 - \zeta^2)^{1/2}$.

We shall consider only the mean square output, that is, the second moment of $X(t)$. As indicated by Eq. (7.22), we have

$$E\{X^2(t)\} = \Gamma_{XX}(t, t) = \int_0^t \int_0^t h(t - u) \, h(t - v) \, \Gamma_{YY}(u - v) \, du \, dv \tag{7.45}$$

which is also the variance $\sigma_X^2(t)$ of $X(t)$ in this case.

The expression given by Eq. (7.45) can be expressed in terms of the power spectral density $S_{YY}(\omega)$ of the s.p. $Y(t)$. The function $S_{YY}(\omega)$ is related to the correlation function $\Gamma_{YY}(\tau)$ by [Eq. (3.62)]

$$\Gamma_{YY}(\tau) = \int_0^\infty S_{YY}(\omega) \cos \omega\tau \, d\omega \tag{7.46}$$

On substituting it into Eq. (7.45), we have

$$\sigma_X^2(t) = \int_0^\infty \int_0^t \int_0^t S_{YY}(\omega) \cos \omega(u - v) \, h(t - u) \, h(t - v) \, du \, dv \, d\omega \tag{7.47}$$

The double integral on t in Eq. (7.47) can be integrated using the expression for $h(t)$ given by Eq. (7.44). The result is

$$\sigma_X{}^2(t) = \int_0^\infty |Z(\omega)|^{-2} S_{YY}(\omega) \left\{ 1 + \exp(-2\omega_0\zeta t)\left[1 + \frac{2\omega_0\zeta}{\omega_1} \sin \omega_1 t \cos \omega_1 t \right. \right.$$

$$- \exp(\omega_0\zeta t)\left(2 \cos \omega_1 t + \frac{2\omega_0\zeta}{\omega_1} \sin \omega_1 t\right) \cos \omega t$$

$$- \frac{2\omega}{\omega_1} \exp(\omega_0\zeta t) \sin \omega_1 t \sin \omega t$$

$$\left. \left. + \frac{1}{\omega_1{}^2} [\omega_0{}^2\zeta^2 - \omega_1{}^2 + \omega^2] \sin^2 \omega_1 t \right] \right\} d\omega \qquad (7.48)$$

where

$$Z(\omega) = \omega_0{}^2 - \omega^2 + 2i\omega\omega_0\zeta \qquad (7.49)$$

Two observations can be made at this point. First, Eq. (7.48) clearly shows that, even in simple cases, the determination of the second moment is in general a tedious one. Assuming that $S_{YY}(\omega)$ is analytic, the integral in Eq. (7.48) can be carried out by means of contour integration, but it is evident that a great deal of work is involved. The second point is that, as seen from the expression given by Eq. (7.48), the variance $\sigma_X{}^2(t)$ is in general a function of t. Hence, a wide-sense stationary input in general gives rise to a nonstationary output. However, for large t, the variance approaches to a constant

$$\sigma_X{}^2(t) \to \int_0^\infty |Z(\omega)|^{-2} S_{YY}(\omega)\, d\omega \qquad (7.50)$$

It can also be shown that $\Gamma_{XX}(t, s) \to \Gamma_{XX}(t - s)$ as $t \to \infty$. Therefore, the output becomes wide-sense stationary in the limit as $t \to \infty$.

Caughey and Stumpf [2] have shown that an approximate evaluation of the integral occurring in Eq. (7.48) is possible if $S_{YY}(\omega)$ is relatively smooth with no sharp peaks and if ζ is small. Under these conditions, the function $|Z(\omega)|^{-2}$ is sharply peaked at $\omega = \omega_0$, and the main contribution to the integral comes from the region around $\omega = \omega_0$. Thus, Eq. (7.48) can be approximated by letting $S_{YY}(\omega) = S_{YY}(\omega_0)$, a constant. It can now be integrated by contour integration with the result

$$\sigma_X{}^2(t) \cong \frac{\pi S_{YY}(\omega_0)}{4\zeta\omega_0{}^3}$$

$$\times \left\{ 1 - \frac{1}{\omega_1{}^2} \exp(-2\omega_0\zeta t)[\omega_1{}^2 + 2(\omega_0\zeta)^2 \sin^2 \omega_1 t + \omega_0\omega_1\zeta \sin 2\omega_1 t] \right\} \qquad (7.51)$$

For the case $\zeta = 0$, Eq. (7.51) reduces to

$$\sigma_X{}^2(t) \cong \frac{\pi}{4} \; \frac{S_{YY}(\omega_0)}{\omega_0{}^3} \; [2\omega_0 t - \sin 2\omega_0 t] \qquad (7.52)$$

Equations (7.51) and (7.52) are plotted in Fig. 7.2. It is shown that the variances take an oscillatory pattern and reach stationary values as t becomes large. For $\zeta = 0.1$, the stationary value is reached in roughly three cycles.

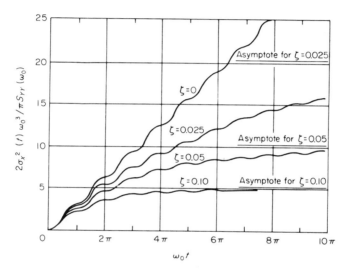

Fig. 7.2. Variance of $X(t)$ for Example 7.2 (from Caughey and Stumpf [2]).

It should be noted that the approximation made above in the evaluation of $\sigma_X{}^2(t)$ is precisely equivalent to saying that the s.p. $Y(t)$ can be approximated by a white noise process with a constant spectral density function $S_{YY}(\omega_0)$. We see here an example where, although the white noise process cannot be realized physically, it leads to considerable mathematical simplifications. Moreover, it gives a good approximation to the actual variance of the output under certain conditions.

In the study of responses of dynamic systems to random excitations, the modeling of random excitations by nonstationary stochastic processes is necessary in many practical cases. Nonstationary inputs lead to nonstationary outputs, and the computation of the output moments is generally quite tedious. A typical situation is considered in the example below.

Example 7.3. Consider the problem of determining the effect on an earthbound structure of an earthquake-type disturbance. In order to describe in some quantitative manner the nature of an earthquake disturbance, Bogdanoff *et al.* [3] have considered simple nonstationary random models for describing ground accelerations due to strong-motion earthquakes. In their analysis, one of the models describing ground accelerations is given by

$$Y(t) = \sum_{j=1}^{n} ta_j \exp(-\alpha_j t) \cos(\omega_j t + \Theta_j), \qquad t \geq 0; \qquad Y(t) = 0, \qquad t < 0$$
$$(7.53)$$

where a_j, α_j, and ω_j are given sets of real positive numbers. The set Θ_j is one of independent random variables uniformly distributed over an interval of length 2π. We easily find that

$$E\{Y(t)\} = 0$$
$$(7.54)$$
$$\Gamma_{YY}(t, s) = \sum_{j=1}^{n} \tfrac{1}{2} tsa_j^2 \exp[-\alpha_j(t + s)] \cos \omega_j(t - s), \qquad t, s \geq 0$$

It is clear from the above that the s.p. $Y(t)$ defined by Eq. (7.53) is nonstationary. It is also evident that it is m.s. continuous.

The plausibility of choosing $Y(t)$ having this particular form stems mainly from the fact that the sample functions generated by Eq. (7.53) possess the usual appearance of earthquake accelograms. Figure 7.3 gives a typical sample function of $Y(t)$ for $n = 20$ together with the corresponding variance. We see that it compares favorably with a typical earthquake accelogram record shown in Fig. 7.4. Both indicate a general trend of gradual increasing and then decreasing variance as a function of time, and both exhibit erratic behavior.

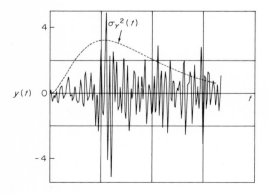

Fig. 7.3. A typical sample function of $Y(t)$ for $n = 20$ (from Bogdanoff *et al.* [3]).

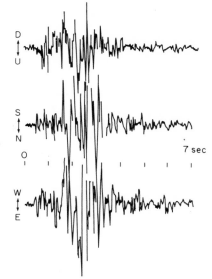

Fig. 7.4. Section of a typical earthquake accelogram record in up–down, North–South, and East–West directions (from Bogdanoff *et al.* [3]).

In the work of Bogdanoff *et al.* [3] a simple structure approximating a linear one-story building is considered. As shown in Fig. 7.5, horizontal displacement of the roof is assumed and it is also assumed that the ground motion due to earthquake disturbances is in the same direction. The mean and the covariance of the roof displacement with respect to the ground are of our chief interest.

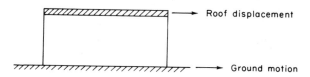

Fig. 7.5. A linear one-story building.

Let us assume that the structure is at rest at $t = 0$ and let $X(t)$, $t \geq 0$, be the relative horizontal displacement of the roof with respect to the ground. Then, based upon an idealized linear model, the relative displacement $X(t)$ is governed by [3]

$$\ddot{X}(t) + 2\zeta\omega_0\dot{X} + \omega_0^2 X = -Y(t), \qquad t \geq 0 \tag{7.55}$$

The m.s. solution of the above is

$$X(t) = -\int_0^t h(t - s)\, Y(s)\, ds \tag{7.56}$$

With $\zeta < 1$, the impulse response $h(t)$ is given by [see Eq. (7.44)]

$$h(t) = (1/\omega_0') \exp(-\zeta\omega_0 t) \sin \omega_0' t, \qquad t \geq 0$$

$$\omega_0' = \omega_0 (1 - \zeta^2)^{1/2} \tag{7.57}$$

The mean and the correlation function of $Y(t)$ are given in Eq. (7.54). We easily find that

$$E\{X(t)\} = 0 \tag{7.58}$$

The correlation function of $X(t)$ is given by

$$\Gamma_{XX}(t, s) = \int_0^s \int_0^t h(t - u) \, h(s - v) \, \Gamma_{YY}(u, v) \, du \, dv \tag{7.59}$$

and, upon substituting Eqs. (7.54) and Eq. (7.57) into Eq. (7.59) and integrating, we obtain

$$\Gamma_{XX}(t, s) = \frac{1}{2\omega_0'^2} \exp[-\zeta\omega_0(t + s)] \sum_{j=1}^{n} \left(\frac{a_j}{b_j}\right)^2 [c_j(t, s) + c_j(s, t)] \tag{7.60}$$

where

$$b_j = (\alpha_j'^2 - \omega_j^2 - \omega_0'^2)^2 + 4\alpha_j'^2\omega_j^2 \tag{7.61}$$

$$\begin{aligned}
c_j(t, s) = {}& \tfrac{1}{2}\omega_0'^2 \exp[-\alpha_j'(t + s)][\beta_j(t)\,\beta_j(s) + 4\omega_j^2\,\gamma_j(t)\,\gamma_j(s)] \cos \omega_j(t - s) \\
& + \omega_0' \exp(-\alpha_j' t)\{[4\alpha_j'\omega_j^2\,\gamma_j(t) \\
& + \beta_j(t)(\alpha_j'^2 - \omega_j^2 - \omega_0'^2)] \cos \omega_j t \sin \omega_0' s \\
& - 2\omega_0'[2\omega_j^2\,\gamma_j(t) + \alpha_j'\,\beta_j(t)] \cos \omega_j t \cos \omega_0' s \\
& + 2\omega_j[(\alpha_j'^2 - \omega_j^2 - \omega_0'^2)\,\gamma_j(t) - \alpha_j\beta_j(t)] \sin \omega_j t \cos \omega_0' s\} \\
& - \alpha_j'\omega_0'(\alpha_j'^2\omega_j^2 - \omega_0'^2) \sin \omega_0'(t + s) \\
& + 2\omega_0'^2(\omega_j^2 + \alpha_j'^2) \cos \omega_0' t \cos \omega_0' s \\
& + \tfrac{1}{2}[4\alpha_j'^2\omega_j^2 + (\alpha_j'^2 - \omega_j^2 - \omega_0'^2)^2] \sin \omega_0' t \sin \omega_0' s
\end{aligned} \tag{7.62}$$

$$\alpha_j' = \alpha_j - \zeta\omega_j, \quad \beta_j(t) = (\alpha_j'^2 - \omega_j^2 - \omega_0'^2)t + 2\alpha_j', \quad \gamma_j(t) = \alpha_j' t + 1 \tag{7.63}$$

The variance $\sigma_X^2(t)$ is easily obtained from Eq. (7.60). It has the form

$$\sigma_X^2(t) = \frac{1}{\omega_0'^2} \exp(-2\zeta\omega_0 t) \sum_{j=1}^{n} \left(\frac{a_j}{b_j}\right)^2 c_j(t, t) \tag{7.64}$$

A practical question may be raised at this point: In what ways are these moment results useful for designers of earthquake-resisting structures? The answer must hinge upon the criteria chosen for determining the structural safety. For example, if one chooses to consider that whenever $|\ddot{X}(t)|$, the absolute value of the relative acceleration, exceeds a certain value λ, the failure occurs, then we know from the Tchebycheff inequality that

$$P\{|\ddot{X}(t)| \geq \lambda\} \leq \sigma_{\ddot{X}}^2(t)/\lambda^2 \tag{7.65}$$

where $\sigma_{\ddot{X}}^2(t)$ can be determined from $\Gamma_{XX}(t, s)$ by [see Eq. (4.133)]

$$\sigma_{\ddot{X}}^2(t) = \Gamma_{\ddot{X}\ddot{X}}(t, t) = \partial^4\Gamma_{XX}(t, s)/\partial t^2\ \partial s^2\ |_{s=t} \tag{7.66}$$

Hence, our moment results can be used to provide an upper bound on the probabilistic failure criterion chosen here. Addition discussions in the same vein are given in Appendix B.

7.2.2. A Special Case

An important special case of the foregoing is one where the linear system under investigation is time-invariant, the input is wide-sense stationary, and only the steady-state output is considered. Under these conditions, we can write

$$\mathbf{X}(t) = \int_{-\infty}^{t} \Phi(t - s)\ \mathbf{Y}(s)\ ds, \tag{7.67}$$

where $\mathbf{Y}(t)$ is wide-sense stationary. Time invariance implies that the coefficient matrix $F(t)$ in Eq. (7.4) is a constant matrix. As indicated by Eq. (6.9), the principal matrix is now only a function of $(t - s)$. The negative infinity in the lower limit of the integral indicates that the solution process is in general a function of all past values of the input process $\mathbf{Y}(t)$. Indeed, the condition $t_0 \to -\infty$ eliminates the possibility of "seeing" the starting transient.

By a change of variable, Eq. (7.67) takes the form

$$\mathbf{X}(t) = \int_{0}^{\infty} \Phi(s)\ \mathbf{Y}(t - s)\ ds \tag{7.68}$$

Since $\mathbf{Y}(t)$ is wide-sense stationary, its mean is a constant and it follows directly from Eq. (7.68) that $\mathbf{X}(t)$ has a constant mean value. Consider the correlation function matrix of $\mathbf{X}(t)$. We have

$$\Gamma_{XX}(t, s) = \int_{0}^{\infty} \int_{0}^{\infty} \Phi(u)\ \Gamma_{YY}(t - u, s - v)\ \Phi^T(v)\ du\ dv$$

$$= \int_{0}^{\infty} \int_{0}^{\infty} \Phi(u)\ \Gamma_{YY}(t - s - u + v)\ \Phi^T(v)\ du\ dv \tag{7.69}$$

In particular, the mean square value of $\mathbf{X}(t)$, $\Gamma_{XX}(t, t)$, is given by

$$\Gamma_{XX}(t, t) = \int_0^\infty \int_0^\infty \Phi(u)\, \Gamma_{YY}(u - v)\, \Phi^{\mathrm{T}}(v)\, du\, dv \qquad (7.70)$$

It is seen from Eq. (7.69) that the correlation function matrix $\Gamma_{XX}(t, s)$ is only a function of $(t - s)$. Thus, we make the important observation that, for constant-coefficient linear equations with solution processes represented by Eq. (7.67), wide-sense stationary inputs lead to wide-sense stationary solutions.

The stationarity in the wide-sense of the solution process in this situation implies the existence of its associated power spectral density. In what follows, we shall derive an important relation between the spectral densities of $\mathbf{Y}(t)$ and $\mathbf{X}(t)$.

Notice first that, since $\Phi(t) = 0$ for $t < 0$ for physically realizable systems, we may replace the lower limit of the integral in Eq. (7.68) by negative infinity. Let us now consider the cross-correlation function matrix $\Gamma_{YX}(s) = E\{\mathbf{Y}(t)\,\mathbf{X}^{\mathrm{T}}(t + s)\}$. Using Eq. (7.68), we have

$$\Gamma_{YX}(s) = E\left\{ \int_{-\infty}^{\infty} \mathbf{Y}(t)\, \mathbf{Y}^{\mathrm{T}}(t + s - u)\, \Phi^{\mathrm{T}}(u)\, du \right\}$$

$$= \int_{-\infty}^{\infty} \Gamma_{YY}(s - u)\, \Phi^{\mathrm{T}}(u)\, du \qquad (7.71)$$

Similarly, we can write

$$\Gamma_{XX}(s) = E\left\{ \int_{-\infty}^{\infty} \Phi(u)\, \mathbf{Y}(t - u)\, \mathbf{X}^{\mathrm{T}}(t + s)\, du \right\}$$

$$= \int_{-\infty}^{\infty} \Phi(u)\, \Gamma_{YX}(s + u)\, du \qquad (7.72)$$

The definitions of power spectral densities and cross-power spectral densities are given in Section 3.2.2. Hence, the Fourier transforms of Eqs. (7.71) and (7.72) yield, respectively,

$$S_{YX}(\omega) = S_{YY}(\omega)\, H^{\mathrm{T}}(i\omega) \qquad (7.73)$$

and

$$S_{XX}(\omega) = H^*(i\omega)\, S_{YX}(\omega) \qquad (7.74)$$

where

$$H(i\omega) = \int_{-\infty}^{\infty} \Phi(\tau) e^{-i\omega\tau}\, d\tau \qquad (7.75)$$

is the well-known system function or frequency-response function associated with the linear system (7.4). $H^*(i\omega)$ denotes its complex conjugate.

Finally, the substitution of Eq. (7.73) into Eq. (7.74) gives the desired result

$$S_{XX}(\omega) = H^*(i\omega)\, S_{YY}(\omega)\, H^T(i\omega) \tag{7.76}$$

This simple relation is one of the primary reasons for the use of spectral densities in the analysis of stationary stochastic processes. We also note that the second moment of $X(t)$ can be computed from Eq. (7.76) by integrating a single integral

$$\Gamma_{XX}(0) = \tfrac{1}{2} \int_{-\infty}^{\infty} S_{XX}(\omega)\, d\omega = \tfrac{1}{2} \int_{-\infty}^{\infty} H^*(i\omega)\, S_{YY}(\omega)\, H^T(i\omega)\, d\omega \tag{7.77}$$

In general, Eq. (7.77) is considerably simpler to use than Eq. (7.70).

Example 7.4. Let us reexamine the problem in Example 7.2; our interest is now only in the determination of the second moment of the steady-state response $X(t)$.

Consider the differential equation (7.42). Let us first mention that, without converting the second-order differential equation into a system of first-order differential equations, the input–output relation for the spectral densities corresponding to Eq. (7.76) can be easily shown to be

$$S_{XX}(\omega) = |\, y(i\omega)\,|^2\, S_{YY}(\omega) \tag{7.78}$$

where

$$y(i\omega) = \int_{-\infty}^{\infty} h(t)\, e^{-i\omega t}\, dt \tag{7.79}$$

is the frequency response function associated with Eq. (7.42). It takes the simple form

$$y(i\omega) = 1/Z(\omega) \tag{7.80}$$

where $Z(\omega)$ is defined by Eq. (7.49).

In terms of the spectral density of the input, we can use Eq. (7.78) and immediately write down the spectral density of the output. We thus have

$$S_{XX}(\omega) = |\, Z(\omega)\,|^{-2}\, S_{YY}(\omega) \tag{7.81}$$

As seen from Eq. (7.77), the second moment of $X(t)$ is given by

$$E\{X^2(t)\} = \Gamma_{XX}(0) = \tfrac{1}{2} \int_{-\infty}^{\infty} S_{XX}(\omega)\, d\omega = \int_{0}^{\infty} |\, Z(\omega)\,|^{-2}\, S_{YY}(\omega)\, d\omega \tag{7.82}$$

We see that this result is in agreement with that given by Eq. (7.50), which is the asymptotic value of the second moment considered in Example 7.2.

7.2.3. Distribution and Density Functions

In Section 4.6, we have considered a technique for determining the joint probability distributions of the m.s. integral of a s.p. $X(t)$ in terms of the joint distributions of $X(t)$. Hence, at least in principle, a procedure exists for determining the joint probability distributions of the solution process as represented by Eq. (7.20). We shall not pursue this further here because it does not in general lead to fruitful results. It is important to point out, however, that we have an exception when the input process $Y(t)$ is Gaussian. According to Theorem 4.6.4, $X(t)$ is also Gaussian in this case and the knowledge of its mean and correlation matrix determine its joint distributions.

In this section, we shall follow an alternate route in determining the joint distributions of the solution process. This technique has the advantage that explicit results can be found for a class of linear as well as nonlinear equations of physical importance. Our development is based upon the work of Bartlett [4] and Pawula [5], and is somewhat parallel to the discussion in Chapter 6 based upon the Liouville's theorem.

Let $f(x, t)$ be the first probability density function of a s.p. $X(t)$, $t \in T$, at t. It then has the general property that

$$f(x, t + \Delta t) = \int_{-\infty}^{\infty} f(x, t + \Delta t \mid x', t) f(x', t) \, dx' \qquad (7.83)$$

where we recall that $f(x, t + \Delta t \mid x', t)$ is the conditional density function of the r.v. $X(t + \Delta t)$ given that $X(t) = x'$. If we denote by $\phi(u, t + \Delta t \mid x', t)$ the conditional characteristic function of the r.v. $\Delta X = X(t + \Delta t) - X(t)$ given that $X(t) = x'$, that is,

$$\phi(u, t + \Delta t \mid x', t) = E\{e^{iu \Delta X} \mid x', t\}$$

$$= \int_{-\infty}^{\infty} e^{iu \Delta x} f(x, t + \Delta t \mid x', t) \, dx, \quad \Delta x = x - x', \quad (7.84)$$

then, by inverse Fourier transform,

$$f(x, t + \Delta t \mid x', t) = (1/2\pi) \int_{-\infty}^{\infty} e^{-iu \Delta x} \phi(u, t + \Delta t \mid x', t) \, du \qquad (7.85)$$

and, upon expanding $\phi(u, t + \Delta t \mid x', t)$ in a Taylor series about $u = 0$, we have

$$f(x, t + \Delta t \mid x', t) = \sum_{n=0}^{\infty} \frac{a_n(x', t)}{2\pi n!} \int_{-\infty}^{\infty} (iu)^n \, e^{-iu\,\Delta x} \, du$$

$$= \sum_{n=0}^{\infty} \frac{(-1)^n}{n!} \, a_n(x', t) \frac{\partial^n}{\partial x^n} \, \delta(\Delta x) \qquad (7.86)$$

where

$$a_n(x', t) = E\{\Delta X^n \mid x', t\}$$

$$= E\{[X(t + \Delta t) - X(t)]^n \mid X(t) = x'\} \qquad (7.87)$$

These expectations are sometimes called the *incremental moments* associated with $X(t)$.

Upon substituting Eq. (7.86) into Eq. (7.83) we get, after integration,

$$f(x, t + \Delta t) = \sum_{n=0}^{\infty} \frac{(-1)^n}{n!} \frac{\partial^n}{\partial x^n} [a_n(x, t) f(x, t)] \qquad (7.88)$$

or

$$f(x, t + \Delta t) - f(x, t) = \sum_{n=1}^{\infty} \frac{(-1)^n}{n!} \frac{\partial^n}{\partial x^n} [a_n(x, t) f(x, t)] \qquad (7.89)$$

Dividing the equation above by Δt and taking the limit as $\Delta t \to 0$ gives

$$\frac{\partial f(x, t)}{\partial t} - \sum_{n=1}^{\infty} \frac{(-1)^n}{n!} \frac{\partial^n}{\partial x^n} [\alpha_n(x, t) f(x, t)] = 0 \qquad (7.90)$$

where

$$\alpha_n(x, t) = \lim_{\Delta t \to 0} (1/\Delta t) E\{[X(t + \Delta t) - X(t)]^n \mid X(t) = x\}, \qquad n = 1, 2, \dots \qquad (7.91)$$

are called the *derivate moments* of the s.p. $X(t)$. We shall refer to Eq. (7.90) as the *kinetic equation* associated with the s.p. $X(t)$.

The kinetic equation is a deterministic partial differential equation satisfied by the first density function $f(x, t)$ of the s.p. $X(t)$. When $X(t)$ represents the solution process of a random differential equation, it provides a mathematical tool for determining the density function provided that the derivate moments can be found and that suitable initial and boundary conditions are obtainable. It is clear that there are inherent difficulties in solving Eq. (7.90) in its general form. This approach is useful, however, when only a few of the derivate moments are nonzero.

Let us now pursue it a bit further and see whether this technique can be extended to the determination of the second density functions. For a second density function, say, $f(x, t; x_1, t_1)$, we may start by writing

$$f(x, t; x_1, t_1) = f(x, t \mid x_1, t_1) f(x_1, t_1) \qquad (7.92)$$

Knowing $f(x_1, t_1)$, we only need to determine the conditional density function $f(x, t \mid x_1, t_1)$, and we now show that this can be done by following the same procedure. As a starting point, we replace Eq. (7.83) by

$$f(x, t+\Delta t \mid x_1, t_1) = \int_{-\infty}^{\infty} f(x, t+\Delta t \mid x', t; x_1, t_1) f(x', t \mid x_1, t_1) \, dx' \qquad (7.93)$$

Following the same steps, we find that $f(x, t \mid x_1, t_1)$ satisfies

$$\frac{\partial f(x, t \mid x_1, t_1)}{\partial t} - \sum_{n=1}^{\infty} \frac{(-1)^n}{n!} \frac{\partial^n}{\partial x^n} [\beta_n(x, t; x_1, t_1) f(x, t \mid x_1, t_1)] = 0 \qquad (7.94)$$

where $\beta_n(x, t; x_1, t_1)$, $n = 1, 2, \ldots$, are now the derivate moments conditional upon *both* $X(t) = x$ and $X(t_1) = x_1$, that is,

$$\beta_n(x, t; x_1, t_1) = \lim_{\Delta t \to 0} \frac{1}{\Delta t} E\{[X(t + \Delta t) - X(t)]^n \mid X(t) = x, \quad X(t_1) = x_1\} \qquad (7.95)$$

It is evident that other similar partial differential equations can be set up for obtaining the general nth density function of the s.p. $X(t)$.

Let us now make several observations regarding the kinetic equation (7.90).

(a) Let

$$\lambda(x, t) = - \sum_{n=1}^{\infty} \frac{(-1)^n}{n!} \frac{\partial^{n-1}}{\partial x^{n-1}} [\alpha_n(x, t) f(x, t)] \qquad (7.96)$$

Eq. (7.90) becomes

$$\frac{\partial f(x, t)}{\partial t} + \frac{\partial \lambda(x, t)}{\partial x} = 0 \qquad (7.97)$$

which has the form of the equation of mass transport or diffusion involving a flow of mass. In our case, Eq. (7.97) describes a flow of probability and it may be interpreted as the *equation of conservation of probability*. The function $\lambda(x, t)$ represents the amount of probability crossing x in the positive direction per unit time.

(b) We have remarked that the applicability of the kinetic equation in its general form is limited by severe mathematical difficulties. Since we will not be able to consider the case where the order of differentiation with respect to the spatial coordinate is infinite, the theorem below shows that the only other case possible is one where the order of differentiation is two or less. In other words, if Eq. (7.90) is of finite order, it must be of order two or less.

Theorem 7.2.1.* If the derivate moment $\alpha_n(x, t)$ exists for all n and is zero for some even n, then

$$\alpha_n(x, t) = 0, \qquad n \geq 3 \tag{7.98}$$

Proof. For $n \geq 3$ and let n be odd,

$$\alpha_n(x, t) = \lim_{\Delta t \to 0} \frac{1}{\Delta t} E\{[X(t + \Delta t) - X(t)]^n \mid X(t) = x\}$$

$$= \lim_{\Delta t \to 0} \frac{1}{\Delta t} E\{[X(t + \Delta t) - X(t)]^{(n-1)/2}$$
$$\times [X(t + \Delta t) - X(t)]^{(n+1)/2} \mid X(t) = x\} \tag{7.99}$$

By means of the Schwarz inequality, we have

$$\alpha_n{}^2(x, t) \leq \lim_{\Delta t \to 0} \frac{1}{\Delta t^2} [E\{[X(t + \Delta t) - X(t)]^{n-1} \mid X(t) = x\}$$
$$\times E\{[X(t + \Delta t) - X(t)]^{n+1} \mid X(t) = x\}]$$
$$= \alpha_{n-1}(x, t)\, \alpha_{n+1}(x, t), \qquad n \text{ odd}, n \geq 3 \tag{7.100}$$

Similarly, we can show that

$$\alpha_n{}^2(x, t) \leq \alpha_{n-2}(x, t)\, \alpha_{n+2}(x, t), \qquad n \text{ even}, n \geq 4 \tag{7.101}$$

Let $\alpha_r(x, t) = 0$ where r is an even integer. Setting $n = r - 1, r + 1$ in Eq. (7.100) and $r - 2, r + 2$ in Eq. (7.101), we have the following four inequalities

$$\begin{aligned} \alpha_{r-2}^2 \leq \alpha_{r-4}\alpha_r, \quad r \geq 6; \qquad \alpha_{r-1}^2 \leq \alpha_{r-2}\alpha_r, \quad r \geq 4 \\ \alpha_{r+1}^2 \leq \alpha_r\alpha_{r+2}, \quad r \geq 2; \qquad \alpha_{r+2}^2 \leq \alpha_r\alpha_{r+4}, \quad r \geq 2 \end{aligned} \tag{7.102}$$

Since $\alpha_r = 0$ and α_n exists for all n, the last two of Eqs. (7.102) imply that $\alpha_n = 0$ for all $n > r$. The first two of Eqs. (7.102) imply that $\alpha_n = 0$ for all $n < r$ and $n \geq 3$. The proof is complete.

* See Pawula [5].

In view of the theorem above, we will be primarily interested in applying the kinetic equation to the study of stochastic processes where $\alpha_n(x, t)$ vanishes for all $n \geq 3$. Physically, this implies that the stochastic process can change only by small amounts in a small time interval; the incremental moments $a_n(x, t)$ defined by Eq. (7.87) will approach zero faster than Δt as $\Delta t \to 0$ for $n \geq 3$.

(c) Under the condition that $\alpha_n(x, t) = 0$, $n \geq 3$, the kinetic equation has the form of the so-called Fokker–Planck equation in the theory of Markov stochastic processes. This connection will be explored in detail in the next section.

(d) Appropriate initial and boundary conditions are needed in the solution of the kinetic equation. It is clear that the initial condition is the prescribed density function $f(x, t_0) = f_0(x)$ at some initial time t_0.

The boundary conditions are not always easy to determine. Let us consider several possible types.

A common boundary condition to be imposed on the density function $f(x, t)$ is that x is allowed to assume all values on the real line. In this case, the boundary conditions describe conditions at $x = \pm\infty$. Integrating Eq. (7.97) with respect to x from $-\infty$ and $+\infty$ and noting that

$$\int_{-\infty}^{\infty} f(x, t)\, dx = 1$$

for all t, the boundary conditions are

$$\lambda(\pm\infty, t) = 0 \tag{7.103}$$

In certain cases, the stronger boundary conditions

$$f(\pm\infty, t) = 0 \tag{7.104}$$

are imposed. Equation (7.104) implies that the values of x remain finite with probability one. It can be interpreted as having an absorbing barrier placed at infinity for the flow of probability.

Generalizing the foregoing remark, we may impose absorbing barriers symmetrically placed about $x = 0$. The conditions for placing absorbing barriers at $x = \pm a$ are

$$f(\pm a, t) = 0 \tag{7.105}$$

In cases where the values of x are restricted in an interval $a \leq x \leq b$ the boundary conditions take the forms

$$\lambda(a, t) = \lambda(b, t) = 0 \tag{7.106}$$

These results are obtained directly by integrating Eq. (7.97) with respect to x from a to b. The conditions described above are ones of having reflecting barriers placed at $x = a$ and $x = b$.

(e) Under the conditions that the derivate moments $\alpha_n(x, t)$ are independent of t, the density function $f(x, t)$ is expected to approach a stationary value, $f_s(x)$, as time increases. In this case, we have

$$\partial f_s(x)/\partial t = 0$$

As seen from the kinetic equation (7.90), the limiting density function satisfies the ordinary differential equation

$$\sum_{n=1}^{\infty} \frac{(-1)^n}{n!} \frac{d^n}{dx^n} [\alpha_n(x) f_s(x)] = 0 \tag{7.107}$$

or, integrating once,

$$\lambda(x) = \text{const} \tag{7.108}$$

The solution $f_s(x)$ of Eq. (7.107) or Eq. (7.108), if it exists, is called the *stationary solution* of the kinetic equation.

(f) Generalization of Eq. (7.90) to the vector-process case can be carried out in a straightforward fashion. Let $\mathbf{X}(t)$ be an m-dimensional vector stochastic process with components $X_j(t)$, $j = 1, 2, \ldots, m$. Then, starting with the vector version of Eq. (7.83),

$$f(\mathbf{x}, t + \Delta t) = \int_{-\infty}^{\infty} f(\mathbf{x}, t + \Delta t \mid \mathbf{x}', t) f(\mathbf{x}', t) \, d\mathbf{x}' \tag{7.109}$$

and following the same steps, we obtain

$$\frac{\partial f(\mathbf{x}, t)}{\partial t} = \sum_{n_1, n_2, \ldots, n_m = 1}^{\infty} \left[\prod_{j=1}^{m} \frac{(-1)^{n_j}}{(n_j)!} \frac{\partial^{n_j}}{\partial x_j^{n_j}} \right] [\alpha_{n_1, n_2, \ldots, n_m}(\mathbf{x}, t) f(\mathbf{x}, t)] \tag{7.110}$$

where

$$\alpha_{n_1, n_2, \ldots, n_m} = \lim_{\Delta t \to 0} \frac{1}{\Delta t} E\left[\prod_{j=1}^{m} [X_j(t + \Delta t) - X_j(t)]^{n_j} \mid \mathbf{X}(t) = \mathbf{x} \right] \tag{7.111}$$

It is clear that all the remarks made above are equally valid with respect to Eq. (7.94) and those satisfied by other conditional density functions.

To see how the kinetic equation (7.90) can be applied to the study of random differential equations, let us consider a simple example.

Example 7.5. Consider the first-order differential equation

$$\dot{X}(t) + \beta X(t) = W(t), \qquad t \geq 0 \tag{7.112}$$

where $W(t)$ is a Gaussian white noise process with zero mean and covariance $2D\ \delta(t - s)$ or, to be precise, the formal derivative of the Wiener process. We wish to determine the first density function $f(x, t)$, $t \geq 0$, of $X(t)$ given that

$$f(x, 0) = f_0(x) \tag{7.113}$$

In order to determine the derivate moments, we write Eq. (7.112) in the difference form

$$\Delta X(t) + \beta\, X(t)\, \Delta t = \Delta B(t) + o(\Delta t) \tag{7.114}$$

where $o(\Delta t)$ stands for terms such that

$$\lim_{\Delta t \to 0} o(\Delta t)/\Delta t = 0 \tag{7.115}$$

and $\Delta B(t) = B(t + \Delta t) - B(t)$ is an increment of the Wiener process.

Consider the conditional characteristic function defined by Eq. (7.84). We have

$$\phi(u, t + \Delta t \mid x', t) = E\{e^{iu\Delta X} \mid x', t\}$$
$$= E\{\exp[iu(-\beta X\, \Delta t + \Delta B + o(\Delta t))] \mid x', t\} \tag{7.116}$$

Let us carry out the conditional expectation term by term. We clearly have

$$E\{\exp[iu(-\beta X\, \Delta t)] \mid x', t\} = \exp(-iu\beta x'\, \Delta t) \tag{7.117}$$

Assuming that $X(0)$ is independent of $\Delta B(t)$, $t \geq 0$, it is seen that the second term gives

$$E\{\exp[iu\, \Delta B(t)] \mid x', t\} = E\{\exp[iu\, \Delta B(t)]\} \tag{7.118}$$

since $\Delta B(t)$ is independent of $X(t)$ (see Section 5.2.2).

Now, $\Delta B(t)$ is a Gaussian random variable with mean zero and variance $2D\ \Delta t$, Eq. (7.118) thus yields

$$E\{\exp[iu\, \Delta B(t)] \mid x', t\} = \exp[-Du^2\, \Delta t] \tag{7.119}$$

Hence, the conditional characteristic function in Eq. (7.116) takes the form

$$\phi(u, t + \Delta t \mid x', t) = \exp[-iu\beta x'\, \Delta t - Du^2\, \Delta t + o(\Delta t)] \tag{7.120}$$

Finally, the definition of the derivate moments leads to

$$\alpha_1(x', t) = \frac{1}{i} \lim_{\Delta t \to 0} \frac{1}{\Delta t} \left. \frac{\partial \phi}{\partial u} \right|_{u=0} = -\beta x'$$

$$\alpha_2(x', t) = (-1) \lim_{\Delta t \to 0} \frac{1}{\Delta t} \left. \frac{\partial^2 \phi}{\partial u^2} \right|_{u=0} = 2D \qquad (7.121)$$

$$\alpha_n(x', t) = 0, \qquad n \geq 3$$

Let us remark that the use of the conditional characteristic function for determining the derivate moments is not necessary. These moments can be found directly from Eq. (7.114) by carrying out appropriate expectations and limiting operations.

Having determined the derivate moments, the partial differential equation (7.90) in this case takes the form

$$\frac{\partial f(x, t)}{\partial t} = \beta \frac{\partial}{\partial x} [x f(x, t)] + D \frac{\partial^2 f(x, t)}{\partial x^2} \qquad (7.122)$$

with initial condition (7.113) and, assuming $f(x, t)$ is sufficiently smooth, with boundary conditions

$$f(\pm\infty, t) = 0 \qquad (7.123)$$

The solution of Eq. (7.122) can be obtained either by transform methods or by using the method of separation of variables. Let us use the method of separation of variables and consider

$$\tau = \beta t \qquad \text{and} \qquad y = ix(\beta/D)^{1/2} \qquad (7.124)$$

as the new variables. Eq. (7.122) takes the form

$$\frac{\partial f(y, \tau)}{\partial \tau} = f(y, \tau) + y \frac{\partial f(y, \tau)}{\partial y} - \frac{\partial^2 f(y, \tau)}{\partial y^2} \qquad (7.125)$$

Let us write

$$f(y, \tau) = Y(y) T(\tau) \qquad (7.126)$$

We find that, following the usual procedure for separation of variables,

$$Y(y) = H_n(y), \qquad T(\tau) = e^{(n+1)\tau} \qquad (7.127)$$

where $H_n(y)$ is the Hermite polynomial of the nth order, that is,

$$H_n(y) = (-1)^n \exp(y^2/2) \frac{d^n}{dy^n} \exp(-y^2/2) \tag{7.128}$$

The solution $f(y, \tau)$ thus takes the form

$$f(y, \tau) = \sum_{n=0}^{\infty} a_n e^{(n+1)\tau} H_n(y) \tag{7.129}$$

where the constants a_n are determined from the initial condition. They are

$$a_n = \frac{1}{n!(2\pi)^{1/2}} \int_{-\infty}^{\infty} H_n(z) f_0(z) \exp(-z^2/2) \, dz \tag{7.130}$$

Hence, we finally have

$$
\begin{aligned}
f(y, \tau) &= \frac{1}{n!(2\pi)^{1/2}} \sum_{n=0}^{\infty} \int_{-\infty}^{\infty} H_n(y) H_n(z) \exp[(n + 1)\tau - z^2/2] f_0(z) \, dz \\
&= \frac{1}{i[2\pi(1 - e^{-2\tau})]^{1/2}} \int_{-\infty}^{\infty} \exp\left[\frac{(y - ze^{-\tau})^2}{2(1 - e^{-2\tau})}\right] f_0(z) \, dz
\end{aligned} \tag{7.131}
$$

where the second expression in the above is obtained by using the identity

$$\sum_{n=0}^{\infty} \frac{1}{n!} H_n(y) H_n(z) e^{n\tau} = [1 - e^{2\tau}]^{-1/2} \exp\left[\frac{2yze^{\tau} - (y^2+z^2)e^{2\tau}}{2(1 - e^{2\tau})}\right] \tag{7.132}$$

For the simple case where the initial condition is

$$f_0(z) = \delta(z - y_0), \qquad y_0 = ix_0(\beta/D)^{1/2} \tag{7.133}$$

The solution is

$$f(y, \tau) = f(y, \tau \mid y_0, 0) = \frac{1}{i[2\pi(1 - e^{-2\tau})]^{1/2}} \exp\left[\frac{(y - y_0e^{-\tau})^2}{2(1 - e^{-2\tau})}\right] \tag{7.134}$$

and, after reverting to the old variables,

$$f(x, t \mid x_0, 0) = [2\pi D(1 - e^{-2\beta t})/\beta]^{-1/2} \exp\left[-\frac{\beta(x - x_0e^{-\beta t})^2}{2D(1 - e^{-2\beta t})}\right] \tag{7.135}$$

The solution is seen to have a Gaussian form with mean $x_0 \exp(-\beta t)$ and variance $D[1 - \exp(-2\beta t)]/\beta$. A plot of the conditional density function $f(x, t \mid x_0, 0)$ is shown in Fig. 7.6 for several values of t; the dotted line gives the mean curve as a function of t.

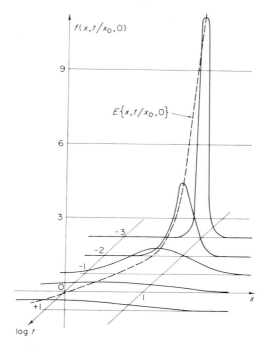

Fig. 7.6. The conditional density function $f(x, t \mid x_0, 0)$, $(D = \beta = 1$ and $x_0 = 1)$.

It is noteworthy that the equation we have just solved is simply the Langevin equation (7.3) we mentioned at the beginning of this chapter. The conditional density function given by Eq. (7.135) is the familiar one associated with the *Ornstein–Uhlenbeck process* [6].

We have used this simple example to illustrate the working details involved in applying the kinetic equation to the study of random differential equations. For this simple example, there is clearly a more direct way of obtaining the solution as indicated by Eq. (7.135). We see from the original differential equation (7.112) that, given that $X(0)$ is a deterministic constant, the solution process $X(t)$ is Gaussian; its mean and variance can be easily found following the development in Section 7.2.1.

It is noted that the stationary solution of $f(x, t \mid x_0, 0)$ can be obtained directly from Eq. (7.135) by letting $t \to \infty$. It has the form

$$f_s(x) = \lim_{t \to \infty} f(x, t \mid x_0, 0) = (2\pi D/\beta)^{-1/2} \exp(-\beta x^2/2D) \qquad (7.136)$$

which is Gaussian with mean zero and variance D/β.

We will defer additional examples to the next section.

7.3. The Case of Markov Processes

We observe that the kinetic equation (7.90) takes a simple form in Example 7.5 and it can be readily solved. This simplicity stems from the fact that, because the governing differential equation is one of the Ito type, the solution process $X(t)$ has the nice property of being Markovian. Indeed, fruitful results can in general be obtained by means of Eq. (7.90) when the stochastic process under consideration is either a Markov process or a *projection* of a Markov process, that is, the stochastic process itself is not Markovian but it is a component of a vector Markov process. For this reason, we wish to focus our attention on Markov processes in this section and clearly identify the role played by the kinetic equation in this situation.

The Markov stochastic process is defined in Section 3.3.2. It is shown that a Markov process $X(t)$ is completely characterized by its second density functions. Therefore, if $f(x_0, t_0)$ is specified, the knowledge of the transition probability density $f(x, t \mid x_0, t_0)$ completely specifies the stochastic process. Furthermore, the transition probability density satisfies the Smoluchowski–Chapman–Kolmogorov equation [Eq. (3.105)].

$$f(x_3, t_3 \mid x_1, t_1) = \int_{-\infty}^{\infty} f(x_3, t_3 \mid x_2, t_2) f(x_2, t_2 \mid x_1, t_1) \, dx_2, \qquad t_1 < t_2 < t_3 \tag{7.137}$$

Consequently, we are primarily concerned with the transition probability density $f(x, t \mid x_0, t_0)$. In what follows, we shall assume that $f(x, t \mid x_0, t_0)$ satisfies the following conditions:

(a) The partial derivatives $\partial f/\partial t$, $\partial f/\partial x$, $\partial^2 f/\partial x^2$, $\partial f/\partial t_0$, $\partial f/\partial x_0$, and $\partial^2 f/\partial x_0^2$ exist and satisfy appropriate regularity conditions.

(b) The process $X(t)$ is continuous; continuity condition is satisfied if the derivate moments $\alpha_1(x, t)$ and $\alpha_2(x, t)$ defined by Eq. (7.91) exist and

$$\alpha_n(x, t) = 0, \qquad n \geq 3 \tag{7.138}$$

A Markov processes satisfying the conditions above is commonly referred to as a *diffusion process*.

7.3.1. The Fokker–Planck Equation and the Kolmogorov Equation

The transition probability density $f(x, t \mid x_0, t_0)$ certainly satisfies Eq. (7.90) and, in view of the assumptions above, it has the simple form

$$\frac{\partial f(x, t \mid x_0, t_0)}{\partial t} = -\frac{\partial}{\partial x} [\alpha_1(x, t)f] + \tfrac{1}{2} \frac{\partial^2}{\partial x^2} [\alpha_2(x, t)f] \tag{7.139}$$

with the initial condition

$$f(x, t_0 \mid x_0, t_0) = \delta(x - x_0) \tag{7.140}$$

In the theory of Markov processes, the parabolic partial differential equation (7.139) is known as the *Fokker–Planck equation*. The name *forward equation* is also used because the transition probability density in it is regarded as a function of the *forward* variables x and t, moving forward in time.

Another equation satisfied by $f(x, t \mid x_0, t_0)$, which is the adjoint of Eq. (7.139), can be derived by regarding the transition probability density as a function of x_0 and t_0 moving backward from t. Consider the Taylor series expansion of $f(x_3, t_3 \mid x_2, t_2)$ about $x_2 = x_1$, that is

$$f(x_3, t_3 \mid x_2, t_2) = f(x_3, t_3 \mid x_1, t_2) + (x_2 - x_1) \left.\frac{\partial f}{\partial x_2}\right|_{x_2 = x_1}$$
$$+ \tfrac{1}{2}(x_2 - x_1)^2 \left.\frac{\partial^2 f}{\partial x_2{}^2}\right|_{x_2 = x_1} + o(x_2 - x_1)^2 \tag{7.141}$$

Upon substituting it into Eq. (7.137) and performing the integration, we get

$$f(x_3, t_3 \mid x_1, t_1) = f(x_3, t_3 \mid x_1, t_2)$$
$$+ \left.\frac{\partial f}{\partial x_2}\right|_{x_2 = x_1} \int_{-\infty}^{\infty} (x_2 - x_1) f(x_2, t_2 \mid x_1, t_1)\, dx_2$$
$$+ \tfrac{1}{2} \left.\frac{\partial^2 f}{\partial x_2{}^2}\right|_{x_2 = x_1} \int_{-\infty}^{\infty} (x_2 - x_1)^2 f(x_2, t_2 \mid x_1, t_1)\, dx_2$$
$$+ o(x_2 - x_1)^2 \tag{7.142}$$

Transposing the term on the left-hand side of Eq. (7.142) and dividing it by $\varDelta t = t_2 - t_1$ yields

$$\frac{1}{\varDelta t} \left[f(x_3, t_3 \mid x_1, t_2) - f(x_3, t_3 \mid x_1, t_1) \right]$$
$$+ \left.\frac{\partial f}{\partial x_2}\right|_{x_2 = x_1} \frac{1}{\varDelta t} \int_{-\infty}^{\infty} (x_2 - x_1) f(x_2, t_2 \mid x_1, t_1)\, dx_2$$
$$+ \tfrac{1}{2} \left.\frac{\partial^2 f}{\partial x_2{}^2}\right|_{x_2 = x_1} \frac{1}{\varDelta t} \int_{-\infty}^{\infty} (x_2 - x_1)^2 f(x_2, t_2 \mid x_1, t_1)\, dx_2$$
$$+ \frac{1}{\varDelta t} o(x_2 - x_1)^2 = 0 \tag{7.143}$$

Let us now take the limit as $\Delta t \to 0$. In view of the condition (7.138), Eq. (7.143) then becomes

$$\frac{\partial f(x_3, t_3 \mid x_1, t_1)}{\partial t_1} + \frac{\partial f}{\partial x_1} \alpha_1(x_1, t_1) + \tfrac{1}{2} \frac{\partial^2 f}{\partial x_1^2} \alpha_2(x_1, t_1) = 0 \qquad (7.144)$$

With an appropriate change of variables, we arrive at what is known as the *Kolmogorov equation* or the *backward equation*

$$\frac{\partial f(x, t \mid x_0, t_0)}{\partial t_0} = - \frac{\partial f}{\partial x_0} \alpha_1(x_0, t_0) - \tfrac{1}{2} \frac{\partial^2 f}{\partial x_0^2} \alpha_2(x_0, t_0) \qquad (7.145)$$

with the same initial condition (7.140).

The transition probability density associated with a Markov process can be determined using either the Fokker–Planck equation or the Kolmogorov equation when appropriate boundary conditions are specified. We note in passing that the normalization condition must be always satisfied, that is

$$\int_{-\infty}^{\infty} f(x, t \mid x_0, t_0) \, dx = 1 \qquad (7.146)$$

The vector versions of Eqs. (7.139) and (7.145) can also be established. Let $X(t)$·be an m-dimensional vector Markov process. The vector version of the kinetic equation is given by Eq. (7.110). Accordingly, the corresponding Fokker–Planck equation has the form

$$\frac{\partial f(\mathbf{x}, t \mid \mathbf{x}_0, t_0)}{\partial t} = - \sum_{j=1}^{m} \frac{\partial}{\partial x_j} [\alpha_j(\mathbf{x}, t)f] + \tfrac{1}{2} \sum_{i,j=1}^{m} \frac{\partial^2}{\partial x_i \, \partial x_j} [\alpha_{ij}(\mathbf{x}, t)f] \qquad (7.147)$$

where

$$\alpha_j(\mathbf{x}, t) = \lim_{\Delta t \to 0} \frac{1}{\Delta t} E\{[X_j(t + \Delta t) - X_j(t)] \mid \mathbf{X}(t) = \mathbf{x}\} \qquad (7.148)$$

$$\alpha_{ij}(\mathbf{x}, t) = \lim_{\Delta t \to 0} \frac{1}{\Delta t} E\{[X_i(t + \Delta t) - X_i(t)]$$
$$\times [X_j(t + \Delta t) - X_j(t)] \mid \mathbf{X}(t) = \mathbf{x}\} \qquad (7.149)$$

Similarly, the vector version of the Kolmogorov equation is

$$\frac{\partial f(\mathbf{x}, t \mid \mathbf{x}_0, t_0)}{\partial t_0} = - \sum_{j=1}^{m} \frac{\partial f}{\partial x_{0j}} \alpha_j(\mathbf{x}_0, t) - \tfrac{1}{2} \sum_{i,j=1}^{m} \frac{\partial^2 f}{\partial x_{0i} \, \partial x_{0j}} \alpha_{ij}(\mathbf{x}_0, t)$$
$$(7.150)$$

where the initial condition for both Eqs. (7.147) and (7.150) is

$$f(\mathbf{x}, t_0 \mid \mathbf{x}_0, t_0) = \prod_{j=1}^{m} \delta(x_j - x_{0j}) \tag{7.151}$$

7.3.2. Applications to Equations of the Ito Type

Theorem 5.2.5 states that the solution process generated by the Ito equation is Markovian. Hence, an immediate application of the Fokker–Planck and Kolmogorov equations is found in the study of solution processes $\mathbf{X}(t)$ governed by

$$d\mathbf{X}(t) = \mathbf{f}(\mathbf{X}(t), t) \, dt + G(\mathbf{X}(t), t) \, d\mathbf{B}(t), \qquad t \geq t_0 \tag{7.152}$$

where $\mathbf{X}(t)$ is the n-dimensional solution process, $G(\mathbf{X}(t), t)$ is an $n \times m$ matrix function, and $\mathbf{B}(t)$ is an m-dimensional vector Wiener process with components $B_j(t)$, $j = 1, 2, \ldots, m$, having the properties

$$E\{\Delta B_j(t)\} = E\{B_j(t + \Delta t) - B_j(t)\} = 0$$
$$E\{\Delta B_i(t) \, \Delta B_j(t)\} = 2D_{ij} \, \Delta t, \qquad t \geq t_0; \quad i, j = 1, 2, \ldots, m \tag{7.153}$$

Let us derive the Fokker-Planck equation associated with the Ito equation given above. We note in passing that, in as much as we are concerned only with differential equations with random nonhomogeneous parts in this chapter, the matrix $G(\mathbf{X}(t), t)$ is not a function of $\mathbf{X}(t)$ in this case. Nevertheless, we will give our derivations for the general case so that the results can be used in Chapters 8 and 9 as well.

According to the development in the preceding section, the transition probability density of $\mathbf{X}(t)$, $f(\mathbf{x}, t \mid \mathbf{x}_0, t_0)$, satisfies the Fokker–Planck equation

$$\frac{\partial f(\mathbf{x}, t \mid \mathbf{x}_0, t_0)}{\partial t} = - \sum_{j=1}^{n} \frac{\partial}{\partial x_j} [\alpha_j(\mathbf{x}, t) f]$$

$$+ \tfrac{1}{2} \sum_{i,j=1}^{n} \frac{\partial^2}{\partial x_i \, \partial x_j} [\alpha_{ij}(\mathbf{x}, t) f] \tag{7.154}$$

with the initial condition

$$f(\mathbf{x}, t_0 \mid \mathbf{x}_0, t_0) = \prod_{j=1}^{n} \delta(x_j - x_{0j}) \tag{7.155}$$

The derivate moments $\alpha_j(\mathbf{x}, t)$ and $\alpha_{ij}(\mathbf{x}, t)$ in Eq. (7.154) can be derived either by means of the conditional characteristic function or, as we shall do here, by direct averaging. In incremental form, Eq. (7.152) leads to

$$\Delta X_j(t) = X_j(t + \Delta t) - X_j(t)$$
$$= f_j(\mathbf{X}(t), t)\, \Delta t + \sum_{k=1}^{m} G_{jk}(\mathbf{X}(t), t)\, \Delta B_k(t) + o(\Delta t) \qquad (7.156)$$

$$\Delta X_i(t)\, \Delta X_j(t) = \sum_{k,l=1}^{m} G_{ik}(\mathbf{X}(t), t)\, G_{jl}(\mathbf{X}(t), t)\, \Delta B_k(t)\, \Delta B_l(t)$$
$$+ f_i(\mathbf{X}(t), t) \sum_{k=1}^{m} G_{jk}(\mathbf{X}(t), t)\, \Delta B_k(t)$$
$$+ f_j(\mathbf{X}(t), t) \sum_{k=1}^{m} G_{ik}(\mathbf{X}(t), t)\, \Delta B_k(t)$$
$$+ f_i(\mathbf{X}(t), t) f_j(\mathbf{X}(t), t)(\Delta t)^2 + o(\Delta t) \qquad (7.157)$$

Using Eqs. (7.153) and noting that $\Delta \mathbf{B}(t)$ is independent of the value of $\mathbf{X}(t)$ at t, we obtain

$$\alpha_j(\mathbf{x}, t) = \lim_{\Delta t \to 0} \frac{1}{\Delta t}\, E\{\Delta X_j(t) \mid \mathbf{X}(t) = \mathbf{x}\} = f_j(\mathbf{x}, t) \qquad (7.158)$$

$$\alpha_{ij}(\mathbf{x}, t) = \lim_{\Delta t \to 0} \frac{1}{\Delta t}\, E\{\Delta X_i(t)\, \Delta X_j(t) \mid \mathbf{X}(t) = \mathbf{x}\}$$
$$= 2 \sum_{k,l=1}^{m} D_{kl} G_{ik}(\mathbf{x}, t)\, G_{jl}(\mathbf{x}, t)$$
$$= 2(GDG^{\mathrm{T}})_{ij}, \qquad i, j = 1, 2, \ldots, n. \qquad (7.159)$$

where D denotes the $m \times m$ matrix whose ijth element is D_{ij}.

Hence, the Fokker–Planck equation associated with Eq. (7.152) has the form

$$\frac{\partial f(\mathbf{x}, t \mid \mathbf{x}_0, t_0)}{\partial t} = - \sum_{j=1}^{n} \frac{\partial}{\partial x_j} [f_j(\mathbf{x}, t) f]$$
$$+ \sum_{i,j=1}^{n} \frac{\partial^2}{\partial x_i \, \partial x_j} [(GDG^{\mathrm{T}})_{ij} f] \qquad (7.160)$$

As was mentioned earlier, we are only concerned with the case where the matrix $G(\mathbf{X}(t), t)$ is not a function of $\mathbf{X}(t)$ in this chapter. Under this condition, the choice of a method of solution of the Fokker–Planck equation with appropriate boundary conditions depends largely on the form of $f_j(\mathbf{x}, t)$. If the original differential equation is nonlinear, the functions $f_j(\mathbf{x}, t)$ will be nonlinear in $\mathbf{x}$ and a general solution of Eq. (7.160) is difficult to obtain.

In linear and constant-coefficient cases, the solution of the Fokker–Planck equation can be solved by the ordinary method of separation of variables. Another method of great utility, particularly in solving multidimensional Fokker–Planck equations for the linear equations, is one of transforming the parabolic partial differential equation (7.160) into a linear, first-order partial differential equation by means of Fourier transforms; the first-order equation is then solved by ordinary means which yields the solution for the conditional characteristic function associated with $f(\mathbf{x}, t \mid \mathbf{x}_0, t_0)$.

In order to illustrate the Fourier transform technique, consider the case where $m = n$, $G = I$ (unit matrix), and

$$f_j(\mathbf{x}, t) = a_j x_j \tag{7.161}$$

the Fokker–Planck equation (7.160) now has the form

$$\frac{\partial f}{\partial t} = - \sum_{j=1}^{n} a_j \frac{\partial}{\partial x_j} [x_j f] + \sum_{i,j=1}^{n} D_{ij} \frac{\partial^2 f}{\partial x_i \, \partial x_j} \tag{7.162}$$

Its solution is required to satisfy the initial condition

$$f(\mathbf{x}, t_0 \mid \mathbf{x}_0, t_0) = \prod_{j=1}^{n} \delta(x_j - x_{j0}) \tag{7.163}$$

and the boundaries conditions

$$f(\mathbf{x}, t \mid \mathbf{x}_0, t_0) \to 0 \tag{7.164}$$

as $x_j \to \pm\infty$ for any j.

Equation (7.162) is a parabolic partial differential equation with $n + 1$ independent variables. In order to transform it into one of the first order, let us consider the characteristic function $\phi(\mathbf{u}, t)$, which is related to $f(\mathbf{x}, t \mid \mathbf{x}_0, t_0)$ by the n-dimensional Fourier transform

$$\phi(\mathbf{u}, t) = \mathscr{F}[f(\mathbf{x}, t \mid \mathbf{x}_0, t_0)] = \int_{-\infty}^{\infty} \cdots \int_{-\infty}^{\infty} \exp(i\mathbf{u}^{\mathsf{T}}\mathbf{x}) f(\mathbf{x}, t \mid \mathbf{x}_0, t_0) \, d\mathbf{x} \tag{7.165}$$

It is easy to show that

$$\mathscr{F}\left(\frac{\partial f}{\partial t}\right) = \frac{\partial \phi}{\partial t}, \quad \mathscr{F}\left[\frac{\partial}{\partial x_j}(x_j f)\right] = -u_j \frac{\partial \phi}{\partial u_j}, \quad \mathscr{F}\left(\frac{\partial^2 f}{\partial x_i \, \partial x_j}\right) = -u_i u_j \phi \tag{7.166}$$

Hence, the Fourier transform of Eq. (7.162) leads to the first-order partial differential equation

$$\frac{\partial \phi}{\partial t} = \sum_{j=1}^{n} a_j u_j \frac{\partial \phi}{\partial u_j} - \sum_{i,j=1}^{n} D_{ij} u_i u_j \phi \tag{7.167}$$

As we have shown in Chapter 6 in connection with the Liouville equation, the solution of Eq. (7.167) can be obtained by considering its associated Lagrange system

$$\frac{dt}{1} = -\frac{du_1}{a_1 u_1} = -\frac{du_2}{a_2 u_2} = \cdots = -\frac{du_n}{a_n u_n} = -\frac{d\phi}{\phi \sum_{i,j=1}^{n} D_{ij} u_i u_j} \tag{7.168}$$

The integrals of the equations

$$dt/1 = -du_j/a_j u_j$$

are given by

$$u_j \exp(a_j t) = c_j (\text{const}), \qquad j = 1, 2, \ldots, n \tag{7.169}$$

Substituting the relations

$$u_j = c_j \exp(-a_j t) \tag{7.170}$$

into the last expression in Eq. (7.168) gives

$$\frac{dt}{1} = -\frac{d\phi}{\phi \sum_{i,j=1}^{n} D_{ij} u_i u_j} = -\frac{d\phi}{\phi \sum_{i,j=1}^{n} D_{ij} c_i c_j \exp[-(a_i + a_j)t]} \tag{7.171}$$

which can be integrated with the result

$$\ln(c\phi) = \sum_{i,j=1}^{n} D_{ij} c_i c_j [a_i + a_j]^{-1} \exp[-(a_i + a_j)t]$$

$$= \sum_{i,j=1}^{n} D_{ij} u_i u_j (a_i + a_j)^{-1} \tag{7.172}$$

or

$$\phi \exp\left[-\sum_{i,j=1}^{n} D_{ij} u_i u_j (a_i + a_j)^{-1}\right] = c_{n+1} \tag{7.173}$$

Therefore, the general solution of Eq. (7.167) has the form

$$\phi(\mathbf{u}, t) = \psi[u_1 \exp(a_1 t), u_2 \exp(a_2 t), \ldots, u_n \exp(a_n t)]$$

$$\times \exp\left[\sum_{i,j=1}^{n} D_{ij} u_i u_j / (a_i + a_j)\right] \tag{7.174}$$

where ψ is some function of the indicated arguments. Now, the initial condition (7.163) leads to

$$\phi(\mathbf{u}, t_0) = \exp\left(i \sum_{j=1}^{n} u_j x_{j0}\right) \qquad (7.175)$$

Hence, the function ψ must take the form such that

$$\psi(u_1, u_2, \ldots, u_n, t_0) = \exp\left[i \sum_{j=1}^{n} u_j x_{j0} - \sum_{i,j=1}^{n} D_{ij} u_i u_j/(a_i + a_j)\right] \qquad (7.176)$$

We finally have for the characteristic function

$$\phi(\mathbf{u}, t) = \exp\left[i \sum_{j=1}^{n} u_j x_{j0} \exp(a_j t)\right.$$

$$\left. + \sum_{i,j=1}^{n} D_{ij} \frac{u_i u_j}{(a_i + a_j)} [1 - \exp(a_i + a_j)t]\right] \qquad (7.177)$$

The solution for $\phi(\mathbf{u}, t)$ clearly defines an n-dimensional Gaussian distribution. Hence, the solution for $f(\mathbf{x}, t \mid \mathbf{x}_0, t_0)$ has the form

$$f(\mathbf{x}, t \mid \mathbf{x}_0, t_0) = (2\pi)^{-n/2} \mid \Lambda \mid^{-1/2} \exp[-\tfrac{1}{2}(\mathbf{x} - \mathbf{m})^{\mathrm{T}} \Lambda^{-1}(\mathbf{x} - \mathbf{m})] \qquad (7.178)$$

where the components of the mean vector $\mathbf{m}$ and the covariance matrix Λ are, respectively,

$$m_j = x_{j0} \exp(a_j t)$$
$$\Lambda_{ij} = -\frac{2D_{ij}}{(a_i + a_j)} [1 - \exp(a_i + a_j)t], \qquad i, j = 1, 2, \ldots, n \qquad (7.179)$$

Example 7.6. It is clear that the results obtained above are directly applicable to the first-order system considered in Example 7.5. In that example,

$$n = 1, \qquad a_1 = -\beta, \qquad D_{11} = D \qquad (7.180)$$

The substitution of the above into Eqs. (7.178) and (7.179) immediately gives Eq. (7.135), the result previously obtained using the method of separation of variables.

Example 7.7. It is more interesting to apply the foregoing to the second-order differential equation

$$\ddot{X}(t) + 2\beta \dot{X}(t) + \omega_0^2 X(t) = W(t), \qquad \beta^2 < \omega_0^2, \quad t \geq t_0 \qquad (7.181)$$

where $W(t)$, $t \geq t_0$, is a Gaussian white noise with zero mean and covariance function $2D \, \delta(t - s)$. Let $X(t) = X_1(t)$, $\dot{X}(t) = X_2(t)$, and

$$\mathbf{X}(t) = \begin{bmatrix} X_1(t) \\ X_2(t) \end{bmatrix}$$

Equation (7.181) can be cast into the Ito form (7.152) with

$$\mathbf{f}(\mathbf{X}(t), t) = \begin{bmatrix} X_2(t) \\ -\omega_0{}^2 X_1(t) - 2\beta X_2(t) \end{bmatrix}$$

$$G(\mathbf{X}(t), t) = \begin{bmatrix} 0 \\ 1 \end{bmatrix}$$

(7.182)

Accordingly, it follows from Eq. (7.158) and (7.159) that

$$\alpha_1 = x_2, \quad \alpha_2 = -\omega_0{}^2 x_1 - 2\beta x_2, \quad \alpha_{11} = \alpha_{12} = \alpha_{21} = 0, \quad \alpha_{22} = 2D \quad (7.183)$$

The results derived above are not directly applicable in this case as the resulting Fokker–Planck equation does not conform with Eq. (7.162). To circumvent this difficulty, we proceed as follows.

Let us write the first two of Eqs. (7.183) in the vector form

$$\begin{bmatrix} \alpha_1 \\ \alpha_2 \end{bmatrix} = T \begin{bmatrix} x_1 \\ x_2 \end{bmatrix} = \begin{bmatrix} 0 & 1 \\ -\omega_0{}^2 & -2\beta \end{bmatrix} \begin{bmatrix} x_1 \\ x_2 \end{bmatrix}$$

(7.184)

The base coordinates x_j, $j = 1, 2$, are first linearly transformed into a new set of coordinates y_j, $j = 1, 2$, through the transformation

$$\begin{bmatrix} y_1 \\ y_2 \end{bmatrix} = C \begin{bmatrix} x_1 \\ x_2 \end{bmatrix}$$

(7.185)

where the transformation matrix C diagonalizes the matrix T. Hence, we seek a solution of the matrix equation (see, for example, Sokolinkoff and Redheffer [7])

$$CTC^{-1} = \begin{bmatrix} \lambda_1 & 0 \\ 0 & \lambda_2 \end{bmatrix}$$

(7.186)

where λ_j, $j = 1, 2$, are the roots of the characteristic equation

$$|T - \lambda I| = 0$$

(7.187)

In our case, they are

$$\lambda_{1,2} = -\beta \pm (\beta^2 - \omega_0{}^2)^{1/2}$$

(7.188)

The elements of the matrix C are then determined from Eq. (7.186) by finding the corresponding characteristic vectors. The result is

$$C = \frac{1}{\lambda_2 - \lambda_1} \begin{bmatrix} \lambda_2 & -1 \\ -\lambda_1 & 1 \end{bmatrix}, \quad C^{-1} = \begin{bmatrix} 1 & 1 \\ \lambda_1 & \lambda_2 \end{bmatrix} \tag{7.189}$$

In terms of the new variables y_j defined by the transformation (7.185), the Fokker–Planck equation now takes the desired form

$$\frac{\partial f(\mathbf{y}, t \mid \mathbf{y}_0, t_0)}{\partial t} = -\sum_{j=1}^{2} \lambda_j \frac{\partial}{\partial y_j} (y_j f) + \sum_{i,j=1}^{2} \sigma_{ij} \frac{\partial^2 f}{\partial y_i \, \partial y_j} \tag{7.190}$$

where

$$[\sigma_{ij}] = C \begin{bmatrix} \alpha_{11} & \alpha_{12} \\ \alpha_{21} & \alpha_{22} \end{bmatrix} C^{\mathrm{T}} \tag{7.191}$$

The associated initial condition is

$$f(\mathbf{y}, t_0 \mid \mathbf{y}_0, t_0) = \delta(y_1 - y_{10})\,\delta(y_2 - y_{20}), \quad \begin{bmatrix} y_{10} \\ y_{20} \end{bmatrix} = C \begin{bmatrix} x_{10} \\ x_{20} \end{bmatrix} \tag{7.192}$$

and the boundary conditions are

$$f(y_1, \pm\infty; t \mid \mathbf{y}_0, t_0) = 0, \quad f(\pm\infty, y_2; t \mid \mathbf{y}_0, t_0) = 0 \tag{7.193}$$

We are now in the position to write down the solution of Eq. (7.190). According to Eqs. (7.178) and (7.179), the transition probability density $f(\mathbf{y}, t \mid \mathbf{y}_0, t_0)$ is bivariate Gaussian with the mean vector

$$\mathbf{m}_Y = [E\{Y_j(t)\}] = [y_{j0} \exp(\lambda_j t)] \tag{7.194a}$$

and the covariance matrix

$$\Lambda_Y = [\mathrm{cov}(Y_i(t)\,Y_j(t))] = \left[-\frac{\sigma_{ij}}{\lambda_i + \lambda_j} [1 - \exp(\lambda_i + \lambda_j)t)] \right], \quad i, j = 1, 2 \tag{7.194b}$$

It is now simple to go back to the original stochastic processes $X_1(t)$ and $X_2(t)$. From Eqs. (7.185) and (7.189),

$$\begin{bmatrix} X_1(t) \\ X_2(t) \end{bmatrix} = C^{-1} \begin{bmatrix} Y_1(t) \\ Y_2(t) \end{bmatrix} = \begin{bmatrix} 1 & 1 \\ \lambda_1 & \lambda_2 \end{bmatrix} \begin{bmatrix} Y_1(t) \\ Y_2(t) \end{bmatrix} \tag{7.195}$$

The linearity of the transform implies that the transition probability density

$f(\mathbf{x}, t \mid \mathbf{x}_0, t_0)$ is also bivariate Gaussian. The mean vector is clearly

$$\mathbf{m}_X = [E\{X_j(t)\}] = C^{-1}\mathbf{m}_Y \qquad (7.196)$$

and the covariance matrix is

$$\Lambda_X = [\mathrm{cov}(X_i(t)\, X_j(t))] = C^{-1}\Lambda_Y (C^{-1})^{\mathrm{T}} \qquad (7.197)$$

Referring back to the original problem, we have found the conditional density function of $X(t)$ and $\dot{X}(t)$, $f(x, \dot{x}; t \mid x_0, \dot{x}_0; t_0)$. Given the initial joint density function $f(x_0, \dot{x}_0; t_0)$, the joint density $f(x, \dot{x}; t)$ or the marginal densities $f(x, t)$ and $f(\dot{x}, t)$ at any time t are obtained by integration. The moments of $X(t)$ and $\dot{X}(t)$ are, of course, also obtainable from these results. Armed with this information, we are able to answer a number of questions concerning the probabilistic behavior of the solution process $X(t)$. For example, the zero and threshold crossing problem as formulated in Appendix B is essentially solved by having the knowledge of $f(x, \dot{x}; t)$ at a given t.

Example 7.7 serves to demonstrate that the technique of Fourier transform can be applied to linear equations of higher order. There is, however, in general a need of performing a coordinate transformation.

Before leaving this section, let us point out that the application of the Fokker–Planck equation is not limited to systems excited by white noise inputs. A case of physical importance is one where the input to a system cannot be approximated by a Gaussian white noise but it can be obtained by passing white noise through a dynamic system (a filtered white noise). Let us consider, for example, a second-order system

$$\ddot{X}(t) = f(X(t), \dot{X}(t), t) + Y(t) \qquad (7.198)$$

where the input process $Y(t)$ is nonwhite. However, if it can be assumed that $Y(t)$ satisfies

$$dY(t) = g(Y(t), t)\, dt + dB(t) \qquad (7.199)$$

then it is clear that the vector process having the components $X(t)$, $\dot{X}(t)$, and $Y(t)$ satisfies a differential equation of the Ito type. The probabilistic behavior of $X(t)$ can thus be analyzed via a three-dimensional Fokker–Planck equation.

As we have indicated earlier, Fokker–Planck approach is equally applicable to a class of nonlinear random differential equations. The nonlinear problem will be taken up in the next section.

7.4. Solution Processes for Nonlinear Equations

As we would expect from our experience in dealing with ordinary non-linear problems, the study of nonlinear stochastic problems is generally difficult. In this section, no attempt will be made to develop a unified nonlinear theory or a systematic method of solving nonlinear random differential equations; instead, we shall only discuss several methods of attack commonly encountered in practice. There are limitations and advantages associated with each of these techniques, and we hope to bring out these pertinent features by means of examples.

7.4.1. Applications of the Fokker–Planck Equation

The groundwork for the use of the Fokker–Planck equation in nonlinear problems has been laid in Section 7.3. In this section, let us explore further this application in connection with the Ito equation

$$dX(t) = f(X(t), t) + G(X(t), t)\, dB(t), \qquad t \geq t_0 \tag{7.200}$$

We are now interested in the case where $G(X(t), t)$ is independent of $X(t)$ and $f(X(t), t)$ is a nonlinear function of $X(t)$.

In the solution of nonlinear equations of this type, the Fokker–Planck equation can be used to give (a) the transition probability density of the solution process, (b) the stationary solution of the transition probability density or the steady-state distribution, and (c) the moments of the solution process.

(a) Transition Probability Densities. As we have mentioned in Section 7.3, the Fokker–Planck equation associated with a nonlinear Ito equation is in general difficult to solve. The method of Fourier transform is not fruitful in general and the method of separation of variables can be applied only in a very limited number of cases. Let us consider one example.

Example 7.8. Consider the first-order equation

$$dX(t) = -\beta \operatorname{sgn}(X(t))\, dt + dB(t), \qquad t \geq 0 \tag{7.201}$$

where

$$\operatorname{sgn}(X) = \begin{cases} 1 & \text{for} \quad X > 0 \\ 0 & \text{for} \quad X = 0 \\ -1 & \text{for} \quad X < 0 \end{cases} \tag{7.202}$$

Equation (7.201) represents a bang-bang servo.

Let

$$E\{dB(t)\} = E\{B(t + dt) - B(t)\} = 0, \qquad E\{[dB(t)]^2\} = 2D \, dt \qquad (7.203)$$

The Fokker–Planck equation is, following Eq. (7.160),

$$\frac{\partial f(x, t \mid x_0, 0)}{\partial t} = \beta \frac{\partial}{\partial x} [\text{sgn}(x) f] + D \frac{\partial^2 f}{\partial x^2} \qquad (7.204)$$

with the initial condition

$$f(x, 0 \mid x_0, 0) = \delta(x - x_0) \qquad (7.205)$$

For boundary conditions, let us use

$$f(\pm x_1, t \mid x_0, 0) = 0 \qquad (7.206)$$

where x_1 is arbitrary. There is, of course, also the normalization condition that must be satisfied.

The solution of Eq. (7.204) together with indicated initial and boundary conditions is discussed by Wishner [8]. Using the method of separation of variables, let

$$f(x, t \mid x_0, 0) = X(x) \, T(t) \qquad (7.207)$$

Upon substituting Eq. (7.207) into Eq. (7.204), we have

$$\frac{1}{DT} \frac{dT(t)}{dt} = -\lambda, \qquad \frac{\beta}{DX} \frac{d}{dx} [\text{sgn}(x) \, X] + \frac{1}{X} \frac{d^2 X}{dx^2} = -\lambda \qquad (7.208)$$

The equation governing the function $T(t)$ is easily solved. It has the solution

$$T(t) = c e^{-\lambda DT} \qquad (7.209)$$

The equation satisfied by $X(x)$ can be written in the form

$$\frac{d^2 X}{dx^2} + \frac{\beta}{D} \frac{dX}{dx} + \lambda X = 0, \qquad x > 0$$

$$\frac{d^2 X}{dx^2} - \frac{\beta}{D} \frac{dX}{dx} + \lambda X = 0, \qquad x < 0 \qquad (7.210)$$

At the origin, we impose the continuity condition

$$X(0^+) = X(0^-) \qquad (7.211)$$

The condition on the derivative of $X(x)$ at the origin may be found by integrating the second of Eqs. (7.208) over a small interval $[-\varepsilon, \varepsilon]$ and then letting $\varepsilon \to 0$. The integration gives

$$\frac{dX}{dx}\bigg|_{-\varepsilon}^{\varepsilon} + \frac{\beta}{D}\,\text{sgn}(x)X\bigg|_{-\varepsilon}^{\varepsilon} + \lambda \int_{-\varepsilon}^{\varepsilon} X(x)\,dx = 0 \qquad (7.212)$$

Upon setting $\varepsilon \to 0$ and using the mean value theorem, the required condition is

$$\dot{X}(0^+) + \frac{\beta}{D}X(0^+) = \dot{X}(0^-) - \frac{\beta}{D}X(0^-) \qquad (7.213)$$

where the dot denotes derivative with respect to x.

The general solution of Eqs. (7.210) is

$$X(x) = \begin{cases} \exp[-\beta x/2D][ae^{\lambda'x} + be^{-\lambda'x}], & x > 0 \\ \exp[\beta x/2D][ce^{\lambda'x} + de^{-\lambda'x}], & x < 0 \end{cases} \qquad (7.214)$$

where

$$\lambda' = \left(\frac{\beta^2}{4D^2} - \lambda\right)^{1/2} \qquad (7.215)$$

and the integration constants a, b, c, and d are to be determined from the conditions (7.211), (7.213), and the conditions

$$X(\pm x_1) = 0 \qquad (7.216)$$

The conditions (7.216) imply that

$$a = -b\exp(-2\lambda'x_1), \qquad d = -c\exp(-2\lambda'x_1) \qquad (7.217)$$

Condition (7.211) gives

$$a + b = c + d \qquad (7.218)$$

and condition (7.213) yields

$$(a - b - c + d)\lambda' = -(\beta/2D)(a + b + c + d) \qquad (7.219)$$

If we substitute Eqs. (7.217) into Eq. (7.218), we get

$$b[1 - \exp(-2\lambda'x_1)] = c[1 - \exp(-2\lambda'x_1)] \qquad (7.220)$$

Two cases must be considered:

Case 1. $1 - \exp(-2\lambda'x_1) \neq 0$. $\qquad\qquad\qquad\qquad\qquad (7.221)$

Case 2. $1 - \exp(-2\lambda'x_1) = 0$. $\qquad\qquad\qquad\qquad\qquad (7.222)$

Treating Case 1 first, it is immediate that Eqs. (7.220) and (7.218) give

$$b = c \quad \text{and} \quad a = d \tag{7.223}$$

Hence, Eq. (7.219) in this case takes the form

$$\lambda' = (\beta/2D) \tanh(\lambda' x_1) \tag{7.224}$$

Together with Eq. (7.215), Eq. (7.224) gives the characteristic values λ in this case, which can be solved by graphical or other approximate means. Hence, for Case 1 the solution (7.214) can be written in the form

$$X^{(1)}(x) = g(\lambda^{(1)}) \exp\left(-\frac{\beta \mid x \mid}{2D}\right) \sin\left[\left(\lambda^{(1)} - \frac{\beta^2}{4D^2}\right)^{1/2}(\mid x \mid - x_1)\right] \tag{7.225}$$

where the superscript (1) is used to denote quantities corresponding to Case 1. It is noted that the functions

$$\sin[(\lambda_k^{(1)} - \beta^2/4D^2)^{1/2}(\mid x \mid - x_1)]$$

form an orthogonal set over the interval $[-x_1, x_1]$. The weight of orthogonality is

$$\begin{aligned}
\omega^{(1)} &= \int_{-x_1}^{x_1} \sin^2[(\lambda_k^{(1)} - \beta^2/4D^2)^{1/2}(\mid x \mid - x_1)] \, dx \\
&= x_1 - \frac{\sin[2x_1(\lambda_k^{(1)} - \beta^2/4D^2)^{1/2}]}{2(\lambda_k^{(1)} - \beta^2/4D^2)^{1/2}}
\end{aligned} \tag{7.226}$$

Let us now consider Case 2. Equation (7.222) can be immediately used to obtain the eigenvalues in this case. They are

$$\lambda_k^{(2)} = \beta^2/4D^2 + (k\pi/x_1)^2 \tag{7.227}$$

We have, from Eqs. (7.217),

$$a = -b, \quad d = -c \tag{7.228}$$

Corresponding to Case 2, the solution (7.214) can be shown to be

$$X^{(2)}(x) = h(\lambda^{(2)}) \exp[-\beta \mid x \mid /2D] \sin[(\lambda^{(2)} - \beta^2/4D^2)^{1/2} x] \tag{7.229}$$

Again the set

$$\sin[(\lambda_k^{(2)} - \beta^2/4D^2)^{1/2} x]$$

is an orthogonal one over the interval $[-x_1, x_1]$, the weight being

$$\omega^{(2)} = \int_{-x_1}^{x_1} \sin^2[(\lambda_k^{(2)} - \beta^2/4D^2)^{1/2}x] \, dx = x_1 \qquad (7.230)$$

In view of Eqs. (7.209), (7.225), and (7.230), the general solution for the transition probability density is thus of the form

$f(x, t \mid x_0, 0)$

$$= \exp\left[-\frac{\beta \mid x \mid}{2D}\right]\left\{\sum_{k=1}^{\infty} g_k \exp(-\lambda_k^{(1)} Dt) \sin\left[\left(\lambda_k^{(1)} - \frac{\beta^2}{4D^2}\right)^{1/2}(\mid x \mid - x_1)\right]\right.$$

$$\left. + \sum_{k=1}^{\infty} h_k \exp(-\lambda_k^{(2)} Dt) \sin\left[\left(\lambda_k^{(2)} - \frac{\beta^2}{4D^2}\right)^{1/2}x\right]\right\} \qquad (7.231)$$

There remains the problem of determining the coefficients g_k and h_k. These are obtained by applying the initial condition (7.205). We first note from Eq. (7.231) that

$$f(x, 0 \mid x_0, 0) \exp\left[\frac{\beta \mid x \mid}{2D}\right] = \sum_{k=1}^{\infty} g_k \sin\left[\left(\lambda_k^{(1)} - \frac{\beta^2}{4D^2}\right)^{1/2}(\mid x \mid - x_1)\right]$$

$$+ \sum_{k=1}^{\infty} h_k \sin\left[\left(\lambda_k^{(2)} - \frac{\beta^2}{4D^2}\right)^{1/2}x\right] \qquad (7.232)$$

Multiplying both sides by

$$\sin[(\lambda_j^{(1)} - \beta^2/4D^2)^{1/2}(\mid x \mid - x_1)]$$

and integrating over $[-x_1, x_1]$ gives

$$\int_{-x_1}^{x_1} f(x, 0 \mid x_0, 0) \exp\left[\frac{\beta \mid x \mid}{2D}\right] \sin\left[\left(\lambda_j^{(1)} - \frac{\beta^2}{4D^2}\right)^{1/2}(\mid x \mid - x_1)\right] dx$$

$$= \sum_{k=1}^{\infty} g_k \omega^{(1)} \delta_{jk} \qquad (7.233)$$

Inserting Eq. (7.205) into Eq. (7.233), we have

$$g_k = \frac{1}{\omega^{(1)}} \exp\left[\frac{\beta \mid x_0 \mid}{2D}\right] \sin\left[\left(\lambda_k^{(1)} - \frac{\beta^2}{4D^2}\right)^{1/2}(\mid x_0 \mid - x_1)\right] \qquad (7.234)$$

The coefficients h_k can be obtained by multiplying both sides of Eq. (7.232) by

$$\sin\left[\left(\lambda_j^{(2)} - \frac{\beta^2}{4D^2}\right)^{1/2}x\right]$$

and integrating over the interval $[-x_1, x_1]$. The result is

$$h_k = \frac{1}{\omega^{(2)}} \exp\left[\frac{\beta \mid x_0 \mid}{2D}\right] \sin\left[\left(\lambda_k^{(2)} - \frac{\beta^2}{4D^2}\right)^{1/2} x_0\right] \quad (7.235)$$

The transition probability density for the bang-bang servo is now completely determined. It has the form given by Eq. (7.231) with g_k given by Eq. (7.234) and h_k by Eq. (7.235). The characteristic values $\lambda_k^{(1)}$ are specified by Eq. (7.224) and $\lambda_k^{(2)}$ are given by Eq. (7.227).

It should be mentioned that, in the determination of the coefficients g_k and h_k, the initial condition is expanded in a series of orthogonal functions as indicated by Eq. (7.232). The validity of this representation requires the convergence of this series expansion and this is by no means assured for our case where the initial condition is a Dirac delta function. In order to avoid this convergence question, the Dirac delta function can be approximated by a narrow rectangular density function centered around x_0. The integrals such as the one in Eq. (7.233) are now more complicated but the results are free of mathematical objections.

As an illustration, the transition density given by Eq. (7.231) is plotted at $t = 1$ and $t = 4$ in Fig. 7.7 with $D = 1$, $\beta = 1$, $x_1 = 10$, and $x_0 = 0$.

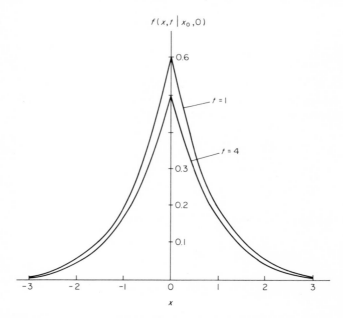

Fig. 7.7. The transition probability density $f(x, t \mid x_0, 0)$ for $D = \beta = 1$, $x_1 = 10$, and $x_0 = 0$ (from Wishner [8]).

For this particular case, the coefficients h_k are identically zero and the results presented in Fig. 7.7 are obtained by truncating the first series in Eq. (7.231) at the tenth term.

(b) The Stationary Solution. A less ambitious goal in the application of the Fokker–Planck equation to nonlinear problems is the determination of its stationary solution. Consider the n-dimensional Fokker–Planck equation as reproduced from Eq. (7.147)

$$\frac{\partial f(\mathbf{x}, t \mid \mathbf{x}_0, t_0)}{\partial t} = -\sum_{j=1}^{n} \frac{\partial}{\partial x_j} [\alpha_j(\mathbf{x})f] + \tfrac{1}{2} \sum_{i,j=1}^{\infty} \frac{\partial^2}{\partial x_i \partial x_j} [\alpha_{ij}(\mathbf{x})f] \quad (7.236)$$

where the derivate moments α_j and α_{ij} are assumed to be independent of time. Following our discussion in Section 7.2.3, a solution of this equation can be facilitated by setting the term on the left-hand side, $\partial f/\partial t$, equal to zero. It implies that the vector Markov process $\mathbf{X}(t)$ approaches a steady state as the transition time $t - t_0$ approaches infinity. The stationary probability density, denoted by $f_s(\mathbf{x})$, is independent of t and the initial conditions, and it satisfies

$$\sum_{j=1}^{n} \frac{\partial}{\partial x_j} [\alpha_j(\mathbf{x})f_s] - \tfrac{1}{2} \sum_{i,j=1}^{n} \frac{\partial^2}{\partial x_i \partial x_j} [\alpha_{ij}(\mathbf{x})f_s] = 0 \quad (7.237)$$

with appropriate boundary conditions and the normalization condition.

It is of practical interest to consider conditions under which a closed-form solution of Eq. (7.237) can be obtained.

In the one-dimensional case, a general solution can be obtained by ordinary methods. Consider Eq. (7.237) with $n = 1$,

$$\frac{d}{dx} [\alpha_1(x)f_s] - \tfrac{1}{2} \frac{d^2}{dx^2} [\alpha_2(x)f_s] = 0 \quad (7.238)$$

We have, upon integrating once,

$$\alpha_1(x) f_s(x) - \tfrac{1}{2} \frac{d}{dx} [\alpha_2(x) f_s(x)] = c_1 = \text{const} \quad (7.239)$$

If we write

$$\alpha_2(x) f_s(x) = h(x) \quad (7.240)$$

Equation (7.239) becomes

$$\frac{dh(x)}{dx} - 2 \frac{\alpha_1(x)}{\alpha_2(x)} h(x) = -2c_1 \quad (7.241)$$

This equation has the general solution

$$h(x) = c_2 \exp\left(2 \int_0^x \frac{\alpha_1(s)}{\alpha_2(s)} \, ds\right) - 2c_1 \int_0^x \exp\left[2 \int_r^x \frac{\alpha_1(s)}{\alpha_2(s)} \, ds\right] dr \quad (7.242)$$

Hence, the general solution for $f_s(x)$ is, upon substituting Eq. (7.242) into Eq. (7.240),

$$f_s(x) = \frac{c_2}{\alpha_2(x)} \exp\left[2 \int_0^x \frac{\alpha_1(s)}{\alpha_2(s)} \, ds\right] - \frac{2c_1}{\alpha_2(x)} \int_0^x \exp\left[2 \int_r^x \frac{\alpha_1(s)}{\alpha_2(s)} \, ds\right] dr$$

$$(7.243)$$

The constants of integration, c_1 and c_2, are determined from normalization and boundary conditions. For example, if we assume that

$$f_s(\pm\infty) = 0 \qquad \text{and} \qquad df_s(\pm\infty)/dx = 0 \qquad (7.244)$$

It is seen from Eqs. (7.240) and (7.241) that $c_1 = 0$ and we have

$$f_s(x) = \frac{c_2}{\alpha_2(x)} \exp\left[2 \int_0^x \frac{\alpha_1(s)}{\alpha_2(s)} \, ds\right] \qquad (7.245)$$

where c_2 is determined by the normalization condition.

The problem is more difficult to solve in the multidimensional case. Instead of attempting to solve the general problem, we give here an approach which is fruitful *provided* that the solution of Eq. (7.237) is of the product form, that is

$$f_s(\mathbf{x}) = \prod_{j=1}^n f_{sj}(x_j) \qquad (7.246)$$

A number of interesting cases falls into this category [9].

If Eq. (7.237) can be written in the form

$$\sum_{j=1}^n L_j(\mathbf{x})[\beta_j(x_j)f_s + \gamma_j(x_j) \, \partial f_s/\partial x_j] = 0 \qquad (7.247)$$

where $L_j(\mathbf{x})$, $j = 1, 2, \ldots, n$, are arbitrary partial differential operators, it is clear that one solution of Eq. (7.247) can be obtained by requiring that $f_s(\mathbf{x})$ satisfies

$$\beta_j(x_j) f_s(\mathbf{x}) + \gamma_j(x_j) \, \partial f_s(\mathbf{x})/\partial x_j = 0 \qquad (7.248)$$

for all j. These first-order homogeneous differential equations then immediately lead to the result

$$f_s(\mathbf{x}) = c \prod_{j=1}^n \exp\left[-\int_0^{x_j} \frac{\beta_j(s)}{\gamma_j(s)} \, ds\right] \qquad (7.249)$$

where c is the normalization constant. It follows from the uniqueness of the stationary solution [10] that Eq. (7.249) is the desired result.

Example 7.9. To illustrate the approach outlined above, let us consider a second-order equation

$$\ddot{X}(t) + 2\beta \dot{X}(t) + k(X) = W(t) \tag{7.250}$$

With

$$X(t) = X_1(t), \qquad \dot{X}(t) = X_2(t), \qquad \text{and} \qquad \mathbf{X}(t) = \begin{bmatrix} X_1(t) \\ X_2(t) \end{bmatrix}$$

it can be represented by

$$d\mathbf{X}(t) = \begin{bmatrix} X_2(t) \\ -k(X_1) - 2\beta X_2(t) \end{bmatrix} dt + \begin{bmatrix} 0 \\ dB(t) \end{bmatrix} \tag{7.251}$$

where

$$E\{dB(t)\} = 0, \qquad E\{[dB(t)]^2\} = 2D\, dt \tag{7.252}$$

The Fokker–Planck equation in this case takes the form

$$\frac{\partial f(\mathbf{x}, t \mid \mathbf{x}_0, t_0)}{\partial t} = -\frac{\partial}{\partial x_1}(x_2 f) + \frac{\partial}{\partial x_2}\{[k(x_1) + 2\beta x_2]f\} + \frac{\partial^2}{\partial x_2{}^2}(Df) \tag{7.253}$$

The stationary density function $f_s(\mathbf{x}) = f_s(x_1, x_2)$ thus satisfies

$$D\frac{\partial^2 f_s}{\partial x_2{}^2} - \frac{\partial}{\partial x_1}(x_2 f_s) + \frac{\partial}{\partial x_2}\{[k(x_1) + 2\beta x_2]f_s\} = 0 \tag{7.254}$$

Upon rearranging the terms in the equation above, we find that it can be written in the form

$$\frac{\partial}{\partial x_2}\left[k(x_1)f_s + \frac{D}{2\beta}\frac{\partial f_s}{\partial x_1}\right] + \left(2\beta \frac{\partial}{\partial x_2} - \frac{\partial}{\partial x_1}\right)\left[x_2 f_s + \frac{D}{2\beta}\frac{\partial f_s}{\partial x_2}\right] = 0 \tag{7.255}$$

which is seen to have the form of Eq. (7.247) with

$$
\begin{aligned}
L_1 &= \partial/\partial x_2 & L_2 &= 2\beta(\partial/\partial x_2) - \partial/\partial x_1 \\
\beta_1(x_1) &= k(x_1) & \beta_2(x_2) &= x_2 \\
\gamma_1(x_1) &= D/2\beta & \gamma_2(x_2) &= D/2\beta
\end{aligned}
\tag{7.256}
$$

The substitution of the equations above into Eq. (7.249) results in

$$f_s(x_1, x_2) = c \exp\left\{-2\beta/D\left[\int_0^{x_1} k(x)\,dx + x_2^2/2\right]\right\} \tag{7.257}$$

It is seen that the marginal density function $f_s(x_2)$ is of the Gaussian form. The component X_1 in general will not follow a Gaussian law because of the nonlinearity in this component in the original differential equation.

Example 7.10. Let us consider a slightly more general second-order equation

$$\ddot{X}(t) + g(\dot{X}) + k(X) = W(t) \tag{7.258}$$

We shall show that the approach developed above does not apply in this case unless the function $g(\dot{X})$ is linear in $\dot{X}$.

Again letting $X(t) = X_1(t)$ and $\dot{X}(t) = X_2(t)$, it is easy to show that the stationary density function $f_s(\mathbf{x}) = f_s(x_1, x_2)$ satisfies the equation

$$D\frac{\partial^2 f_s}{\partial x_2^2} - \frac{\partial}{\partial x_1}(x_2 f_s) + \frac{\partial}{\partial x_2}[(k(x_1) + g(x_2))f_s] = 0 \tag{7.259}$$

Integrating it with respect to x_2 gives

$$k(x_1) + g(x_2) = \frac{1}{f_s}\left[\int_0^{x_2} \frac{\partial}{\partial x_1}[xf_s(x_1, x)]\,dx - D\frac{\partial f_s}{\partial x_2} + c(x_1)\right] \tag{7.260}$$

where $c(x_1)$ is an integration constant.

Let us assume that the solution to the equation above takes the product form

$$f_s(x_1, x_2) = f_1(x_1)f_2(x_2) \tag{7.261}$$

Equation (7.260) then becomes

$$k(x_1) + g(x_2) = \frac{1}{f_1(x_1)}\frac{df_1(x_1)}{dx_1}\left[\frac{1}{f_2(x_2)}\int_0^{x_2} xf_2(x)\,dx\right]$$
$$- \frac{D}{f_2(x_2)}\frac{df_2(x_2)}{dx_2} + \frac{c(x_1)}{f_1(x_1)f_2(x_2)} \tag{7.262}$$

In order that Eq. (7.262) be admissible, each of the terms on the right-hand side must be either a function of only x_1 or a function of only x_2. Hence, under general conditions we must have

$$c(x_1) = 0, \qquad [1/f_2(x_2)] \int_0^{x_2} x f_2(x)\, dx = -c_1 = \text{const}$$

$$g(x_2) = \frac{-D}{f_2(x_2)} \frac{df_2(x_2)}{dx_2}$$

(7.263)

The second of Eqs. (7.263) then implies that

$$f_2(x_2) = c_2 \exp(-x_2{}^2/2c_1)$$

(7.264)

and the substitution of the above into the third of Eqs. (7.263) shows that $g(x_2)$ must have the form

$$g(x_2) = c_1 D x_2$$

(7.265)

Therefore, the stationary density function in the separable form of Eq. (7.261) is permissible only when the function $g(x_2)$ is linear in x_2. A stationary solution of Eq. (7.260) in its general form has not been found.

While the transition probability density is in general difficult to obtain, we have seen that the stationary solution can be found for a class of problems of practical interest. The differential equation considered in Example 7.9, for instance, represents oscillators with nonlinear spring characteristics, control systems with nonlinear transfer properties, simple structures with nonlinear material properties, and other physically interesting phenomena.

(c) **Moments.** Recently, several attempts have been made in determining in some approximate way the time-dependent moments associated with the solution processes of nonlinear differential equations of the Ito type. The procedure for doing this begins with the derivation of a set of differential equations satisfied by these moments.

Let us consider in general terms the Ito equation

$$d\mathbf{X}(t) = \mathbf{f}(\mathbf{X}(t), t)\, dt + G(\mathbf{X}(t), t)\, d\mathbf{B}(t), \qquad t \geq t_0$$

(7.266)

with its associated Fokker–Planck equation [Eq. (7.160)]

$$\frac{\partial f(\mathbf{x}, t \mid \mathbf{x}_0, t_0)}{\partial t} = -\sum_{j=1}^{n} \frac{\partial}{\partial x_j} [f_j(\mathbf{x}, t) f] + \sum_{i,j=1}^{n} \frac{\partial^2}{\partial x_i\, \partial x_j} [(GDG^{\mathrm{T}})_{ij} f]$$

(7.267)

The differential equations satisfied by the moments associated with the solution process $\mathbf{X}(t)$ can be established with the help of the Fokker–

Planck equation (7.267). Consider the expectation

$$E\{X_1^{k_1}(t)\, X_2^{k_2}(t)\, \cdots\, X_n^{k_n}(t)\} = m_{k_1 k_2 \ldots k_n}(t) \tag{7.268}$$

By definition, this expectation is given by

$$m_{k_1 k_2 \ldots k_n}(t) = \int_{-\infty}^{\infty} \cdots \int_{-\infty}^{\infty} h(\mathbf{x})\, f(\mathbf{x}_0, t_0)\, f(\mathbf{x}, t \mid \mathbf{x}_0, t_0)\, d\mathbf{x}\, d\mathbf{x}_0 \tag{7.269}$$

where

$$h(\mathbf{X}) = X_1^{k_1}(t)\, X_2^{k_2}(t)\, \cdots\, X_n^{k_n}(t) \tag{7.270}$$

The time derivative of Eq. (7.269) gives

$$\dot{m}_{k_1 k_2 \ldots k_n}(t) = \int_{-\infty}^{\infty} \cdots \int_{-\infty}^{\infty} h(\mathbf{x})\, f(\mathbf{x}_0, t_0)\, \frac{\partial f(\mathbf{x}, t \mid \mathbf{x}_0, t_0)}{\partial t}\, d\mathbf{x}\, d\mathbf{x}_0 \tag{7.271}$$

Now, if we replace $\partial f/\partial t$ in the equation above by the right-hand side of Eq. (7.267) and integrate the right-hand side of Eq. (7.271) by parts over the entire phase space $-\infty \le x_j$, $x_{0j} \le \infty$ for all j, the resulting equation is a first-order ordinary differential equation containing only moments of the solution process. Therefore, for different values of k_1, k_2, $\ldots$, a set of moment equations can be generated. This procedure is essentially due to Bogdanoff and Kozin [11].

Cumming [12] gives a more direct method of generating these moment equations. Consider an arbitrary function of $\mathbf{X}(t)$ and t, $h(\mathbf{X}, t)$, whose partial derivatives $\partial^2 h/\partial X_i\, \partial X_j$ and $\partial h/\partial t$ are jointly continuous and bounded over any finite interval of $\mathbf{X}$ and t. If we use δ as a finite forward increment operator over the time increment δt, we have

$$\delta h = h(\mathbf{X} + \delta\mathbf{X}, t + \delta t) - h(\mathbf{X}, t) \tag{7.272}$$

The Taylor series expansion of δh then gives

$$\delta h = \sum_{j=1}^{n} \delta X_j\, \partial h/\partial X_j + \tfrac{1}{2} \sum_{i,j=1}^{n} \delta X_i\, \delta X_j\, \partial^2 h/\partial X_i\, \partial X_j + \delta t\, \partial h/\partial t$$
$$+ o(\delta\mathbf{X}\, \delta\mathbf{X}^{\mathrm{T}}) + o(\delta t) \tag{7.273}$$

Referring back to the general Ito equation (7.266), we have shown that [Eqs. (7.158) and (7.159)]

$$E\{\delta X_j(t) \mid \mathbf{X}\} = f_j(\mathbf{X}, t)\, \delta t + o(\delta t)$$
$$E\{\delta X_i(t)\, \delta X_j(t) \mid \mathbf{X}\} = 2(GDG^{\mathrm{T}})_{ij}\, \delta t + o(\delta t) \tag{7.274}$$

The conditional expectation of Eq. (7.273) given $\mathbf{X}$ thus has the form

$$E\{\delta h \mid \mathbf{X}\} = \sum_{j=1}^{n} f_j(\mathbf{X}, t)(\partial h/\partial X_j)\, \delta t + \sum_{i,j=1}^{n} (GDG^{\mathrm{T}})_{ij}(\partial^2 h/\partial X_i\, \partial X_j)\, \delta t$$
$$+ (\partial h/\partial t)\, \delta t + o(\delta t) \tag{7.275}$$

We observe that the expectation $E\{\delta h \mid \mathbf{X}\}$ is regarded here as a random variable. It has the form $E\{Y \mid \mathbf{X}\}$ which takes the value $E\{Y \mid \mathbf{x}\}$ according to

$$E\{Y \mid \mathbf{X} = \mathbf{x}\} = \int_{-\infty}^{\infty} yf(y \mid \mathbf{x})\, dy \tag{7.276}$$

Hence, the expectation of $E\{Y \mid \mathbf{X}\}$ is

$$E\{E\{Y \mid \mathbf{X}\}\} = \int_{-\infty}^{\infty} \cdots \int_{-\infty}^{\infty} \left[\int_{-\infty}^{\infty} y f(y \mid \mathbf{x})\, dy \right] f(\mathbf{x})\, d\mathbf{x} = E\{Y\} \tag{7.277}$$

In view of the above, the expectation of Eq. (7.275) gives

$$E\{\delta h\} = \sum_{j=1}^{n} E\{f_j(\mathbf{X}, t)(\partial h/\partial X_j)\}\, \delta t$$
$$+ \sum_{i,j=1}^{n} E\{(GDG^{\mathrm{T}})_{ij}(\partial^2 h/\partial X_i\, \partial X_j)\}\, \delta t + E\{\partial h/\partial t\}\, \delta t + o(\delta t) \tag{7.278}$$

provided that the indicated expectations exist. Finally, upon dividing it by δt, passing to the limit $\delta t \to 0$, and interchanging differentiation and expectation, we obtain the ordinary differential equation

$$\frac{dE\{h\}}{dt} = \sum_{j=1}^{n} E\{f_j\, \partial h/\partial X_j\} + \sum_{i,j=1}^{n} E\{(GDG^{\mathrm{T}})_{ij}\, \partial^2 h/\partial X_i\, \partial X_j\}$$
$$+ E\{\partial h/\partial t\} \tag{7.279}$$

This is the moment equation for $E\{X_1^{k_1}(t)X_2^{k_2}(t) \cdots X_n^{k_n}(t)\}$ if we set

$$h(\mathbf{X}, t) = X_1^{k_1}X_2^{k_2} \cdots X_n^{k_n} \tag{7.280}$$

In order to see what this moment equation entails in a given situation, let us consider the simple first-order nonlinear Ito equation

$$dX(t) = -[X(t) + aX^3(t)]\, dt + dB(t) \tag{7.281}$$

Let $h(X, t) = X^k$. We have $n = 1$, $f_1 = -(X + aX^3)$, $G = 1$, $\partial h/\partial X = kX^{k-1}$, $\partial^2 h/\partial X^2 = k(k-1)X^{k-2}$, and $\partial h/\partial t = 0$. The substitution of these quantities into Eq. (7.279) yields the moment equation

$$\dot{m}_k(t) = -k(m_k + am_{k+2}) + Dk(k-1)m_{k-2} \tag{7.282}$$

where

$$m_k(t) = E\{X^k(t)\}, \qquad k = 1, 2, \ldots \tag{7.283}$$

It is seen that the moment equations are relatively easy to establish; however, as we see from Eq. (7.282), a difficulty arises because the moment equation for $m_k(t)$ contains moments of orders higher than k. Hence, this situation leads to an infinite hierarchy of differential equations for the moments. Indeed, we see upon little reflection that this drawback is present whenever the governing differential equation is nonlinear.

Hence, in order to obtain moment solutions, some closure approximations must be used to truncate this hierarchy. Sancho [13, 14] and Wilcox and Bellman [15] have considered a number of truncation schemes. These schemes range from simply dropping the higher-order moments at a certain truncation point to approximating these higher-order moments as functions of lower-order ones in some optimal way.

As an illustration, consider the first two moment equations defined by Eq. (7.282). They are

$$\dot{m}_1(t) = -m_1(t) - am_3(t), \qquad \dot{m}_2(t) = 2[D - m_2(t) - am_4(t)] \tag{7.284}$$

We wish to approximate $m_3(t)$ and $m_4(t)$ in some sense so that the first two moments can be determined from the equations above.

It is not desirable to simply set $m_3(t) = m_4(t) = 0$, since this procedure is equivalent to setting $a = 0$ and thus reduces the problem to the linear case.

Another possible approach is to use the approximations

$$m_3(t) \cong a_0 + a_1 m_1 + a_2 m_2, \qquad m_4(t) \cong b_0 + b_1 m_1 + b_2 m_2 \tag{7.285}$$

where the coefficients a_0, a_1, a_2, b_0, b_1, and b_2 are allowed to be independent of time. A natural way of choosing these coefficients is to follow a mean-square averaging procedure in which we minimize the integrals

$$\int_{-\infty}^{\infty} f(x, t)(x^3 - a_0 - a_1 x - a_2 x^2)^2 \, dx$$
$$\int_{-\infty}^{\infty} f(x, t)(x^4 - b_0 - b_1 x - b_2 x^2)^2 \, dx \tag{7.286}$$

Now, since $f(x, t)$ is unavailable, the next best way is to replace $f(x, t)$ by $f_s(x)$, the stationary density function, which can be found in this case.

Another possible approximation scheme is to drop the third- and fourth-order central moments. Since

$$\mu_3 = m_3 - 3m_1m_2 - 2m_1^3, \qquad \mu_4 = m_4 - 4m_1m_3 + 6m_1^2m_2 - 3m_1^4 \quad (7.287)$$

this procedure means that we set $\mu_3 = \mu_4 = 0$ and determine the moments m_3 and m_4 as functions of m_1 and m_2.

It is clear that all the procedures outlined above lack sound mathematical basis. Unless there is a physical basis for choosing a certain truncation procedure, the merits of these approximation schemes are difficult to assess in general.

In closing, we remark that, when a truncation procedure is used, it is also important to determine whether the procedure preserves the moment properties which are known to exist for exact solutions. For example, variances must be positive. A step toward answering this question is given by Bellman and Richardson [16].

7.4.2. Perturbation Techniques

The perturbation approach is well developed in the study of deterministic nonlinear problems when the nonlinearity is considered small. For nonlinear random differential equations with random inputs, the same technique can be applied fruitfully under certain restrictions.

Consider a second-order system governed by the differential equation

$$\ddot{X}(t) + 2\beta\dot{X}(t) + \omega_0^2[X(t) + \varepsilon g(X)] = Y(t), \quad (7.288)$$

where the nonlinear function $g(X)$ is assumed to be differentiable with respect to X up to a suitable order and $Y(t)$ is a random input to the system. We shall be concerned with approximate solutions of Eq. (7.288) when the parameter ε is small.

The perturbation method is based on the assumption that the solution process $X(t)$ can be expanded in the powers of the parameter ε. Let us write

$$X(t) = X_0(t) + \varepsilon X_1(t) + \varepsilon^2 X_2(t) + \cdots \quad (7.289)$$

Substituting Eq. (7.289) into Eq. (7.288) and equating terms of the same power of ε gives

$$\ddot{X}_0(t) + 2\beta \dot{X}_0(t) + \omega_0{}^2 X_0(t) = Y(t)$$
$$\ddot{X}_1(t) + 2\beta \dot{X}_1(t) + \omega_0{}^2 X_1(t) = -\omega_0{}^2 g(X_0) \qquad (7.290)$$
$$\ddot{X}_2(t) + 2\beta \dot{X}_2(t) + \omega_0{}^2 X_2(t) = -\omega_0{}^2 X_1(t)\, g'(X_0)$$
$$\vdots$$

where $g'(X_0)$ stands for the derivative of $g(X)$ evaluated at $X(t) = X_0(t)$.

Equations (7.290) indicate that each term in the expansion of $X(t)$ satisfies a linear differential equation with a random input. Consequently, the non-linear problem is reduced to one of solving a system of linear random differential equations of the type discussed in Section 7.2.

We see that the application of the perturbation method to solving non-linear differential equations is straightforward. From the computational viewpoint, however, the statistical properties of the higher order terms in Eq. (7.289) are difficult to obtain because the inhomogeneous parts rapidly become exceedingly complex as a function of the input process $Y(t)$.

Another observation to be made is that, while the convergence of the series representation (7.289) can be established for sufficiently small perturbation parameter ε in the deterministic situation (see, for example, Stoker [17]), no proof is available to show that the s.p. $X(t)$ represented by Eq. (7.289) is convergent in mean square or in any other sense. Rigorously speaking, the validity of the perturbation procedure thus requires the restrictive assumption that the solution process $X(t)$ must be such that each of its sample functions can be represented by a convergent series in the powers of ε.

Example 7.11. Consider a specific problem where the function $g(X)$ in Eq. (7.288) is of the form

$$g(X) = X^3 \qquad (7.291)$$

This form corresponds to the familiar case of the Duffing oscillator. We further assume that the input process is stationary.

With the solution process represented by Eq. (7.289), let us determine the second moment of $X(t)$ up to the first order term. It is given by

$$E\{X^2(t)\} = E\{X_0{}^2(t)\} + 2\varepsilon E\{X_0(t)\, X_1(t)\} \qquad (7.292)$$

This problem has been considered by Crandall [18].

Let us consider only the steady-state solution. Based upon the linear theory, the solution for $X_0(t)$ from the first of Eqs. (7.290) is

$$X_0(t) = \int_{-\infty}^{t} h(t-s) Y(s) \, ds = \int_{0}^{\infty} h(s) Y(t-s) \, ds \qquad (7.293)$$

where, for the underdamped case $(\beta^2/\omega_0^2 < 1)$,

$$h(t) = [1/(\omega_0^2 - \beta^2)^{1/2}] e^{-\beta t} \sin(\omega_0^2 - \beta^2)^{1/2} t \qquad (7.294)$$

Hence,

$$E\{X_0^2(t)\} = \int_{0}^{\infty} \int_{0}^{\infty} h(u) h(v) \Gamma_{YY}(u-v) \, du \, dv \qquad (7.295)$$

The first term of Eq. (7.292) can thus be determined from the knowledge of the correlation function $\Gamma_{YY}(\tau)$ of the s.p. $Y(t)$.

Let us turn now to the second term. The process $X_1(t)$ is governed by the second of Eqs. (7.290) and, with $g(X)$ specified by Eq. (7.291), we have

$$X_1(t) = \int_{-\infty}^{t} h(t-s)[-\omega_0^2 X_0^3(s)] \, ds = -\omega_0^2 \int_{0}^{\infty} h(s) X_0^3(t-s) \, ds \qquad (7.296)$$

Therefore, the expectation $E\{X_0(t) X_1(t)\}$ can be written in the form

$$E\{X_0(t) X_1(t)\} = -\omega_0^2 \int_{0}^{\infty} h(s) E\{X_0(t) X_0^3(t-s)\} \, ds \qquad (7.297)$$

In terms of the statistics of the input process $Y(t)$, we substitute Eq. (7.293) into Eq. (7.297) and obtain

$$E\{X_0(t) X_1(t)\}$$

$$= -\omega_0^2 \int_{0}^{\infty} h(s) \, ds \int_{0}^{\infty} h(s_1) \, ds_1 \int_{0}^{\infty} h(s_2) \, ds_2 \int_{0}^{\infty} h(s_3) \, ds_3 \int_{0}^{\infty} h(s_4) \, ds_4$$

$$\times E\{Y(t-s_1) Y(t-s-s_2) Y(t-s-s_3) Y(t-s-s_4)\} \qquad (7.298)$$

It is seen that the evaluation of the expectation $E\{X_0(t) X_1(t)\}$ requires the knowledge of the fourth-order moments of the input process.

If we further assume that the input process is Gaussian with zero mean, the fourth-order moments of $Y(t)$ are expressible in terms of its correlation functions. Using the identity for a stationary Gaussian process (Problem 2.18)

$$E\{Y(t_1) Y(t_2) Y(t_3) Y(t_4)\} = \Gamma_{YY}(t_1 - t_2) \Gamma_{YY}(t_3 - t_4)$$
$$+ \Gamma_{YY}(t_1 - t_3) \Gamma_{YY}(t_2 - t_4)$$
$$+ \Gamma_{YY}(t_1 - t_4) \Gamma_{YY}(t_2 - t_3) \qquad (7.299)$$

The fourth-order moment of $Y(t)$ in Eq. (7.176) thus takes the form

$$E\{Y(t - s_1)\, Y(t - s - s_2)\, Y(t - s - s_3)\, Y(t - s - s_4)\}$$
$$= \Gamma_{YY}(s - s_1 + s_2)\, \Gamma_{YY}(s_3 - s_4) + \Gamma_{YY}(s - s_1 + s_3)\, \Gamma_{YY}(s_2 - s_4)$$
$$+ \Gamma_{YY}(s - s_1 + s_4)\, \Gamma_{YY}(s_2 - s_3) \qquad (7.300)$$

The fivefold integral in Eq. (7.298) becomes partially separable upon the substitution of Eq. (7.300) into it. It can be put in the convenient form

$$E\{X_0(t)\, X_1(t)\} = -3\omega_0^2\, \Gamma_{X_0}(0) \int_0^\infty h(s)\, \Gamma_{X_0 X_0}(s)\, ds \qquad (7.301)$$

where $\Gamma_{X_0 X_0}(s)$ is the correlation function of $X_0(t)$ and it is given by

$$\Gamma_{X_0 X_0}(s) = \int_0^\infty \int_0^\infty h(s_1)\, h(s_2)\, \Gamma_{YY}(s - s_1 + s_2)\, ds_1\, ds_2 \qquad (7.302)$$

Finally, the substitution of Eqs. (7.295), (7.301), and (7.302) into Eq. (7.292) gives the mean square value of $X(t)$

$$E\{X^2(t)\} = \Gamma_{X_0 X_0}(0)\left[1 - 6\varepsilon\omega_0^2 \int_0^\infty h(s)\, \Gamma_{X_0 X_0}(s)\, ds\right] \qquad (7.303)$$

with $\Gamma_{X_0 X_0}(s)$ given by Eq. (7.302).

Other useful statistical properties can be found following essentially the same procedure outlined above. For example, the correlation function of $X(t)$ is, to the first order of ε,

$$\Gamma_{XX}(\tau) = E\{X(t)\, X(t + \tau)\}$$
$$= E\{X_0(t)\, X_0(t + \tau)\} + \varepsilon[E\{X_0(t)\, X_1(t + \tau)\} + E\{X_1(t)\, X_0(t + \tau)\}]$$
$$\qquad (7.304)$$

The procedure for evaluating it is straightforward but the computational labor is considerable.

7.4.3. Equivalent Linearization

The technique of equivalent linearization was first introduced by Krylov and Bogoliubov [19] in connection with deterministic nonlinear problems. The underlying idea in this approach is that, given a nonlinear differential equation, an "equivalent" linear equation is constructed so that the behavior of the linear system approximates that of the nonlinear system in some sense. Once the "equivalent" linear equation is established, the properties of the solution process can be easily analyzed by means of the linear theory and the results should be approximations to the solution properties of the original nonlinear equation.

Extensions of the equivalent linearization technique to nonlinear random problems have been applied to a number of problems in control and mechanical vibrations. The advantage of this technique over the perturbation method is that it is not restricted to problems with small nonlinearities. On the other hand, a number of difficulties developes in the process of derivation and they can be overcome satisfactorily only for a certain class of nonlinear problems.

Consider a second-order system described by the differential equation

$$\ddot{X}(t) + g[X(t), \dot{X}(t)] = Y(t), \qquad t \in T \tag{7.305}$$

In the method of equivalent linearization, Eq. (7.305) is approximated by an "equivalent" linear equation

$$\ddot{X}(t) + \beta_e \dot{X}(t) + k_e X(t) = Y(t) \tag{7.306}$$

where the parameters β_e and k_e are to be selected so that the linear equation above produces a solution which "best" approximates that of the original nonlinear equation. The central problem, of course, is the determination of these parameters. With the knowledge of β_e and k_e, the approximate properties of the solution process $X(t)$ can be obtained from Eq. (7.306) by means of the linear theory.

Adding the terms $(\beta_e \dot{X}(t) + k_e X(t))$ to both sides of Eq. (7.305) and rearranging gives

$$\ddot{X}(t) + \beta_e \dot{X}(t) + k_e X(t) = Y(t) + N(t) \tag{7.307}$$

where

$$N(t) = \beta_e \dot{X}(t) + k_e X(t) - g[X(t), \dot{X}(t)] \tag{7.308}$$

The quantity $N(t)$ can be considered as the error term in the approximation procedure. In our case, it is a stochastic process. In order to minimize the approximate error, a common criterion is to minimize the mean-squared value of the error process $N(t)$. Hence, we require that the parameters β_e and k_e be chosen such that

$$E\{N^2(t)\} = E\{[\beta_e \dot{X} + k_e X - g(X, \dot{X})]^2\} \tag{7.309}$$

is minimized for $t \in T$.

The first- and the second-order derivatives of $E\{N^2(t)\}$ with respect to β_e and k_e are

$$\frac{\partial}{\partial\beta_e} E\{N^2\} = 2E\{\beta_e\dot{X}^2 + k_e X\dot{X} - \dot{X}g(X, \dot{X})\}$$

$$\frac{\partial}{\partial k_e} E\{N^2\} = 2E\{k_e X^2 + \beta_e X\dot{X} - Xg(X, \dot{X})\}$$

$$\frac{\partial^2}{\partial\beta_e{}^2} E\{N^2\} = 2E\{\dot{X}^2\} > 0 \qquad\qquad (7.310)$$

$$\frac{\partial^2}{\partial k_e{}^2} E\{N^2\} = 2E\{X^2\} > 0$$

$$\frac{\partial^2}{\partial\beta_e{}^2} E\{N^2\} \frac{\partial^2}{\partial k_e{}^2} E\{N^2\} - \left[\frac{\partial^2}{\partial k_e\,\partial\beta_e} E\{N^2\}\right]^2$$
$$= 4[E\{X^2\}\, E\{\dot{X}^2\} - E^2\{X\dot{X}\}] \geq 0$$

It is seen from the above that the conditions

$$\frac{\partial}{\partial\beta_e} E\{N^2\} = 0 \qquad \text{and} \qquad \frac{\partial}{\partial k_e} E\{N^2\} = 0 \qquad (7.311)$$

lead to the minimization of $E\{N^2(t)\}$ and, therefore, the desired parameters β_e and k_e are the solutions of the pair of algebraic equations

$$\beta_e E\{\dot{X}^2\} + k_e E\{X\dot{X}\} - E\{\dot{X}g(X, \dot{X})\} = 0$$
$$k_e E\{X^2\} + \beta_e E\{X\dot{X}\} - E\{Xg(X, \dot{X})\} = 0 \qquad (7.312)$$

In this form, the solution for β_e and k_e requires the knowledge of the indicated expectations and, unfortunately, these expectations are not easily obtainable because they require in general the knowledge of the (unknown) joint density function $f(x, t; \dot{x}, t)$. Two approximations are possible at this point [20, 21]. We may replace $f(x, t; \dot{x}, t)$ by the stationary density function $f_s(x, \dot{x})$ computed from the original nonlinear equation (7.305). This can be made available in some cases by, for example, solving the associated Fokker–Planck equation when applicable. The second approach is to determine $f_s(x, \dot{x})$ approximately by using the linearized equation (7.306). We note that the expectations are now implicit functions of the parameters β_e and k_e. Hence, Eqs. (7.312) become nonlinear in β_e and k_e under this approximation.

In the case where the input process $Y(t)$ is a stationary Gaussian process with zero mean, the second alternative leads to the result that $X(t)$ and $\dot{X}(t)$

are stationary independent Gaussian processes with the joint density function (See Problem 4.13)

$$f_s(x, \dot{x}) = \frac{1}{2\pi\sigma_X\sigma_{\dot{X}}} \exp\left[-\tfrac{1}{2}\left(\frac{x^2}{\sigma_X{}^2} + \frac{\dot{x}^2}{\sigma_{\dot{X}}{}^2}\right)\right] \qquad (7.313)$$

where the variances $\sigma_X{}^2$ and $\sigma_{\dot{X}}{}^2$ are functions of β_e and k_e in general.

Example 7.12. Consider the equation

$$\ddot{X}(t) + 2\beta\dot{X}(t) + \omega_0{}^2[X(t) + aX^3(t)] = Y(t) \qquad (7.314)$$

where $Y(t)$ is a stationary Gaussian process with zero mean and spectral density function $S_{YY}(\omega)$.

The equivalent linear equation has the form

$$\ddot{X}(t) + 2\beta_e\dot{X}(t) + \omega_e{}^2X(t) = Y(t) \qquad (7.315)$$

and β_e and $\omega_e{}^2$ satisfy, as seen from Eqs. (7.312),

$$2(\beta_e - \beta)\, E\{\dot{X}^2\} + (\omega_e{}^2 - \omega_0{}^2)\, E\{X\dot{X}\} - a\omega_0{}^2\, E\{\dot{X}X^3\} = 0$$
$$(\omega_e{}^2 - \omega_0{}^2)\, E\{X^2\} + 2(\beta_e - \beta)\, E\{X\dot{X}\} - a\omega_0{}^2\, E\{X^4\} = 0 \qquad (7.316)$$

Following the second approach, the stochastic processes $X(t)$ and $\dot{X}(t)$ are stationary, independent, and with zero means. Hence,

$$E\{X\dot{X}\} = 0, \qquad E\{\dot{X}X^3\} = 0, \qquad E\{X^2\} = \sigma_X{}^2$$
$$E\{\dot{X}^2\} = \sigma_{\dot{X}}{}^2, \qquad E\{X^4\} = 3\sigma_X{}^4, \qquad (7.317)$$

The first of Eqs. (7.316) gives immediately

$$\beta_e = \beta \qquad (7.318)$$

and the second gives

$$\omega_e{}^2 = \omega_0{}^2[1 + 3a\sigma_X{}^2] \qquad (7.319)$$

The variance $\sigma_X{}^2$ can be computed from the spectral density function $S_{YY}(\omega)$ of $Y(t)$ by

$$\sigma_X{}^2 = \int_0^\infty S_{XX}(\omega)\, d\omega = \int_0^\infty |y(i\omega)|^2\, S_{YY}(\omega)\, d\omega \qquad (7.320)$$

where $y(i\omega)$ is the frequency response associated with the equivalent linear system described by Eq. (7.315). It has the form

$$y(i\omega) = [\omega_e{}^2 - \omega^2 + 2i\beta_e\omega]^{-1}$$
$$= [\omega_e{}^2 - \omega^2 + 2i\beta\omega]^{-1} \qquad (7.321)$$

The substitution of Eqs. (7.320) and (7.321) into Eq. (7.319) results in an algebraic equation with ω_e^2 as the only unknown quantity. The evaluation of ω_e^2 can still be difficult since Eq. (7.319) is in general highly nonlinear.

In this example, a further simplification is justified if the parameter a in Eq. (7.314) is assumed small. To facilitate the computation of ω_e, we approximate the frequency response function $y(i\omega)$ by

$$y_0(i\omega) = [\omega_0^2 - \omega^2 + 2i\beta\omega]^{-1} \tag{7.322}$$

where $y_0(i\omega)$ is simply the frequency response of the linear system by setting $a = 0$ in Eq. (7.314). The equivalent linear equation is thus of the form

$$\ddot{X}(t) + 2\beta\dot{X}(t) + \omega_0^2[1 + 3a\sigma_{X_0}^2] X(t) = Y(t) \tag{7.323}$$

where $\sigma_{X_0}^2$ is a known constant and it is equal to

$$\sigma_{X_0}^2 = \int_0^\infty |y_0(i\omega)|^2 S_{YY}(\omega)\, d\omega = \int_0^\infty S_{YY}(\omega)[(\omega_0^2 - \omega^2)^2 + 4\omega^2\beta^2]^{-1}\, d\omega \tag{7.324}$$

References

1. E. A. Coddington and N. Levinson, *Theory of Ordinary Differential Equations.* McGraw-Hill, New York, 1955.
2. T. K. Caughey and H. J. Stumpf, Transient response of a dynamic system under random excitation. *J. Appl. Mech.* **28**, 563–566 (1961).
3. J. L. Bogdanoff, J. E. Goldberg, and M. C. Bernard, Response of a simple structure to a random earthquake-type disturbance. *Bull. Seism. Soc. Amer.* **51**, 293–310 (1961).
4. M. S. Bartlett, *An Introduction to Stochastic Processes*, 2nd ed. Cambridge Univ. Press, London and New York, 1966.
5. R. F. Pawula, Generalizations and extensions of the Fokker-Planck-Kolmogorov equations. *IEEE Trans. Information Theory* **IT-13**, 33–41 (1967).
6. G. E. Uhlenbeck and L. S. Ornstein, On the theory of the Brownian motion. *Phys. Rev.* **36**, 823–841 (1930). Reprinted in *Selected Papers on Noise and Stochastic Processes* (N. Wax, ed.). Dover, New York, 1954.
7. I. S. Sokolinkoff and R. M. Redheffer, *Mathematics of Physics and Modern Engineering*, Chapter 4. McGraw-Hill, New York, 1958.
8. R. P. Wishner, On Markov processes in control systems. Rep. R-116. Coordinated Sci. Lab., Univ. of Illinois, Urbana, Illinois, 1960.
9. S. C. Liu, Solutions of Fokker-Planck equation with applications in nonlinear random vibration. *Bell System Tech. J.* **48**, 2031–2051 (1969).
10. A. H. Gray, Uniqueness of steady-state solutions to the Fokker-Planck equation. *J. Math. Phys.* **4**, 644–647 (1965).

11. J. L. Bogdanoff and F. Kozin, Moments of the output of linear random systems. *J. Acoust. Soc. Amer.* **34**, 1063–1068 (1962).

12. I. G. Cumming, Derivation of the moments of a continuous stochastic system. *Internat. J. Control* **5**, 85–90 (1967).

13. N. G. F. Sancho, Technique for finding the moment equations of a nonlinear stochastic system. *J. Math. Phys.* **11**, 771–774 (1970).

14. N. G. F. Sancho, On the approximate moment equations of a nonlinear stochastic differential equation. *J. Math. Anal. Appl.* **29**, 384–391 (1970).

15. R. M. Wilcox and R. Bellman, Truncation and preservation of moment properties for Fokker-Planck moment equations. *J. Math. Anal. Appl.* **32**, 532–542 (1970).

16. R. Bellman and J. M. Richardson, Closure and preservation of moment properties. *J. Math. Anal. Appl.* **23**, 639–644 (1968).

17. J. J. Stoker, *Nonlinear Vibrations.* Wiley (Interscience), New York, 1950.

18. S. H. Crandall, Perturbation techniques for random vibration of nonlinear systems. *J. Acoust. Soc. Amer.* **35**, 1700–1705 (1961).

19. N. Krylov and N. Bogoliubov, *Introduction to Nonlinear Mechanics: Asymptotic and Approximate Methods* (Ann. Math. Studies No. 11). Princeton Univ. Press, Princeton, New Jersey, 1947.

20. R. C. Booten, Jr., Nonlinear control systems with random inputs. *IRE Trans. Circuit Theory* **CT-1**, 9–18 (1954).

21. T. K. Caughey, Equivalent linearization techniques. *J. Acoust. Soc. Amer.* **35**, 1706–1711 (1961).

22. H. W. Liepmann, On the application of statistical concepts to the buffeting problem. *J. Aerosp. Sci.* **19**, 793–800 (1952).

PROBLEMS

In Problems 7.1–7.5, we consider solution processes in the steady state only.

7.1. Consider the output process $X(t)$ of a simple RC filter as shown in the figure below. Assuming that the input process $Y(t)$ is wide-sense stationary with mean zero and correlation function

$$\Gamma_{YY}(\tau) = ae^{-b|\tau|}$$

determine the mean square value of the output process $X(t)$ using Eq. (7.70), and then using Eq. (7.77). Compare the computational labor involved in these two procedures.

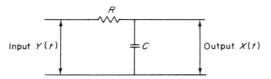

7.2. A "band-limited" white noise $Y(t)$ is one whose power spectral density has the form shown in the accompanying figure. Let $Y(t)$ be the input process to a system described by

$$\ddot{X}(t) + 2\beta\omega_0\dot{X}(t) + \omega_0{}^2X(t) = Y(t)$$

Compute the mean square response of the system. Under what conditions can the response be approximated by one subject to an ideal white-noise excitation?

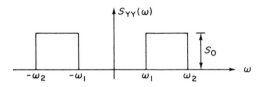

7.3. The differential equation in Problem 7.2 is an idealized equation of motion of a lifting surface in response to atmospheric turbulence in a steady flight [22]. The input in this case is the random lift due to air turbulence. Based upon a two-dimensional air-foil theory, the input process can be described as a wide-sense stationary process with the power spectral density

$$S_{YY}(\omega) = \left(1 + \frac{\pi\omega C}{u}\right)^{-1}\left\{\frac{1 + 3(L\omega/u)^2}{[1 + (L\omega/u)^2]^2}\right\}$$

where u is the airfoil forward speed, C is the chord length of the airfoil, and L is the scale of turbulence. Determine $E\{X^2(t)\}$.

7.4. Let $Y(t)$ and $X(t)$ be, respectively, the input and output stochastic processes of a constant-parameter linear system. The input process $Y(t)$ is assumed to be wide-sense stationary.

(a) Show that

$$\Gamma_{YX}(\tau) = \int_0^\infty h(t)\,\Gamma_{YY}(\tau - t)\,dt$$

(b) Show that

$$\Gamma_{YX}(\tau) = S_0 h(\tau)$$

if

$$S_{YY}(\omega) = S_0$$

A useful application of the second result is that the impulse response function of a linear system can be estimated based upon the scheme shown below. The correlator is a device which delays the input $Y(t)$ by τ time units, multiplies the delayed input by $X(t)$, and performs an integration of the product.

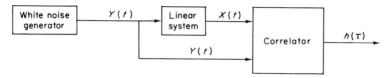

7.5. Consider the relation

$$X(t) = (1/2T) \int_{t-T}^{t+T} Y(\tau)\, d\tau$$

(a) Determine the impulse response $h(t)$ of a linear system which gives $X(t)$ as the output with input process $Y(t)$.

(b) Show that

$$S_{XX}(\omega) = \frac{\sin^2 T\omega}{(T\omega)^2}\, S_{YY}(\omega)$$

7.6. The impulse response $h(t)$ of a linear system is given by

$$h(t) = (1/T) \exp(-t/T), \qquad t > 0$$
$$h(t) = 0, \qquad\qquad\qquad t \le 0$$

Let the input process $Y(t)$ be a white noise process with $\Gamma_{YY}(\tau) = \pi S_0 \delta(\tau)$. Determine $\Gamma_{XX}(s, t)$, the correlation function of the output, and plot $\Gamma_{XX}(s, t)$ at $s = t$.

7.7. Verify the formula given by Eq. (4.134) using Eqs. (7.73), (7.74), and (7.76).

7.8. Following the development in Section 7.3.1, derive the Fokker–Planck equation and the Kolmogorov equation for

(a) the Wiener process. Its transition density function is

$$f(x, t \mid x_0, t_0) = [4\pi D(t - t_0)]^{-1/2} \exp\left[-\frac{1}{4D(t - t_0)}(x - x_0)^2\right]$$

(b) the Ornstein–Uhlenbeck process. Its transition density function is given by Eq. (7.135). Check the resulting Fokker–Planck equation with Eq. (7.122).

7.9. In reference to Example 7.5, obtain the stationary solution $f_s(x)$ given by Eq. (7.136) directly from the Fokker–Planck equation (7.122).

7.10. Consider the second-order system

$$a^2\ddot{X}(t) + \dot{X}(t) + f(X) = W(t), \qquad t \geq t_0$$

where $W(t)$ is a Gaussian white-noise process with mean zero and covariance $\Gamma_{WW}(t, s) = 2D\,\delta(t - s)$

(a) Set up the Fokker–Planck equation.
(b) Obtain the stationary solution $f_s(x, \dot{x})$ from the Fokker–Planck equation.
(c) Compare the marginal density $f_s(x)$ obtained from (b) with the stationary density of $X(t)$ described by the first-order equation

$$\dot{X}(t) + f(X) = W(t), \qquad t \geq t_0$$

7.11. Consider the second-order equation

$$\ddot{X}(t) + 2\beta\dot{X}(t) + \omega_0^2(X(t) + aX^3(t)) = W(t)$$

where a is a small parameter and $W(t)$ is a white Gaussian process as defined in Problem 7.10.

(a) Use Eq. (7.323) as the equivalent linear equation and determine the power spectral density function of the solution process $X(t)$.
(b) Compare the result obtained above with that obtained by the perturbation technique and show that they are the same up to the first order of a.
(c) In the method of equivalent linearization, determine the coefficients k_e and β_e from Eqs. (7.312), where the expectations contained in these equations are computed from the stationary density function associated with the original nonlinear equation. Compare the resulting equivalent linear equation with Eq. (7.323).

7.12. Consider the first-order nonlinear system represented by Eq. (7.201).
(a) Determine the stationary solution of its associated Fokker-Planck equation.
(b) Comment on the accuracy of the technique of equivalent linearization by comparing the stationary solution obtained in (a) with one obtained by means of equivalent linearization.

Chapter 8

Differential Equations with Random Coefficients

Intensive studies of differential equations with random coefficients have been undertaken only recently, and they have already exerted a profound influence on the analysis of a wide variety of problems in engineering and science. The earlier work in this area goes back to the investigation of the propagation of sound in an ocean of randomly varying refraction index [1]. Recently, a great number of physical problems which have been investigated falls into this class. In control and communication areas, for example, the behavior of systems with randomly varying transfer characteristics or with random parametric excitations is described by differential equations with stochastic processes as coefficients. The same type of problems arises in chemical kinetics, economic systems, physiology, drug administration, and many other areas. The study of systems and structures having imprecise parameter values or inherent imperfections also leads to differential equations of this type.

Mathematically speaking, the study of this class of differential equations is the most interesting of the three classes, and it is certainly the most difficult. By virtue of the randomness in the coefficients, the scope of the mean square theory is considerably enlarged. We not only need to develop methods of analyzing properties of the stochastic solution process; but the eigenvalue problem, for example, now becomes a random problem and it needs new mathematical tools. The concept of stability of differential equations, which plays such an important role in physical applications, also takes on an added probabilistic dimension here. New concepts of stability in some probabilistic sense have also attracted considerable attention on the part of the engineers and scientists.

Because of the fact that this class of differential equations is the stochastic counterpart of deterministic differential equations in the most complete sense, the term "random differential equations" in general refers to differential equations with random coefficients, having either deterministic or random inhomogeneous parts and initial conditions.

This chapter is devoted to the study of solution processes associated with this class of differential equations. Stability of random differential equations will be discussed in Chapter 9. The random eigenvalue problems are touched upon in some of the examples herein but will not be treated in detail; the reader is referred to Boyce [2] for a review of this topic.

8.1. The Mean Square Theory

The situation is somewhat cloudy with respect to mean square properties of this class of differential equations. For a general discussion of existence and uniqueness of m.s. solutions, it is difficult to improve upon the classical Picard-type theorem as given in Theorem 5.1.2 which, as we have demonstrated in Chapter 5, has severe limitations when applied to differential equations with random coefficients.

For linear differential equations with random coefficients, an existence and uniqueness theorem under less restrictive conditions is given by Strand [3, 4]. This result is given below without proof.

Consider a system of differential equations having the form

$$\dot{\mathbf{X}}(t) = A(t)\,\mathbf{X}(t) + \mathbf{Y}(t), \qquad t \in T = [t_0, a]$$
$$\mathbf{X}(t_0) = \mathbf{X}_0$$

(8.1)

where the coefficient matrix $A(t)$, the inhomogeneous term $\mathbf{Y}(t)$, and the initial condition $\mathbf{X}_0$ are all regarded as random quantities.

Theorem 8.1.1. Let $A(t)$ be an $n \times n$ matrix of m.s. integrable processes. Let $\mathbf{Y}: T \to L_2^n$ be m.s. integrable and let $\mathbf{X}_0 \in L_2^n$. Then Eq. (8.1) has a unique m.s. solution on T if

(a) $\displaystyle\sum_{k=1}^{\infty} \int_{t_0}^{a} \int_{t_0}^{s_k} \cdots \int_{t_0}^{s_2} \| A(s_k)\, A(s_{k-1}) \cdots A(s_2)\, \mathbf{Y}(s_1) \|_n \, ds_1 \cdots ds_k < \infty$

(b) $\displaystyle\sum_{k=1}^{\infty} \int_{t_0}^{a} \int_{t_0}^{s_k} \cdots \int_{t_0}^{s_2} \| A(s_k) \cdots A(s_1)\mathbf{X}_0 \|_n \, ds_1 \cdots ds_k < \infty$

Conditions (a) and (b) essentially limit the growth rates of the moments associated with the coefficient matrix $A(t)$. These are, therefore, much more natural conditions as compared with the Lipschitz condition given in Theorem 5.1.2. We note, however, that Conditions (a) and (b) are not simple conditions imposed on the random matrix $A(t)$, and it is in general difficult to verify them in a given situation.

In the interest of applications, we will not belabor this point but assume that m.s. solutions exist in what follows.

8.2. First-Order Linear Equations

One of the simplest differential equations possessing an explicit solution representation is a first-order linear differential equation. Let us study in the section the scalar equation

$$\dot{X}(t) + A(t)\, X(t) = Y(t), \qquad X(t_0) = X_0 \tag{8.2}$$

where the coefficient process $A(t)$, together with $Y(t)$ and X_0, is considered as a stochastic process. In what follows, we shall assume that $A(t)$ possesses well-behaved sample functions so that one can make sense of the sample equations as ordinary differential equations. This is a model of, for example, an RC circuit with a randomly varying open-loop gain. Rosenbloom [5] and Tikhonov [6] have given a thorough analysis of equations of this type.

Following the deterministic theory, the m.s. solution of $X(t)$ can be represented in the form

$$X(t) = X_0 \exp\left[- \int_{t_0}^{t} A(s)\, ds \right] + \int_{t_0}^{t} Y(u) \exp\left[- \int_{u}^{t} A(s)\, ds \right] du \tag{8.3}$$

It is immediately clear from the above that the coefficient process $A(t)$ enters the solution in a complex way. The knowledge of the density functions of $A(t)$ is in general necessary for determining even the simple moments of the solution process. Furthermore, the joint probabilistic behavior of $A(t)$, $Y(t)$, and X_0 is also required.

We stress again that Eq. (8.3) may not be valid in the situation where $A(t)$ does not have good sample behavior. As we shall see in Section 8.4, white noise coefficients will invalidate the results presented here.

Example 8.1. Let us show some working details by determining the mean and the correlation function of $X(t)$ assuming that $X_0 \equiv 0$ and that the processes $A(t)$ and $Y(t)$ are stationary and correlated Gaussian processes with known means and known correlation functions.

In this case, we have

$$X(t) = \int_{t_0}^{t} Y(u) \exp\left[- \int_{u}^{t} A(s)\, ds \right] du \qquad (8.4)$$

In calculating the moments of $X(t)$, it is pointed out by Tikhonov [6] that a significant simplification results if we introduce the process

$$Z(t) = \int_{t_0}^{t} \exp\left[-\alpha Y(u) - \int_{u}^{t} A(s)\, ds \right] du \qquad (8.5)$$

We can then write

$$X(t) = -\left[\frac{\partial}{\partial \alpha} Z(t) \right]_{\alpha=0} \qquad (8.6)$$

It is simpler to use Eq. (8.6) because the expectation of the integrand in Eq. (8.5) is closely related to the joint characteristic function of two random variables, which can be easily written down in the Gaussian case.

Consider first the mean of $X(t)$. It is given by

$$E\{X(t)\} = -\left[\frac{\partial}{\partial \alpha} E\{Z(t)\} \right]_{\alpha=0} \qquad (8.7)$$

where

$$E\{Z(t)\} = \int_{t_0}^{t} E\left\{ \exp\left[-\alpha Y(u) - \int_{u}^{t} A(s)\, ds \right] \right\} du \qquad (8.8)$$

Let

$$U(u) = \int_{u}^{t} A(s)\, ds$$

and let $\phi(u_1, u_2)$ denote the joint characteristic function of the two Gaussian r.v.'s $Y(u)$ and $U(u)$. We recognize that the expectation to be evaluated in Eq. (8.8) is simply $\phi(i\alpha, i)$. Hence, in terms of the joint statistical properties of $A(t)$ and $Y(t)$ and after carrying out the necessary computations, we obtain

$$E\left\{ \exp\left[-\alpha Y(u) - \int_{u}^{t} A(s)\, ds \right] \right\}$$

$$= \phi(i\alpha, i) = \exp\left\{ -[\alpha m_Y + m_A(t - u)] \right.$$

$$\left. + \tfrac{1}{2}\left[\alpha^2 \sigma_Y^2 + 2\alpha \sigma_Y \sigma_U(t, u) \int_{u}^{t} \varrho_{AY}(u - s)\, ds + \sigma_U^2(t, u) \right] \right\} \qquad (8.9)$$

where

$$m_Y = E\{Y(t)\}, \qquad m_A = E\{A(t)\}$$

$$\sigma_Y^2 \varrho_{YY}(\tau) = E\{[Y(t) - m_Y][Y(t + \tau) - m_Y]\}$$

$$\sigma_A^2 \varrho_{AA}(\tau) = E\{[A(t) - m_A][A(t + \tau) - m_A]\}$$

$$\sigma_A \sigma_Y \varrho_{AY}(\tau) = E\{[A(t) - m_A][Y(t + \tau) - m_Y]\}$$

$$m_U = E\{U(u)\} = m_A(t - u)$$

$$\sigma_U^2(t, u) = E\{(U(u) - m_U)^2\} = \sigma_A^2 \int_u^t \int_u^t \varrho_{AA}(s_2 - s_1)\, ds_1\, ds_2$$

The mean of $X(t)$ is obtained upon substituting Eq. (8.9) into Eqs. (8.8) and (8.7) and carrying out the indicated integration and differentiation. It can be represented in the form

$$E\{X(t)\} = \int_{t_0}^t \left[m_Y - \sigma_A \sigma_Y \int_{t_0}^t ds_1 \int_{t_0}^t ds_2 \int_u^t \varrho_{AA}(s_2 - s_1)\, \varrho_{AY}(u - s_3)\, ds_3 \right]$$

$$\times \exp\left[-m_A(t - u) + \tfrac{1}{2}\sigma_A^2 \int_u^t \int_u^t \varrho_{AA}(s_2 - s_1)\, ds_1\, ds_2 \right] du \quad (8.10)$$

The correlation function is given by

$$\Gamma_{XX}(t_1, t_2) = E\left\{ \left[\frac{\partial}{\partial \alpha} Z(t_1) \right]_{\alpha=0} \left[\frac{\partial}{\partial \alpha} Z(t_2) \right]_{\alpha=0} \right\}$$

$$= \left[\frac{\partial^2}{\partial \alpha_1 \partial \alpha_2} E\{Z(t_1) Z(t_2)\} \right]_{\alpha_1 = \alpha_2 = 0} \quad (8.11)$$

The expectation involved in the calculation of the correlation function can be identified with a four-dimensional characteristic function of a Gaussian process. It can thus be obtained in a straightforward but cumbersome way. The final expression is

$$\Gamma_{XX}(t_1, t_2) = \int_{t_0}^t \int_{t_0}^t \left\{ \sigma_Y^2 \varrho_{YY}(u - v) \right.$$

$$+ \left[m_Y - \sigma_Y \sigma_U(t_1, u) \int_u^{t_1} \varrho_{AY}(u - s_1)\, ds_1 - \sigma_Y \sigma_U(t_2, u) \int_v^{t_2} \varrho_{AY}(u - s_2)\, ds_2 \right]$$

$$\times \left[m_Y - \sigma_Y \sigma_U(t_1, v) \int_u^{t_1} \varrho_{AY}(v - s_1)\, ds_1 - \sigma_Y \sigma_U(t_2, v) \int_v^{t_2} \varrho_{AY}(v - s_2)\, ds_2 \right]$$

$$\times \exp\left[-m_A(t_1 + t_2 - u - v) + \tfrac{1}{2}\sigma_U^2(t_1, u) + \tfrac{1}{2}\sigma_U^2(t_2, v) \right.$$

$$\left. + \sigma_U(t_1, u)\, \sigma_U(t_2, v) \int_u^{t_1} \int_v^{t_2} \varrho_{AA}(s_2 - s_1)\, ds_1\, ds_2 \right] \Bigg\} du\, dv \quad (8.12)$$

8.3. Differential Equations with Random Constant Coefficients

The performance of a dynamical system is a function of the values of the parameters which constitute the system. The value of a parameter is experimentally determined and, in the performance analysis, it is usually taken to be the mean value of a set of experimental observations. In reality, of course, we recognize that this set of observations represents a distribution for the parameter values. Some spread of this distribution always exists as a result of inherent inaccuracies in the measurements and of tolerance problems in the manufacturing process. Therefore, with increasing emphasis placed upon the accuracy and reliability in performance predictions of physical systems and structures, a more rational approach to the analysis would be to consider systems with random variables as coefficients, and to study the resulting statistical performance properties. The dynamics of simple systems have been studied from this point of view [7–9]. In recent years, this approach has been used in the analysis of a number of problems of engineering interest [10–14].

It is worth mentioning that problems of this type have also received considerable attention in physics. One example is found in the study of lattice dynamics [15]. For thermodynamic reasons, the physicist has been interested in the dynamics of crystal lattices in which there is a small percentage of impurity atoms. Thus, randomness in the parameters arises from the fact that atom masses may take two or more values.

Another area in which equations with random constant coefficients have applications is one of mathematical modeling of biomedical systems. In constructing compartmental models of pharmacokinetics, for example, uncertainties in our knowledge of the rate constants are inevitable. They can be attributed to scatter of experimental kinetic data, environmental effects, and parameter variations in human patients involved. These inherent uncertainties again lead to the study of differential equations with random variables as coefficients [16].

Motivated by the considerations above, we study in this section random differential equations of the type

$$\dot{\mathbf{X}}(t) = \mathbf{f}(\mathbf{X}(t), \mathbf{A}, t), \qquad t \geq t_0; \qquad \mathbf{X}(t_0) = \mathbf{X}_0 \qquad (8.13)$$

The random vector $\mathbf{A}$ (constant in t) enters the equation through the coefficients, and we assume that the joint probability distribution of $\mathbf{A}$ and the initial condition $\mathbf{X}_0$ is specified.

In order to determine the statistical properties of the solution process $\mathbf{X}(t)$ governed by Eq. (8.13), it is instructive to first draw a parallel between this class of equations and the differential equations considered in Chapter 6. Indeed, we can easily show that Eq. (8.13) can be cast in the form of differential equations with only random initial conditions. Hence, all mathematical tools developed in Chapter 6 are applicable in this case.

Let us define the vector process $\mathbf{Z}(t)$ by

$$\mathbf{Z}(t) = \begin{bmatrix} \mathbf{X}(t) \\ \mathbf{A} \end{bmatrix} \tag{8.14}$$

Then, the augmented version of Eq. (8.13) is

$$\dot{\mathbf{Z}}(t) = \mathbf{k}(\mathbf{Z}(t), t), \qquad t \geq t_0; \qquad \mathbf{Z}(t_0) = \mathbf{Z}_0 \tag{8.15}$$

where

$$\mathbf{k} = \begin{bmatrix} \mathbf{f} \\ \mathbf{0} \end{bmatrix} \quad \text{and} \quad \mathbf{Z}_0 = \begin{bmatrix} \mathbf{X}_0 \\ \mathbf{A} \end{bmatrix} \tag{8.16}$$

In terms of $\mathbf{Z}(t)$, Eq. (8.15) describes a vector differential equation where randomness enters only through the initial condition. Hence, it is generally possible to determine the statistical properties of the solution process $\mathbf{Z}(t)$, and consequently that of $\mathbf{X}(t)$, by means of methods developed in Chapter 6. Let us note, however, if an explicit solution of $\mathbf{X}(t)$ as a function of $\mathbf{A}$ and $\mathbf{X}_0$ can be obtained from Eq. (8.13), it can be used to find the statistical properties of $\mathbf{X}(t)$ directly.

Example 8.2. Let us consider a two-compartment model in chemical kinetics or pharmacokinetics as shown in Fig. 8.1. The quantities $X_1(t)$ and $X_2(t)$ represent the chemical concentrations in the compartments and A_1 and A_2 represent the rate constants.

Fig. 8.1. A two-compartment model.

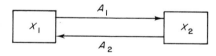

This model is described by

$$\dot{\mathbf{X}}(t) = \begin{bmatrix} \dot{X}_1(t) \\ \dot{X}_2(t) \end{bmatrix} = \begin{bmatrix} -A_1 & A_2 \\ A_1 & -A_2 \end{bmatrix} \begin{bmatrix} X_1(t) \\ X_2(t) \end{bmatrix}, \qquad t \geq 0 \tag{8.17}$$

with initial condition

$$\mathbf{X}(0) = \mathbf{X}_0 = \begin{bmatrix} X_{10} \\ X_{20} \end{bmatrix} \tag{8.18}$$

Let us assume that A_1, A_2, X_{10}, and X_{20} are random variables with a prescribed joint density function. The concentrations $X_1(t)$ and $X_2(t)$ are now stochastic processes whose statistical properties are to be determined.

For this simple case, the solution of Eq. (8.17) can be obtained in an elementary way. We have

$$\mathbf{X}(t) = \mathbf{h}(\mathbf{X}_0, \mathbf{A}, t)$$

$$= \frac{1}{A_1 + A_2} \begin{bmatrix} A_2 + A_1 \exp[-(A_1 + A_2)t] & A_2\{1 - \exp[-(A_1 + A_2)t]\} \\ A_1\{1 - \exp[-(A_1 + A_2)t]\} & A_1 + A_2 \exp[-(A_1 + A_2)t] \end{bmatrix} \mathbf{X}_0$$

$$(8.19)$$

This relation can be used immediately for finding some of the statistical properties of $\mathbf{X}(t)$. The mean of $\mathbf{X}(t)$, for example, can be found by evaluating the integral

$$E\{\mathbf{X}(t)\} = \int_{-\infty}^{\infty} \mathbf{h}(\mathbf{x}_0, \mathbf{a}, t) f_0(\mathbf{x}_0, \mathbf{a}) \, d\mathbf{x}_0 \, d\mathbf{a} \qquad (8.20)$$

where $f_0(\mathbf{x}_0, \mathbf{a})$ is the joint density function of the random variables X_{10}, X_{20}, A_1, and A_2.

Numerical results presented below are obtained by assuming that $X_{10} \equiv 1$ and $X_{20} \equiv 0$. The rate constants A_1 and A_2 are assumed to have a truncated bivariate Gaussian distribution with mean one and variance approximately 0.1 for A_1, and mean 0.5 and variance approximately 0.04 for A_2. In order to insure positivity of the rate constants, the bivariate distribution is restricted to taking values only in the first quadrant of the a_1a_2-plane. In Fig. 8.2, the means of $X_1(t)$ are shown for two cases: the case where A_1 and A_2 are uncorrelated ($\varrho = 0$) and the case where A_1 and A_2 are perfectly correlated ($\varrho = 1$). These curves are compared with the deterministic result for $X_1(t)$ when the mean values of A_1 and A_2 are used. It shows that, in an average sense, the concentrations with randomness in the rate constants taken into account are in general different from that obtained from deterministic procedures where the rate constants are first averaged.

Figure 8.3 gives the corresponding standard deviations, $\sigma_{X_1}(t)$, of $X_1(t)$ as a function of t. Asymptotically, both the means and the standard deviations take stationary values as $X_1(t)$ approaches a random variable $A_2/(A_1 + A_2)$.

We remark that, as indicated by the governing differential equations (8.17), $X_1(t)$ and $X_2(t)$ possess the property that

$$X_1(t) + X_2(t) = 1 \qquad (8.21)$$

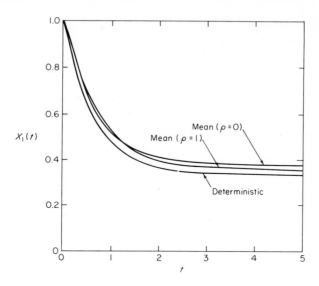

Fig. 8.2. Mean of $X_1(t)$ (Example 8.2) (from Soong [16]).

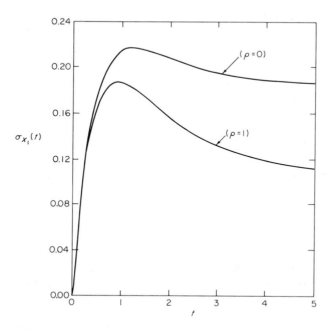

Fig. 8.3. Standard deviation of $X_1(t)$, $\sigma_{X_1}(t)$ (Example 8.2) (from Soong [16]).

with probability one for all t. Hence, the mean and the standard deviation of $X_2(t)$ are given by

$$E\{X_2(t)\} = 1 - E\{X_1(t)\}, \qquad \sigma_{X_2}(t) = \sigma_{X_1}(t) \tag{8.22}$$

The determination of the density functions for $\mathbf{X}(t)$ is of course more difficult. An effective way for obtaining this information is to make use of the augmented equation (8.14) and using the techniques developed in Chapter 6. For this example, the result obtained following this procedure is the joint density function of $X_1(t)$, $X_2(t)$, A_1, and A_2. Marginal density functions such as $f(x_1, t)$ of $X_1(t)$ or $f(x_2, t)$ of $X_2(t)$ are then found by evaluating appropriate integrals.

Let us pursue this a bit and see how the Liouville equation can be applied in this situation.

8.3.1. Applications of the Liouville Equation

In Chapter 6, a Liouville's theorem was developed for determining the joint density function of the components of the solution process at a given time t. In view of the equivalence established between these two classes of problems, the same theorem is applicable here and, as we shall see, it is also of considerable utility.

Let the dimensions of the vectors $\mathbf{X}(t)$ and $\mathbf{A}$ in Eq. (8.13) be, respectively, n and m. The vector process $\mathbf{Z}(t)$ defined by Eq. (8.14) then has $(n + m)$ components. It follows from the Liouville's theorem (Theorem 6.2.2) in Chapter 6 that the joint density function $f(\mathbf{z}, t)$ satisfies

$$\partial f(\mathbf{z}, t)/\partial t + \sum_{j=1}^{n+m} \partial k_j f/\partial z_j = 0 \tag{8.23}$$

The substitution of Eqs. (8.16) into Eq. (8.23) immediately gives

$$\partial f(\mathbf{x}, t; \mathbf{a})/\partial t + \sum_{j=1}^{n} \partial f_j f/\partial x_j = 0 \tag{8.24}$$

Equation (8.24) is satisfied by the joint density function of the solution process $\mathbf{X}(t)$ together with the random coefficient vector $\mathbf{A}$. The initial condition associated with Eq. (8.24) is the joint density function of $\mathbf{X}_0$ and $\mathbf{A}$, which we denote by $f_0(\mathbf{x}_0, \mathbf{a})$.

The general solution of Eq. (8.24) is readily obtained by examining the associated Lagrange system. Following the procedure developed in Chapter

6 [Eq. (6.60)], it is easy to see that $f(\mathbf{x}, t; \mathbf{a})$ is given by

$$f(\mathbf{x}, t; \mathbf{a}) = f_0(\mathbf{x}_0, \mathbf{a}) \exp\left\{-\int_{t_0}^{t} \nabla \cdot \mathbf{f}[\mathbf{x} = \mathbf{h}(\mathbf{x}_0, \mathbf{a}, \tau), \mathbf{a}, \tau]\, d\tau\right\}\Bigg|_{\mathbf{x}_0 = \mathbf{h}^{-1}(\mathbf{x}, \mathbf{a}, t)}$$
(8.25)

Example 8.3. To illustrate the application of the Liouville equation (8.24), let us reconsider Example 8.2. We now wish to determine the joint density function $f(\mathbf{x}, t) = f(x_1, t; x_2, t)$ of $X_1(t)$ and $X_2(t)$ at a given time t. As seen from Eq. (8.17), we have

$$\mathbf{f}(\mathbf{x}, \mathbf{a}, t) = \begin{bmatrix} -a_1 x_1 + a_2 x_2 \\ a_1 x_1 - a_2 x_2 \end{bmatrix}$$
(8.26)

Thus

$$\nabla \cdot \mathbf{f} = -(a_1 + a_2)$$
(8.27)

From Eq. (8.19), it is easy to obtain

$$\mathbf{x}_0 = \mathbf{h}^{-1}(\mathbf{x}, \mathbf{a}, t) = \frac{1}{a_1 + a_2}\begin{bmatrix} a_2 + a_1 \exp(a_1 + a_2)t & a_2(1 - \exp(a_1 + a_2)t) \\ a_1(1 - \exp(a_1 + a_2)t) & a_1 + a_2 \exp(a_1 + a_2)t \end{bmatrix}\mathbf{x}$$
(8.28)

We can now write down the joint density function of $\mathbf{X}(t)$ and $\mathbf{A}$ in terms of $f_0(\mathbf{x}_0, \mathbf{a})$. The substitutions of Eqs. (8.27) and (8.28) into Eq. (8.25) yield

$$f(\mathbf{x}, t; \mathbf{a}) = f_0(\mathbf{x}_0, \mathbf{a}) \exp(a_1 + a_2)t \,|_{\mathbf{x}_0 = \mathbf{h}^{-1}(\mathbf{x}, \mathbf{a}, t)}$$

$$= \exp[(a_1 + a_2)t] f_0\left\{\frac{1}{a_1 + a_2} [(a_2 + a_1 \exp(a_1 + a_2)t)x_1\right.$$

$$+ a_2(1 - \exp(a_1 + a_2)t)x_2], \frac{1}{a_1 + a_2} [a_1(1 - \exp(a_1 + a_2)t)x_1$$

$$\left. + (a_1 + a_2 \exp(a_1 + a_2)t)x_2], a_1, a_2\right\}$$
(8.29)

The joint density function $f(\mathbf{x}, t)$ is obtained from the above by integration with respect to $\mathbf{a}$, that is

$$f(\mathbf{x}, t) = \int_{-\infty}^{\infty} \int_{-\infty}^{\infty} f(\mathbf{x}, t; \mathbf{a})\, da_1\, da_2$$
(8.30)

We note that the knowledge of this joint density function can be quite useful. In this example, it may be of practical interest to determine the probability that chemical concentrations in the compartments do not exceed certain bounds at a given time t. This probability is then given by

$$P = \int_\Omega \int f(\mathbf{x}, t) \, d\mathbf{x} \tag{8.31}$$

where Ω is some appropriate region determined by the prescribed bounds.

Let us remark that the method outlined above is also applicable, under certain conditions, to the case where the random coefficients are time dependent. In the example above, if the rate constants $\mathbf{A}(t)$ can be represented by

$$\mathbf{A}(t) = \mathbf{r}(\mathbf{B}, t) \tag{8.32}$$

where $\mathbf{B}$ is a finite random (constant) vector and $\mathbf{r}$ has a known function form, a Liouville type equation can again be derived which is satisfied by the joint density function of $\mathbf{X}(t)$ and $\mathbf{B}$. In other words, the analysis given here is also valid for the case where the coefficient processes are stochastic processes with a finite degree of randomness. Some work along this line is given by Soong and Chuang [17].

8.4. Differential Equations of the Ito Type

The Ito equation, studied extensively in Sections 5.2, 7.3, and 7.4 has the form

$$d\mathbf{X}(t) = \mathbf{f}(\mathbf{X}(t), t) \, dt + G(\mathbf{X}(t), t) \, d\mathbf{B}(t), \qquad t \geq t_0 \tag{8.33}$$

In general, since the matrix $G(\mathbf{X}(t), t)$ is a function of the vector solution process $\mathbf{X}(t)$, the Ito equation constitutes a special class of differential equations with random coefficients. As we have pointed out in Chapter 5, this class of random differential equations have stimulated considerable interest on the part of engineers in the study of filtering, control, and communication systems. (See References 6–12 cited in Chapter 5).

In Chapter 7, some mathematical tools were developed for the general case but were applied only to solution of Ito equations with nonrandom coefficients. We now see how these tools can be used to study some of the random-coefficient systems.

Let us first recall some of the main results derived in Sections 7.3.2 and 7.4.1.

Consider Eq. (8.33) where $\mathbf{X}(t)$ is the n-dimensional solution process and $\mathbf{B}(t)$ is an m-dimensional vector Wiener process whose components $B_j(t)$ have the properties

$$E\{\Delta B_j(t)\} = E\{B_j(t + \Delta t) - B_j(t)\} = 0$$
$$E\{\Delta B_i(t)\,\Delta B_j(t)\} = 2D_{ij}\,\Delta t; \qquad t \geq t_0, \quad i, j = 1, 2, \ldots, m \tag{8.34}$$

Let $f(\mathbf{x}, t \mid \mathbf{x}_0, t_0)$ be the transition density of the solution process $\mathbf{X}(t)$. The main result given in Section 7.3.2 is that the associated Fokker–Planck equation has the form [Eq. (7.160)]

$$\frac{\partial f(\mathbf{x}, t \mid \mathbf{x}_0, t)}{\partial t} = -\sum_{j=1}^{n} \frac{\partial}{\partial x_j}\,[f_j(\mathbf{x}, t)f] + \sum_{i,j=1}^{n} \frac{\partial^2}{\partial x_i\,\partial x_j}\,[(GDG^{\mathrm{T}})_{ij}f] \tag{8.35}$$

We have shown in Section 7.4.1 that it is also possible to derive a set of differential equations satisfied by the moments of the solution process. These moment equations are of the form [Eq. (7.279)]

$$\frac{d}{dt}\,E\{h(\mathbf{X}, t)\}$$
$$= \sum_{j=1}^{n} E\{f_j\,\partial h/\partial X_j\} + \sum_{i,j=1}^{n} E\{(GDG^{\mathrm{T}})_{ij}\,\partial^2 h/\partial X_i\,\partial X_j\} + E\{\partial h/\partial t\} \tag{8.36}$$

where $h(\mathbf{X}, t)$ is an arbitrary function of $\mathbf{X}(t)$ and t. The indicated partial derivatives and expectations are assumed to exist.

Hence, considerable machinery exists for treating random-coefficient Ito equations. It is a simple matter to establish either the Fokker–Planck equation or the moment equations for a given problem. The solutions of the Fokker–Planck equations, however, are difficult to obtain in general. Except for trivial cases, no solution of practical interest has been found for this class of Fokker–Planck equations.

The moment equations, on the other hand, have considerable utility here, and they are the subject of numerous investigations [18–21]. Let us consider several simple cases.

Example 8.4. Consider the first-order equation

$$\dot{X}(t) + [a + W_1(t)]\,X(t) = 0, \qquad t \geq 0, \quad X(0) = X_0 \tag{8.37}$$

where $W(t)$ is a Gaussian white noise with zero mean and covariance $2D_{11}\delta(t - s)$.

In this case, $n = m = 1$, $f_1 = -aX(t)$, and $(GDG^{\mathrm{T}})_{11} = -D_{11}\,X^2(t)$. The substitution of these quantities into Eq. (8.36) establishes the general moment equation. Specializing $h(X, t) = X(t)$, we arrive at the mean equation

$$\dot{m}_1(t) = -am_1(t), \qquad t \geq 0, \quad m_1(0) = E\{X_0\} \tag{8.38}$$

where $m_1 = E\{X(t)\}$. It is a simple first-order linear differential equation having as its solution

$$m_1(t) = E\{X_0\} \, e^{-at} \tag{8.39}$$

Moment equations for higher order moments can be established in a similar fashion. Using the notation $m_j(t) = E\{X^j(t)\}$, it follows from the general moment equation, Eq. (8.36), that the jth moment of $X(t)$ satisfies

$$\dot{m}_j(t) = -j[a - (j-1)D_{11}]m_j(t), \qquad j = 1, 2, \ldots \tag{8.40}$$

which has the general solution

$$m_j(t) = E\{X_0^j\} \exp\{-j[a - (j-1) D_{11}]t\} \tag{8.41}$$

For this simple example, we see that all moment equations separate, that is, the equation for the moment of a certain order contains only the moment of that order. Each of these moment equations can be readily solved.

Example 8.5. A first-order linear system is characterized by

$$\dot{X}(t) + [a + W_1(t)] X(t) = W_2(t), \qquad t \geq 0 \tag{8.42}$$

This case corresponds to Eq. (8.33) with $n = 1$, $m = 2$, $f_1 = -aX(t)$, and $G = [-X(t) \quad 1]$. The moment equations based upon Eq. (8.36) can again be easily established. They are given by

$$\dot{m}_1(t) = -am_1(t)$$
$$\dot{m}_j(t) = -j[a - (j-1) D_{11}] m_j(t) - 2D_{12} j(j-1) m_{j-1}(t)$$
$$+ j(j-1) D_{22}m_{j-2}(t), \qquad j = 2, 3, \ldots \tag{8.43}$$

Comparing the results of this example with that of the preceding one, we see that the effect of the inhomogeneous term $W_2(t)$ in the differential equation on the moment equations is that the moment equations become inhomogeneous. Furthermore, they are no longer uncoupled. However, they can be solved successively, starting from the moment equation of the first order.

Example 8.6. The differential equation describing a second-order system takes the form

$$\ddot{X}(t) + a_2 \dot{X}(t) + [a_1 + W_1(t)] X(t) = W_2(t), \qquad t \geq 0 \qquad (8.44)$$

Let $X(t) = X_1(t)$ and $\dot{X}(t) = X_2(t)$. We rewrite Eq. (8.44) in the form

$$\dot{X}_1(t) = X_2(t)$$
$$\dot{X}_2(t) = -a_2 X_2(t) - [a_1 + W_1(t)] X_1(t) + W_2(t) \qquad (8.45)$$

These two equations constitute a two-dimensional Ito equation in the form of Eq. (8.33) with

$$\mathbf{f} = \begin{bmatrix} X_2(t) \\ -a_1 X_1(t) - a_2 X_2(t) \end{bmatrix} \quad \text{and} \quad G = \begin{bmatrix} 0 & 0 \\ -X_1(t) & 1 \end{bmatrix} \qquad (8.46)$$

The moment equations can now be established using Eq. (8.36). Let $m_{jk}(t) = E\{X_1{}^j(t)\, X_2{}^k(t)\}$. We find that the first-order moments satisfy

$$\dot{m}_{10}(t) = m_{01}(t)$$
$$\dot{m}_{01}(t) = -a_1 m_{10}(t) - a_2 m_{01}(t) \qquad (8.47)$$

The second-order moment equations are found to be

$$\dot{m}_{20} = 2m_{11}$$
$$\dot{m}_{11} = -a_1 m_{20} - a_2 m_{11} + m_{02} \qquad (8.48)$$
$$\dot{m}_{02} = 2(D_{11} m_{20} - a_1 m_{11} - a_2 m_{02} + D_{22} - 2D_{12} m_{10})$$

Higher-order moment equations can be found in a similar way. We can show that moment equations of any order contain only moments of the same or lower order. Hence, the moments can be solved successively. The first-order equations (8.47) are first solved; Eqs. (8.48) can now be solved for the second moments, and so on.

Example 8.7. Let us now consider a slightly different situation. A second-order linear equation is of the form

$$\ddot{X}(t) + a_2 \dot{X}(t) + [a_1 + Y(t)] X(t) = 0, \qquad t \geq 0 \qquad (8.49)$$

where we assume that the s.p. $Y(t)$, $t \geq 0$, is a filtered Gaussian white noise; specifically, it is assumed to satisfy

$$\dot{Y}(t) + Y(t) = W_1(t) \qquad (8.50)$$

Let $X(t) = X_1(t)$, $\dot{X}(t) = X_2(t)$, and $Y(t) = X_3(t)$. Equations (8.49) and (8.50) can be cast into the Ito form with

$$\mathbf{f} = \begin{bmatrix} X_2(t) \\ -a_1 X_1(t) - a_2 X_2(t) - X_1(t)\, X_3(t) \\ -X_3(t) \end{bmatrix} \quad \text{and} \quad G = \begin{bmatrix} 0 \\ 0 \\ 1 \end{bmatrix} \quad (8.51)$$

Proceeding as before, we can derive the moment equations associated with the solution process. Adopting the notation

$$m_{ijk}(t) = E\{X_1{}^i(t)\, X_2{}^j(t)\, X_3{}^k(t)\},$$

the first-order moment equations are

$$\begin{aligned}
\dot{m}_{100} &= m_{010} \\
\dot{m}_{010} &= -a_1 m_{100} - a_2 m_{010} - m_{101} \\
\dot{m}_{001} &= -m_{001}
\end{aligned} \qquad (8.52)$$

We see from the above that the first-moment solutions, except for m_{001}, require the knowledge of the second moment m_{101}. Consider now the second-order moment equations. They are

$$\begin{aligned}
\dot{m}_{200} &= 2m_{110} \\
\dot{m}_{110} &= -a_1 m_{200} - a_2 m_{110} + m_{020} - m_{201} \\
\dot{m}_{101} &= -m_{101} + m_{011} \\
\dot{m}_{020} &= -2(a_1 m_{110} + a_2 m_{020} + m_{111}) \\
\dot{m}_{011} &= -a_1 m_{101} - (a_2 + 1)m_{011} - m_{102} \\
\dot{m}_{002} &= -2m_{002} + 2D_{11}
\end{aligned} \qquad (8.53)$$

We again see that the second-order moment equations contain third-order moments. Therefore, for differential equations whose coefficients are filtered white-noise processes, the moment equations are not separable in general. In order to obtain moment solutions, some form of truncation procedures similar to those discussed in Section 7.4.1 now becomes necessary.

Example 8.8. Let us now attempt to apply this technique to nonlinear differential equations. Consider a simple nonlinear system characterized by

$$\dot{X}(t) + [a + W_1(t)]\, X^3(t) = 0, \qquad t \geq 0 \qquad (8.54)$$

It is simple to show that the first moment equation is of the form

$$\dot{m}_1(t) = -am_3(t) \tag{8.55}$$

which contains $m_3(t)$, the third-order moment. In general, it is expected that the moment equations associated with nonlinear differential equations will also be nonseparable. Hence, truncation approximations are again needed here.

8.5. Perturbation Techniques

In terms of the mathematical properties of the random coefficients, we have covered in Sections 8.3 and 8.4 essentially two extremes of the spectrum. The random constant case implies no dependence on the part of the coefficients with respect to the independent variable, and the Gaussian white noise case corresponds to, roughly speaking, completely erratic behavior as a function of the independent variable. By and large, the nature of the physical processes generally gives rise to random coefficients which fall somewhere between these two extremes. Unfortunately, no general methods are available for obtaining exact solutions for these cases and we must resort to approximate means.

One of the most powerful approximation techniques is the perturbation approach. Perturbation methods are applicable to those cases where the random parametric variations are 'weak' or 'small' and these cases cover a class of problems of great physical importance. Consider, for example, the transverse vibration of a linear elastic beam. The governing differential equations are

$$d^2Y/dx^2 = M/EI, \qquad d^2M/dx^2 = \omega^2\mu Y \tag{8.56}$$

where Y is the transverse displacement, M the bending moment, EI the beam stiffnes, ω the beam frequency, μ the beam mass per unit length, and x is the beam coordinate. Under ideal conditions, the vibration of a uniform beam is analyzed by means of Eqs. (8.56) where the parameters indicated above are regarded as constants. However, due to inherent imperfections in geometry and in material properties, these coefficients or parameters exhibit variations as functions of the beam coordinate, which are difficult to model deterministically. To take into account these imperfections, we face the problem of solving a system of random-coefficient differential equations where the randomness can be considered small as compared with the deterministic parametric values.

8.5.1. A Direct Approach

To facilitate our development here, it is convenient to write the basic vector differential equation in the form

$$L(\mathbf{A}(t), t)\, \mathbf{X}(t) = \mathbf{Y}(t), \qquad t \geq 0 \tag{8.57}$$

where $L(\mathbf{A}(t), t)$ is an $n \times n$ square matrix whose elements are random differential operators. As before, $\mathbf{A}(t)$ represents a vector random coefficient process, $\mathbf{Y}(t)$ is an n-dimensional random input process. The initial conditions associated with the vector solution process $\mathbf{X}(t)$ are assumed to be specified.

In this subsection, we consider a direct perturbation scheme for solving Eq. (8.57) when the coefficient process $\mathbf{A}(t)$ can be expressed in the form

$$\mathbf{A}(t) = \mathbf{a}_0(t) + \varepsilon \mathbf{A}_1(t) + \varepsilon^2 \mathbf{A}_2(t) + \cdots \tag{8.58}$$

where ε is a small parameter. The leading coefficient $\mathbf{a}_0(t)$ is deterministic and it represents the unperturbed part. The coefficients $\mathbf{A}_1(t), \mathbf{A}_2(t), \ldots$ are stochastic processes, representing the perturbed part.

It is seen from Eqs. (8.57) and (8.58) that the differential operator $L(\mathbf{A}(t), t)$ can be similarly put in the form

$$L(\mathbf{A}(t), t) = L_0(\mathbf{a}_0(t), t) + \varepsilon L_1(t) + \varepsilon^2 L_2(t) + \cdots \tag{8.59}$$

where $L_0(t)$ is the corresponding deterministic operator. The differential operators $L_1(t), L_2(t), \ldots$ are in general functions of $\mathbf{A}_1(t), \mathbf{A}_2(t), \ldots,$ and, therefore, they are stochastic operators. The substitution of Eq. (8.59) into Eq. (8.57) then gives

$$[L_0(t) + \varepsilon L_1(t) + \varepsilon^2 L_2(t) + \cdots]\, \mathbf{X}(t) = \mathbf{Y}(t), \qquad t \geq 0 \tag{8.60}$$

In the direct application of the perturbation scheme, we seek a solution of Eq. (8.60) in the form

$$\mathbf{X}(t) = \mathbf{X}_0(t) + \varepsilon \mathbf{X}_1(t) + \varepsilon^2 \mathbf{X}_2(t) + \cdots \tag{8.61}$$

Upon substituting this solution representation into Eq. (8.60) and equating terms of the same order of ε, we immediately have a system of differential equations

$$L_0(t)\,\mathbf{X}_0(t) = \mathbf{Y}(t)$$
$$L_0(t)\,\mathbf{X}_1(t) = -L_1(t)\,\mathbf{X}_0(t)$$
$$\vdots \tag{8.62}$$
$$L_0(t)\,\mathbf{X}_j(t) = -[L_1(t)\,\mathbf{X}_{j-1}(t) + L_2(t)\,\mathbf{X}_{j-2}(t) + \cdots + L_j(t)\,\mathbf{X}_0(t)],$$
$$j = 1, 2, \ldots$$

It is observed that the differential equations given above are random differential equations with random inputs only, and these inputs are specified by solving Eqs. (8.62) successively. Thus, the perturbation technique transforms random coefficient problems into those with random inputs and the techniques of treating random input problems are well documented in Chapter 7.

In the case where the differential operator $L_0(t)$ is linear with constant coefficients, its inverse operator, $L_0^{-1}(t)$, is well defined. It is an integral operator whose kernel is the weighting function matrix associated with the linear system characterized by $L_0(t)$. Thus, operating on an arbitrary vector function $\mathbf{g}(t)$, $L_0^{-1}(t)$ takes the form

$$L_0^{-1}(t)\,\mathbf{g}(t) = \int_0^t \Phi(t - \tau)\,\mathbf{g}(t)\,d\tau \tag{8.63}$$

where $\Phi(t)$ is the principal matrix associated with $L_0(t)$, satisfying

$$L_0(t)\,\Phi(t) = I\,\delta(t), \qquad t \geq 0 \tag{8.64}$$

where I is the identity matrix.

Consider only particular solutions, the solutions of Eqs. (8.62) in this case can be written in the forms

$$\mathbf{X}_0(t) = L_0^{-1}(t)\,\mathbf{Y}(t)$$
$$\mathbf{X}_j(t) = -L_0^{-1}(t) \sum_{i=1}^{j} L_i(t)\,\mathbf{X}_{j-i}(t), \qquad j = 1, 2, \ldots \tag{8.65}$$

Hence, the solution for $\mathbf{X}(t)$ as expressed by Eq. (8.61) is formally established for this case. It can be written as

$$\mathbf{X}(t) = [I - \varepsilon L_0^{-1}(t)\,L_1(t) - \varepsilon^2 L_0^{-1}(t)(-L_1(t)\,L_0^{-1}(t)\,L_1(t)$$
$$+ L_2(t)) + \cdots]\,L_0^{-1}(t)\,\mathbf{Y}(t) \tag{8.66}$$

While Eq. (8.66) gives $X(t)$ as an explicit function of the coefficient process $A(t)$ and the input $Y(t)$, it is in general difficult to perform calculations beyond that of the simple moments of $X(t)$. Moreover, this can be done with only the first few terms in the expansion as the evaluation of higher-order terms becomes exceedingly complex. For some applications of this approach to physical problems, the reader is referred to the literature [22–24].

8.5.2. The Moment Equations

If we are only concerned with the moments of the solution process, another approach which has attracted considerable attention is the development of moment equations similar to those developed in connection with the Ito equation. This approach is generally superior to the direct approach discussed earlier as the moment equations are easier to handle. Moreover, the forms of the moment equations themselves often shed some light on the solution properties. Consider, for example, the mean of the solution process. It is constructive to establish the deterministic equation satisfied by the mean; not only because the mean can be determined from it, but also that the structure of the mean equation itself reveals some of the properties of solution process in an average sense. The parameters in the mean equation are sometimes called *effective* parameters associated with the random differential equation.

Let us now give a derivation of the mean equation using the perturbation approach. The results described below are due to Keller [25] in his investigation of waves propagating in a random medium. The method is also used by Karal and Keller [26], Chen and Tien [27], Jokipii [28], and Liu [29] in the determination of effective propagation constants for a variety of wave propagation problems.

Consider the perturbation solution of $X(t)$ as expressed by Eq. (8.66). Replacing $L_0^{-1}(t) Y(t)$ by $X_0(t)$ and taking expectation gives, assuming statistical independence between $Y(t)$ and $A(t)$,

$$\langle X \rangle = \langle X_0 \rangle - \varepsilon L_0^{-1} \langle L_1 \rangle \langle X_0 \rangle + \varepsilon^2 L_0^{-1} (\langle L_1 L_0^{-1} L_1 \rangle + \langle L_2 \rangle) \langle X_0 \rangle + O(\varepsilon^3)$$

$$(8.67)$$

where the symbol $\langle \ \rangle$ here stands for mathematical expectation for convenience. The time dependence is not explicitly stated for the sake of brevity.

Upon rearranging, we get from Eq. (8.67)

$$\langle \mathbf{X}_0 \rangle = \langle \mathbf{X} \rangle + \varepsilon L_0^{-1} \langle L_1 \rangle \langle \mathbf{X}_0 \rangle + O(\varepsilon^2)$$
$$= \langle \mathbf{X} \rangle + \varepsilon L_0^{-1} \langle L_1 \rangle \langle \mathbf{X} \rangle + O(\varepsilon^2) \tag{8.68}$$

The substitution of this expression into Eq. (8.67) yields

$$\langle \mathbf{X} \rangle = \langle \mathbf{X}_0 \rangle - \varepsilon L_0^{-1} \langle L_1 \rangle (I + \varepsilon L_0^{-1} \langle L_1 \rangle) \langle \mathbf{X} \rangle$$
$$+ \varepsilon^2 L_0^{-1} (\langle L_1 L_0^{-1} L_1 \rangle + \langle L_2 \rangle) \langle \mathbf{X} \rangle + O(\varepsilon^3)$$
$$= \langle \mathbf{X}_0 \rangle - \varepsilon L_0^{-1} \langle L_1 \rangle \langle \mathbf{X} \rangle$$
$$- \varepsilon^2 L_0^{-1} (\langle L_1 \rangle L_0^{-1} \langle L_1 \rangle - \langle L_1 L_0^{-1} L_1 \rangle + \langle L_2 \rangle) \langle \mathbf{X} \rangle + O(\varepsilon^3) \tag{8.69}$$

Transposing all except the first term on the right-hand side, we have

$$[I + \varepsilon L_0^{-1} \langle L_1 \rangle + \varepsilon^2 L_0^{-1} (\langle L_1 \rangle L_0^{-1} \langle L_1 \rangle - \langle L_1 L_0^{-1} L_1 \rangle + \langle L_2 \rangle) + O(\varepsilon^3)] \langle \mathbf{X} \rangle$$
$$= \langle \mathbf{X}_0 \rangle = L_0^{-1} \langle \mathbf{Y} \rangle \tag{8.70}$$

Finally, operating the equation above by L_0 gives

$$[L_0 + \varepsilon \langle L_1 \rangle + \varepsilon^2 (\langle L_1 \rangle L_0^{-1} \langle L_1 \rangle - \langle L_1 L_0^{-1} L_1 \rangle + \langle L_2 \rangle) + O(\varepsilon^3)] \langle \mathbf{X} \rangle = \langle \mathbf{Y} \rangle \tag{8.71}$$

Up to the second-order terms, this is the desired deterministic equation satisfied by the mean of the solution process. Higher-order terms in the mean equation can be obtained following essentially the same procedure. It is left to the reader to show that the mean equation including terms of the third order in ε is

$$\{L_0 + \varepsilon \langle L_1 \rangle + \varepsilon^2 [\langle L_1 \rangle L_0^{-1} \langle L_1 \rangle - \langle L_1 L_0^{-1} L_1 \rangle + \langle L_2 \rangle]$$
$$+ \varepsilon^3 [\langle L_1 L_0^{-1} L_1 L_0^{-1} L_1 \rangle - \langle L_1 L_0^{-1} L_2 \rangle - \langle L_2 L_0^{-1} L_1 \rangle + \langle L_3 \rangle$$
$$- (\langle L_1 L_0^{-1} L_1 \rangle - \langle L_2 \rangle) L_0^{-1} \langle L_1 \rangle + \langle L_1 \rangle L_0^{-1} \langle L_1 \rangle L_0^{-1} \langle L_1 \rangle$$
$$- \langle L_1 \rangle L_0^{-1} (\langle L_1 L_0^{-1} L_1 \rangle - \langle L_2 \rangle)] + O(\varepsilon^4)\} \langle \mathbf{X} \rangle$$
$$= \langle \mathbf{Y} \rangle \tag{8.72}$$

We remark that it is more convenient to set $\langle L_1 \rangle = 0$. There is no loss of generality in doing this since the mean part of the operator $L_1(t)$ can always be grouped with deterministic operator $L_0(t)$. The mean equations (8.71) and (8.72) simplify considerably under the condition $\langle L_1 \rangle = 0$.

Deterministic equations satisfied by higher-order moments can be derived in a similar, although much more cumbersome, fashion. Let us give a derivation of a correlation equation for a less general case [30].

We consider a scalar equation

$$[L_0(t) + \varepsilon L_1(t) + \varepsilon^2 L_2(t) + O(\varepsilon^3)] X(t) = y(t), \qquad t \geq 0 \quad (8.73)$$

where the operators $L_0(t)$, $L_1(t)$, $L_2(t)$, ... are as defined before, only they are now scalars. The forcing function $y(t)$ is assumed to be deterministic.

It is seen from Eq. (8.66) that we can write

$$L_0(t_i) X(t_i) = [L_0(t_i) - \varepsilon L_1(t_i) - \varepsilon^2(-L_1(t_i) L_0^{-1}(t_i) L_1(t_i)$$
$$+ L_2(t_i)) + \cdots] L_0^{-1}(t_i) y(t_i) \qquad (8.74)$$

at time t_i. Multiplying Eq. (8.74) at t_1 by itself at t_2 and taking expectation, we have up to the second-order terms,

$$L_0(t_1) L_0(t_2)\langle X(t_1) X(t_2)\rangle = J(t_1, t_2) y(t_1) y(t_2) + O(\varepsilon^3) \quad (8.75)$$

where $J(t_1, t_2)$ is an integro-differential operator having the form

$$J(t_1, t_2) = 1 - \varepsilon \sum_{j=1}^{2} \langle L_1(t_j)\rangle L_0^{-1}(t_j)$$

$$+ \varepsilon^2\left[\sum_{j=1}^{2} (\langle L_1(t_j) L_0^{-1}(t_j) L_1(t_j)\rangle L_0^{-1}(t_j) - \langle L_2(t_j)\rangle L_0^{-1}(t_j))\right.$$

$$\left. + L_0^{-1}(t_1) L_0^{-1}(t_2)\langle L_1(t_1) L_1(t_2)\rangle\right] \qquad (8.76)$$

It is not difficult to determine the inverse operator $J^{-1}(t_1, t_2)$. Upon operating Eq. (8.75) by $J^{-1}(t_1, t_2)$ and assuming that $\langle L_1(t)\rangle = 0$, we have

$$\{L_0(t_1) L_0(t_2) - \varepsilon^2[L_0(t_1)\langle L_1(t_2) L_0^{-1}(t_2) L_1(t_2)\rangle$$
$$+ L_0(t_2)\langle L_1(t_1) L_0^{-1}(t_1) L_1(t_1)\rangle + \langle L_1(t_1) L_1(t_2)\rangle$$
$$- L_0(t_1)\langle L_2(t_2)\rangle - L_0(t_2)\langle L_2(t_1)\rangle + O(\varepsilon^3)\}\langle X(t_1) X(t_2)\rangle$$
$$= y(t_1) y(t_2) \qquad (8.77)$$

This is one form of the correlation equation. Another useful form of the correlation equation is one which is independent of the forcing term

$y(t)$ but related to the mean solution $\langle X(t)\rangle$. Consider Eq. (8.67). We may write

$$
\begin{aligned}
x_0 &= \langle X\rangle + [\varepsilon L_0^{-1}\langle L_1\rangle - \varepsilon^2 L_0^{-1}(\langle L_1 L_0^{-1} L_1\rangle + \langle L_2\rangle)]\, x_0 + O(\varepsilon^3)\\
&= \{1 + \varepsilon L_0^{-1}\langle L_1\rangle + \varepsilon^2 [L_0^{-1}\langle L_1\rangle\, L_0^{-1}\langle L_1\rangle - L_0^{-1}\langle L_1 L_0^{-1} L_1\rangle\\
&\quad + L_0^{-1}\langle L_2\rangle\}\langle X\rangle + O(\varepsilon^3)
\end{aligned}
\tag{8.78}
$$

We note from the above that since

$$
x_0(t) = L_0^{-1}(t)\, y(t)
$$

and since both $L_0^{-1}(t)$ and $y(t)$ are nonrandom, $x_0(t)$ is a deterministic function. Now, we see from Eq. (8.66) that the solution $X(t)$ can be written in the form

$$
X(t) = [I - \varepsilon L_0^{-1} L_1 - \varepsilon^2 L_0^{-1}(-L_1 L_0^{-1} L_1 + L_2) + O(\varepsilon^3)]\, x_0 \tag{8.79}
$$

By forming the product $X(t_1) X(t_2)$ from Eq. (8.79) above, replacing x_0 in the resulting equation by Eq. (8.78), and taking expectations, we obtain, upon simplifying,

$$
\begin{aligned}
\langle X(t_1) X(t_2)\rangle &= \{1 + \varepsilon^2 L_0^{-1}(t_1)\, L_0^{-1}(t_2)[\langle L_1(t_1)\, L_1(t_2)\rangle\\
&\quad - \langle L_1(t_1)\rangle\langle L_1(t_2)\rangle]\}\langle X(t_1)\rangle\langle X(t_2)\rangle + O(\varepsilon^3)
\end{aligned}
\tag{8.80}
$$

It is noted that this equation, together with the knowledge of the mean solution as obtained from the mean equation, gives a direct evaluation of the correlation function.

To illustrate the utilities of these moment equations, let us consider two examples.

Example 8.9.* Consider the problem of a scalar wave propagating in a medium whose properties vary in a random way as functions of the spatial coordinates. A one-dimensional wave $U(x, t)$ satisfies

$$
\frac{\partial^2 U}{\partial x^2} = \frac{1}{C^2}\, \frac{\partial^2 U}{\partial t^2} \tag{8.81}
$$

where x and t stand for position and time, respectively. For time harmonic wave with angular frequency Ω, we have the reduced wave equation

$$
d^2 U(x)/dx^2 + K^2 U(x) = 0 \tag{8.82}
$$

* See Chen and Soong [30].

where $K = \Omega/C$ is the propagation constant characterizing the medium in which the wave travels. In the case of a medium whose properties undergo small random variations, the propagation constant K can be written as

$$K = k_0 N(x) \tag{8.83}$$

where

$$N(x) = 1 + \varepsilon V(x) \tag{8.84}$$

is the refraction index. The stochastic process $V(x)$ represents the random variations of the medium properties. In what follows, we shall assume that $V(x)$ is stationary with mean zero.

In this case, Eq. (8.82) takes the form

$$[L_0(x) + \varepsilon L_1(x) + \varepsilon^2 L_2(x)]\, U(x) = 0 \tag{8.85}$$

where

$$L_0(x) = d^2/dx^2 + k_0^2, \qquad L_1(x) = 2k_0^2 V(x), \qquad L_2(x) = k_0^2 V^2(x) \tag{8.86}$$

The inverse operator $L_0^{-1}(x)$ is an integral operator whose kernel is the Green's function

$$g(x - x') = -i/2k_0 \exp(ik_0 \mid x - x' \mid) \tag{8.87}$$

the range of integration being $(-\infty, \infty)$ in this example.

Let us consider the mean of the solution process $U(x)$ up to the second-order terms in ε. Noting that

$$\langle L_1(x) \rangle = 2k_0^2 \langle V(t) \rangle = 0 \tag{8.88}$$

it is seen from Eq. (8.71) that the mean solution $\langle U(x) \rangle$ satisfies

$$[L_0 - \varepsilon^2(\langle L_1 L_0^{-1} L_1 \rangle - \langle L_2 \rangle) + O(\varepsilon^3)]\langle U \rangle = 0 \tag{8.89}$$

The mean equation in this case thus takes the form

$$\frac{d^2}{dx^2}\langle U \rangle + k_0^2 \langle U \rangle - \varepsilon^2 k_0^2 \left[4k_0^2 \int_{-\infty}^{\infty} g(x - x')\, \Gamma_{VV}(x - x')\langle U(x') \rangle \, dx' \right.$$
$$\left. - \Gamma_{VV}(0)\langle U \rangle \right] + O(\varepsilon^3) = 0 \tag{8.90}$$

The behavior of the mean wave is governed by the equation above. Let us write its solution in the form

$$\langle U(x) \rangle = u_0 e^{ik'x} \tag{8.91}$$

The parameter k' can be interpreted as the propagation constant associated with the mean wave, which is in general different from the propagation constant k_0 defined earlier for the unperturbed wave motion. In this sense, the constant k' is usually regarded as the "effective" propagation constant associated with the random wave.

It is of interest to determine this effective propagation constant. Upon substituting Eq. (8.91) into Eq. (8.90), we find that k' must satisfy

$$(k_0^2 - k'^2) + \varepsilon^2 \left[2ik_0^3\sigma^2 \left(\frac{1}{\alpha - i(k_0 - k')} + \frac{1}{\alpha - i(k_0 + k')} \right) + k_0^2\sigma^2 \right]$$

$$+ O(\varepsilon^3) = 0 \tag{8.92}$$

where we have assumed that

$$\Gamma_{VV}(x - x') = \sigma^2 \exp(-\alpha \mid x - x' \mid) \tag{8.93}$$

Solving for k' from Eq. (8.92), we have

$$k' = \mu + i\nu + O(\varepsilon^3) \tag{8.94}$$

where

$$\mu = k_0[1 + \tfrac{1}{2}\varepsilon^2\sigma^2\alpha^2/(\alpha^2 + 4k_0^2)]$$
$$\nu = 2\varepsilon^2\sigma^2k_0^2(\alpha^2 + 2k_0^2)/[\alpha(\alpha^2 + 4k_0^2)] \tag{8.95}$$

In the above, the real part μ represents the propagation frequency. Since $\mu > k_0$, the phase velocity of the mean wave is smaller in the random medium than that in the unperturbed homogeneous medium. The imaginary part is the "effective" attenuation coefficient, which is positive in this case.

The second-moment properties of the random wave can be obtained by means of either Eq. (8.77) or Eq. (8.80). Without carrying out the details, the result for the variance of $U(x)$, $\sigma_U^2(x)$, is

$$\sigma_U^2(x) = \tfrac{1}{4}\varepsilon^2\sigma^2u_0^2k_0^2F(k_0, \alpha, \mu, \nu, x) \tag{8.96}$$

where

$$F = \phi_1(\mu) + \phi_1(-\mu) + [\phi_2(\mu) + \phi_2(-\mu)]$$
$$\times [\phi_3(\mu) + \phi_3(-\mu)] + [\phi_4(\mu) + \phi_4(-\mu)]$$
$$\times \{\tfrac{3}{4}e^{-2\nu x}[\phi_5(\mu) + \phi_5(-\mu)] + \tfrac{1}{4}e^{-(\alpha+\nu)x}[\phi_2(\mu) + \phi_2(-\mu)]\} \tag{8.97}$$

with

$$\phi_1(\mu) = \tfrac{1}{8}[(\alpha - \nu)^2 + (k_0 - \mu)^2]^{-1}\Big\{\frac{\alpha - \nu}{\nu} + \frac{1}{\nu^2 + \mu^2}\,[\nu(\alpha - \nu) - \mu(k_0 - \mu)]\Big\}$$

$$+ [\nu^2 + (k_0 - \mu)^2]^{-1}\{(k_0 - \mu)(2\nu - \alpha)\sin 2k_0 x$$

$$- [(k_0 - \mu)^2 + \nu(\alpha - \nu)]\cos 2k_0 x\}$$

$$+ (\nu^2 + k_0{}^2)^{-1}\{[\nu(k_0 - \mu) - k_0(\alpha - \nu)]\sin 2k_0 x$$

$$- [k_0(k_0 - \mu) - \nu(\alpha - \nu)]\cos 2k_0 x\}$$

$$\phi_2(\mu) = [(\alpha - \nu)^2 + (k_0 - \mu)^2]^{-1}[(k_0 - \mu)\cos k_0 x - (\alpha - \nu)\sin k_0 x]$$

$$\phi_3(\mu) = \tfrac{1}{2}[(\alpha + \nu)^2 + (k_0 - \mu)^2]^{-1}\{e^{-(\alpha + \nu)x}[(k_0 - \mu)\cos \mu x$$

$$- (\alpha + \nu)\sin \mu x + (\alpha + \nu)\sin k_0 x + (k_0 - \mu)\cos k_0 x]\}$$

$$\phi_4(\mu) = [(\alpha + \nu)^2 + (k_0 + \mu)^2]^{-1}[(\alpha + \nu)\sin \mu x + (k_0 + \mu)\cos \mu x]$$

$$\phi_5(\mu) = [(\alpha - \nu)^2 + (k_0 - \mu)^2]^{-1}[(\alpha - \nu)\sin \mu x + (k_0 - \mu)\cos \mu x]$$

(8.98)

Some properties of the mean and the variance for the random wave are illustrated in Figs. 8.4 and 8.5. Figure 8.4 shows a typical attenuated mean wave together with the standard deviation $\sigma_U(x)$. The standard deviation is seen to take an oscillatory pattern, oscillating approximately $\pi/2$ radians out of phase from the mean wave. The dependence of $\sigma_U(x)$ on the parameter α, the correlation length of $V(x)$ in the refraction index, is shown in

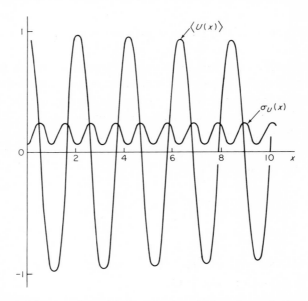

Fig. 8.4. Wave motion with $k_0 = 3$, $\varepsilon\sigma = 0.05$, and $\alpha = \tfrac{1}{3}$ (from [30]).

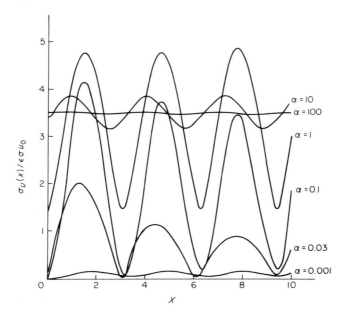

Fig. 8.5. Effect of correlation length on $\sigma_U(x)$ ($k_0 = 1$ and $\varepsilon\sigma = 0.05$) (from [30]).

Fig. 8.5. It is seen that the standard deviation curves in general attenuate with increasing x owing to random phase mixing. This attenuation effect becomes less pronounced as α approaches either extreme of the range of values plotted. These two extremes correspond to the cases where, in the limit, the randomness in the refraction index becomes independent of x at one end and becomes completely random at the other.

Example 8.10. The treatment of vector differential equations can be carried out in a similar fashion. Consider a damped linear oscillator having n degrees of freedom as shown in Fig. 8.6. The equations of motion are

$$M_j\ddot{X}_j = -K_j(X_j - X_{j-1}) - K_{j+1}(X_j - X_{j+1}) - C_j(\dot{X}_j - \dot{X}_{j-1})$$
$$-C_{j+1}(\dot{X}_j - \dot{X}_{j+1}), \quad j = 1, 2, \ldots, n-1, \quad X_0 = 0 \quad (8.99)$$
$$M_n\ddot{X}_n = Y(t) - K_n(X_n - X_{n-1}) - C_n(\dot{X}_n - \dot{X}_{n-1})$$

Fig. 8.6. The linear system.

We assume that the parameters of the system can be represented by

$$\left.\begin{array}{l} M_j(t) = m + \varepsilon M_j'(t), \\ C_j(t) = c + \varepsilon C_j'(t), \\ K_j(t) = k + \varepsilon K_j'(t), \end{array}\right\} j = 1, 2, \ldots, n \qquad (8.100)$$

Equations (8.99) can then be put in the vector form

$$[L_0 + \varepsilon L_1]\, \mathbf{X}(t) = \mathbf{Y}(t) \qquad (8.101)$$

where

$$\mathbf{X}(t) = \begin{bmatrix} X_1(t) \\ \vdots \\ X_n(t) \end{bmatrix} \quad \text{and} \quad \mathbf{Y}(t) = \begin{bmatrix} 0 \\ 0 \\ \vdots \\ 0 \\ Y(t) \end{bmatrix} \qquad (8.102)$$

The matrix $L_1(t)$ is an $n \times n$ symmetric Jacobi matrix whose elements are differential operators. Its nonzero elements are

$$(L_1)_{jj} = M_j'\, \frac{d^2}{dt^2} + (C_j' + C_{j+1}')\, \frac{d}{dt} + (K_j' + K_{j+1}'), \qquad j = 1, 2, \ldots, n-1$$
$$(8.103)$$
$$(L_1)_{nn} = M_n'\, \frac{d^2}{dt^2} + C_n'\, \frac{d}{dt} + K_n'$$

along the diagonal, and

$$(L_1)_{j(j-1)} = (L_1)_{(j-1)j} = -C_j'\, \frac{d}{dt} - K_j', \qquad j = 2, 3, \ldots, n \quad (8.104)$$

along the two contiguous diagonals. The matrix $L_0(t)$ has the same form as $L_1(t)$ with M_j', C_j', and K_j' replaced by m, c, and k, respectively.

If we assume that the means of $M_j'(t)$, $C_j'(t)$, and $K_j'(t)$ are zero, the mean equation for $\mathbf{X}(t)$ takes the form

$$[L_0 - \varepsilon^2 \langle L_1 L_0^{-1} L_1 \rangle] \langle \mathbf{X} \rangle = \langle \mathbf{Y} \rangle \qquad (8.105)$$

In the determination of the "effective" natural frequencies, we let $\langle \mathbf{Y} \rangle = 0$ and assume

$$\langle \mathbf{X} \rangle = A e^{i\lambda t} \qquad (8.106)$$

The operators L_0 and $\langle L_1 L_0^{-1} L_1 \rangle$ in Eq. (8.105) are well defined. With the mean solution expressed in the form above, we can formally write

$$L_0(t)\langle \mathbf{X} \rangle = R(i\lambda)\langle \mathbf{X} \rangle \tag{8.107}$$

and

$$\langle L_1 L_0^{-1} L_1 \rangle \langle \mathbf{X} \rangle = S(i\lambda)\langle \mathbf{X} \rangle \tag{8.108}$$

Upon substituting Eqs. (8.107) and (8.108) into Eq. (8.105), we require that, for nontrivial solution,

$$| R(i\lambda) - \varepsilon^2 S(i\lambda) | = 0 \tag{8.109}$$

The equation above is an algebraic equation containing the unknown parameter λ. The roots of this equation give the effective natural frequencies of the system.

In order to show some working details, let us consider the case where $n = 2$, $c = 0$, $C_j'(t) = K_j'(t) = 0$, $j = 1, 2, \ldots$, with probability one, and $M_j'(t)$, $j = 1, 2, \ldots$, are stationary stochastic processes with means zero and correlation functions

$$\langle M_j'(t)\, M_k'(t') \rangle = \sigma^2 \delta_{jk} e^{-\alpha |t-t'|}. \tag{8.110}$$

For this case, the differential operators take the forms

$$L_0(t) = \begin{bmatrix} m\, \dfrac{d^2}{dt^2} + 2k & -k \\[2ex] -k & m\, \dfrac{d^2}{dt^2} + k \end{bmatrix} \tag{8.111}$$

$$L_1(t) = \begin{bmatrix} M_1'\, \dfrac{d^2}{dt^2} & 0 \\[2ex] 0 & M_2'\, \dfrac{d^2}{dt^2} \end{bmatrix} \tag{8.112}$$

The inverse operator $L_0^{-1}(t)$ is specified by the principal matrix $\Phi(t)$ whose elements are given by

$$\Phi_{jk}(t) = c_{jk} \sin \omega_{01} t + d_{jk} \cos \omega_{02} t, \qquad j, k = 1, 2 \tag{8.113}$$

where

$$
\begin{aligned}
c_{jk} &= [2/5(km)^{1/2}] \sin(\, j\pi/5) \sin(k\pi/5)/\sin(\pi/10) \\
d_{jk} &= [2/5(km)^{1/2}] \sin(3j\pi/5) \sin(3k\pi/5)/\sin(3\pi/10) \\
\omega_{01} &= 2(k/m)^{1/2} \sin(\pi/10) \\
\omega_{02} &= 2(k/m)^{1/2} \sin(3\pi/10)
\end{aligned}
\tag{8.114}
$$

The matrices $R(i\lambda)$ and $S(i\lambda)$ can now be determined. They are

$$R(i\lambda) = \begin{bmatrix} -m\lambda^2 + 2k & -k \\ -k & -m\lambda^2 + k \end{bmatrix} \tag{8.115}$$

$$S(i\lambda) = \begin{bmatrix} \dfrac{\sigma^2\lambda^2 c_{11}\omega_{01}^3}{(\alpha+i\lambda)^2 + \omega_{01}^2} + \dfrac{\sigma^2\lambda^2 d_{11}\omega_{02}^3}{(\alpha+i\lambda)^2 + \omega_{02}^2} & 0 \\ 0 & \dfrac{\sigma^2\lambda^2 c_{22}\omega_{01}^3}{(\alpha+i\lambda)^2 + \omega_{01}^2} + \dfrac{\sigma^2\lambda^2 d_{22}\omega_{02}^3}{(\alpha+i\lambda)^2 + \omega_{02}^2} \end{bmatrix}$$

$$\tag{8.116}$$

The desired effective frequencies are found by substituting Eqs. (8.115) and (8.116) into Eq. (8.109) and solving for λ. The results are

$$\lambda_1 = \omega_{01} + \omega_{01}\,\frac{\varepsilon^2\sigma^2}{2\sqrt{5}\,k}\left\{\left(\omega_{01}^2 - 2\,\frac{k}{m}\right)\left[\frac{c_{22}\omega_{01}^3\alpha^2}{p_1} + \frac{d_{22}\omega_{02}^3(\alpha - \omega_{01}^2 + \omega_{02}^2)}{q_1}\right]\right.$$

$$+ \left(\omega_{01}^2 - \frac{k}{m}\right)\left[\frac{c_{11}\omega_{01}^3\alpha^2}{p_1} + \frac{d_{11}\omega_{02}^3(\alpha - \omega_{01}^2 + \omega_{02}^2)}{q_1}\right]$$

$$- i\left(\omega_{01}^2 - 2\,\frac{k}{m}\right)\left[\frac{2\alpha c_{22}\omega_{01}^4}{p_1} + \frac{2\alpha d_{22}\omega_{01}\omega_{02}^3}{q_1}\right]$$

$$- i\left(\omega_{01}^2 - \frac{k}{m}\right)\left[\frac{2\alpha c_{11}\omega_{01}^4}{p_1} + \frac{2\alpha d_{11}\omega_{01}\omega_{02}^3}{q_1}\right]\right\} \tag{8.117}$$

$$\lambda_2 = \omega_{02} + \omega_{02}\,\frac{\varepsilon^2\sigma^2}{2\sqrt{5}\,k}\left\{-\left(\omega_{02}^2 - 2\,\frac{k}{m}\right)\left[\frac{c_{22}\omega_{01}^3(\alpha^2 + \omega_{01}^2 - \omega_{02}^2)}{p_2}\right.\right.$$

$$+ \frac{d_{22}\omega_{02}^3\alpha^2}{q_2}\bigg]$$

$$- \left(\omega_{02}^2 - \frac{k}{m}\right)\left[\frac{c_{11}\omega_{01}^3(\alpha^2 + \omega_{01}^2 - \omega_{02}^2)}{p_2} + \frac{d_{11}\omega_{02}^3\alpha^2}{q_2}\right]$$

$$+ i\left(\omega_{02}^2 - 2\,\frac{k}{m}\right)\left[\frac{2\alpha c_{22}\omega_{01}^3\omega_{02}}{p_2} + \frac{2\alpha d_{22}\omega_{02}^4}{q_2}\right]$$

$$+ i\left(\omega_{02}^2 - \frac{k}{m}\right)\left[\frac{2\alpha c_{11}\omega_{01}^3\omega_{02}}{p_2} + \frac{2\alpha d_{11}\omega_{02}^4}{q_2}\right]\right\} \tag{8.118}$$

where

$$p_1 = \alpha^4 + 4\alpha^2\omega_{01}^2, \qquad q_1 = (\alpha^2 - \omega_{01}^2 + \omega_{02}^2)^2 + 4\alpha^2\omega_{01}^2$$

$$p_2 = (\alpha^2 + \omega_{01}^2 - \omega_{02}^2)^2 + 4\alpha^2\omega_{02}^2, \qquad q_2 = \alpha^4 + 4\alpha^2\omega_{02}^2 \tag{8.119}$$

It is seen from the results above that

$$\text{Re}[\lambda_i] < \omega_{0i}, \qquad \text{Im}[\lambda_i] > 0, \qquad i = 1, 2,$$

indicating that the "effective" natural frequencies associated with the mean oscillation are lower than the unperturbed natural frequencies. The imaginary parts of the results show that damping is present in the mean motion. We note that for small α, or long correlation length, the imaginary parts of λ_i can become large and hence the damping effect can be prominent. The effective frequencies approach the unperturbed natural frequencies as α becomes large. Figure 8.7. illustrates the behavior of λ_i as a function of α for $\varepsilon\sigma = 1$ and $k = m = 1$.

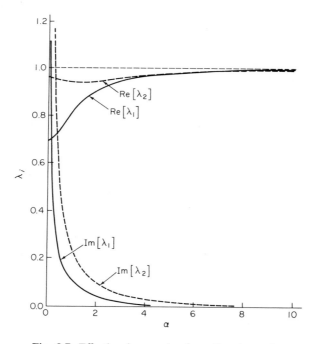

Fig. 8.7. Effective frequencies λ_i as functions of α.

8.6. Truncated Hierarchy Techniques

Let us consider the simple but important scalar differential equation

$$\frac{d^2X(t)}{dt^2} + A(t)\,X(t) = 0, \qquad t \geq 0 \tag{8.120}$$

Suppose that no restrictions are imposed on the stochastic coefficient process $A(t)$; it is rather discouraging to note that we have not so far developed a technique to cope with this situation.

A number of attempts have been made in determining the solution properties of differential equations similar to Eq. (8.120) in some approximate way. We outline in this section some approximate methods generally called truncated hierarchy techniques. Extensively used in statistical mechanics and statistical physics, they have been studied by Richardson [31] as possible tools for solving random linear differential equations. Some applications of these techniques can be found in the engineering literature [32, 33].

In order to develop various hierarchy techniques relatively free of mathematical clutter, we consider the simple scalar random differential equation

$$[L_0(t) + aA(t)] X(t) = 0, \qquad t \geq 0 \tag{8.121}$$

with deterministic initial conditions

$$X(0) = x_0, \qquad \dot{X}(0) = \dot{x}_0, \ldots \tag{8.122}$$

The differential operator $L_0(t)$ in Eq. (8.121) is deterministic and the statistical properties of the stochastic process $A(t)$ are assumed to be specified. For simplicity, we shall assume that the mean of $A(t)$ is zero without loss of generality. The parameter a is introduced in Eq. (8.121) in order to facilitate comparison with the perturbation method; the value of a is, of course, not necessarily small.

In what follows, our primary concern is the determination of the mean of solution process $X(t)$. Extensions to the calculation of higher-order moments can be made following essentially the same procedure.

The hierarchy considered here is constructed by successively multiplying Eq. (8.121) by the coefficient process $A(t)$ and averaging; the independent variable t is given a different value at each multiplication step. The first member of the hierarchy is simply the averaged Eq. (8.121), that is

$$L_0(t)\langle X(t)\rangle + a\langle A(t) X(t)\rangle = 0 \tag{8.123}$$

The hierarchy is now generated by multiplying Eq. (8.121) by factors $A(t_1)$, $A(t_1) A(t_2)$, ..., and averaging. The result is

$$L_0(t)\langle A(t_1) X(t)\rangle + a\langle A(t_1) A(t) X(t)\rangle = 0$$
$$L_0(t)\langle A(t_1) A(t_2) X(t)\rangle + a\langle A(t_1) A(t_2) A(t) X(t)\rangle = 0 \tag{8.124}$$

$$\vdots$$

It is clear that, in order to solve for $\langle X(t) \rangle$, we will not be able to obtain a finite sequence of equations with a consistent number of unknowns. As we have encountered in Example 8.7 and in Section 7.4.1, it is necessary that we consider various truncation procedures where, by a suitable approximation, the truncation leads to a finite number of equations with a consistent number of dependent variables. Two possible schemes are outlined below.

8.6.1. Correlation Discard

The new dependent variable $\langle A(t_1) A(t_2) \cdots A(t_{n-1}) A(t) X(t) \rangle$ appears in the nth member of the hierarchy. In the method of correlation discard, we approximate this term by

$$\langle A(t_1) A(t_2) \cdots A(t_{n-1}) A(t) \rangle \langle X(t) \rangle$$

There is no real justification for using this approximation since the solution process $X(t)$ is statistically dependent upon the coefficient process $A(t)$. Bourret [34], however, gives this type of approximations a "local independence" interpretation which provides a partial justification.

Since the average $\langle A(t_1) \cdots A(t_{n-1}) A(t) \rangle$ is now specified, this approximation reduces the number of unknowns in the first n members of the hierarchy by one, thus obtaining a closed set of n equations with n unknowns.

We note that this approximation procedure is equivalent to setting

$$\langle A(t_1) \cdots A(t_{n-1}) A(t) \Delta X(t) \rangle = 0 \tag{8.125}$$

where

$$\Delta X(t) = X(t) - \langle X(t) \rangle \tag{8.126}$$

It is instructive to investigate the effect of this approximation procedure on the mean solution when the correlation discard is applied at some small values of n. For this purpose, let us write Eq. (8.121) in the form

$$[L_0(t) + aL_1(t)] X(t) = 0, \qquad t \geq 0 \tag{8.127}$$

where

$$L_1(t) = A(t) \tag{8.128}$$

and

$$\langle L_1(t) \rangle = \langle A(t) \rangle = 0 \tag{8.129}$$

For $n = 1$, the mean value $\langle X(t) \rangle$ is given by

$$L_0(t) \langle X(t) \rangle = 0 \tag{8.130}$$

with initial conditions

$$\langle X(t) \rangle |_{t=0} = x_0, \qquad \langle \dot{X}(t) \rangle |_{t=0} = \dot{x}_0, \ldots \tag{8.131}$$

For $n = 2$, we have

$$L_0(t)\langle X(t) \rangle + a\langle L_1(t) X(t) \rangle = 0$$
$$L_0(t)\langle L_1(t_1) X(t) \rangle + a\langle L_1(t_1) L_1(t) \rangle \langle X(t) \rangle = 0 \tag{8.132}$$

with initial conditions

$$\langle X(t) \rangle |_{t=0} = x_0, \qquad\qquad \langle \dot{X}(t) \rangle |_{t=0} = \dot{x}_0, \ldots$$
$$\langle L_1(t_1) X(t) \rangle |_{t=0} = 0, \qquad \frac{d}{dt} \langle L_1(t_1) X(t) \rangle |_{t=0} = 0, \ldots \tag{8.133}$$

An equation satisfied by $\langle X(t) \rangle$ can be easily established from above. The second of Eq. (8.132) leads to

$$\langle L_1(t) X(t) \rangle = -aL_0^{-1}(t)\langle L_1(t_1) L_1(t) \rangle \langle X(t) \rangle |_{t_1=t}$$
$$= -a\langle L_1(t) L_0^{-1}(t) L_1(t) \rangle \langle X(t) \rangle \tag{8.134}$$

The substitution of the equation above into the first of Eq. (8.132) then gives

$$[L_0(t) - a^2\langle L_1(t) L_0^{-1}(t) L_1(t) \rangle]\langle X(t) \rangle = 0 \tag{8.135}$$

A close look at Eqs. (8.130) and (8.135) reveals a remarkable feature associated with this approximation procedure. For $n = 1$ and 2, the resulting equations satisfied by the mean are identical to Eq. (8.71), obtained by using the perturbation method with a as the perturbation parameter. This comparison indicates that, at least for $n = 1$ and 2, the method of correlation discard renders poor approximation for cases when a is not small. Similar remarks are advanced by Adomian [35].

The solution for $\langle X(t) \rangle$ with truncation executed at $n = 3$ or higher is in general difficult to obtain. It is also difficult to assess the validities of these higher-order approximations.

8.6.2. Cumulant Discard

Rather than neglecting the correlation function associated with the random variables $A(t_1), A(t_2), \ldots, A(t_{n-1}), A(t),$ and $\Delta X(t)$ as was done above, this procedure suggests that we neglect the corresponding cumulant. (See Problem 2.10 for definition.) Specifically, we set

$$i^{-(n+1)} \left[\frac{\partial^{n+1}}{\partial \lambda_1 \cdots \partial \lambda_{n-1} \, \partial \lambda \, \partial \mu} \log \left\langle \exp \left[i \sum_{j=1}^{n-1} \lambda_j \, A(t_j) + i\lambda \, A(t) \right. \right. \right.$$

$$\left. \left. \left. + i\mu \, \Delta X(t) \right] \right\rangle \right] \Bigg|_{\lambda_1, \ldots, \lambda_{n-1}, \lambda, \mu = 0} = 0 \qquad (8.136)$$

The result of this approximation is that the new unknown $\langle A(t_1) \, A(t_2) \cdots A(t_{n-1}) \, A(t) \, X(t) \rangle$ in the nth member of the hierarchy is approximated by a linear combination of the lower order averages. Since the cumulants and the correlation functions are identical for $n = 1$ and 2, the methods of cumulant discard and correlation discard give identical results for these two values of n and, as we have shown above, they are the same as the perturbation method.

The differences of these procedures begin to show when $n = 3$ or higher. For $n = 3$, the method of cumulant discard involves the neglect of the term

$$\langle A(t_1) \, A(t_2) \, A(t) \, \Delta X(t) \rangle - \langle A(t_1) \, A(t_2) \rangle \langle A(t) \, \Delta X(t) \rangle$$
$$- \langle A(t_1) \, A(t) \rangle \langle A(t_2) \, \Delta X(t) \rangle - \langle A(t_2) \, A(t) \rangle \langle A(t_1) \, \Delta X(t) \rangle$$

while the method of correlation discard simply neglects the term

$$\langle A(t_1) \, A(t_2) \, A(t) \, \Delta X(t) \rangle$$

Again, because of the lack of a mathematical basis, the merits of these procedures for $n = 3$ or higher cannot be evaluated in general terms.

In closing and without giving details, we mention that several other approximate techniques have been proposed as methods of analyzing differential equations with random coefficients. Richardson [31] has considered a scheme in which the solution process is represented by the output of a linear system driven by the coefficient process; the specification of the linear system is such that it best satisfies the random differential equation in the mean square sense. An iterative method employing a series solution representation was considered by Adomian [35]. Recently, Bellman *et al.* [36] examined the feasibility of obtaining solutions of random differential equations as limiting solution processes of appropriate random difference equations. These methods need to be developed further and, indeed, this whole area is still relatively unplowed at this moment.

References

1. P. G. Bergman, Propagation of radiation in a medium with random inhomogeneities. *Phys. Rev.* **70**, 486–489 (1946).
2. W. E. Boyce, Random eigenvalue problems. In *Probabilistic Methods in Applied Mathematics* (A. T. Bharucha-Reid, ed.), Vol. 1, pp. 1–73. Academic Press, New York, 1968.
3. J. L. Strand, *Stochastic Ordinary Differential Equations*. Ph. D. Thesis, Univ. of California, Berkeley, California, 1968.
4. J. L. Strand, Random ordinary differential equations. *J. Differential Equations* **7**, 538–553 (1970).
5. A. Rosenbloom, Analysis of linear systems with randomly time-varying parameters. *Proc. Symp. Information Networks, Polytech. Inst. of Brooklyn, 1954*, pp. 145–153.
6. V. I. Tikhonov, Fluctuation action in the simplest parametric systems. *Automat. Remote Control* **19**, 705–711 (1958).
7. P. F. Chenea and J. L. Bogdanoff, Impedance of some disordered systems. *Colloq. Mech. Impedance Methods*, pp. 125–128. ASME, New York, 1958.
8. J. L. Bogdanoff and P. F. Chenea, Dynamics of some disordered linear systems. *Internat. J. Mech. Sci.* **3**, 157–169 (1961).
9. F. Kozin, On the probability densities of the output of some random systems. *J. Appl. Mech.* **28**, 161–165 (1961).
10. T. T. Soong and J. L. Bogdanoff, On the natural frequencies of a disordered linear chain of N degrees of freedom. *Internat. J. Mech. Sci.* **5**, 237–265 (1963).
11. T. T. Soong and J. L. Bogdanoff, On the impulsive admittance and frequency response of a disordered linear chain of N degrees of freedom. *Internat. J. Mech. Sci.* **6**, 225–237 (1964).
12. T. T. Soong and F. A. Cozzarelli, Effect of random temperature distributions on creep in circular plates. *Internat. J. Non-linear Mech.* **2**, 27–38 (1967).
13. J. D. Collins and W. T. Thomson, The eigenvalue problem for structural systems with statistical properties. *AIAA J.* **7**, 642–648 (1969).
14. F. A. Cozzarelli and W. N. Huang, Effect of random material parameters on nonlinear steady creep solutions. *Internat. J. Solids Structures* **7**, 1477–1494 (1971).
15. A. Maradudin, E. Montroll, and G. Weiss, *Lattice Vibration in the Harmonic Approximation*. Academic Press, New York, 1963.
16. T. T. Soong, Pharmacokinetics with uncertainties in rate constants. *Math. Biosci.* **12**, 235–244 (1971).
17. T. T. Soong and S. N. Chuang, Solutions of a class of random differential equations. *SIAM J. Appl., Math.* **24** (4) (1973).
18. J. L. Bogdanoff and F. Kozin, Moments of the output of linear random systems. *J. Acoust. Soc. Amer.* **34**, 1063–1068 (1962).
19. T. K. Caughey and J. K. Dienes, The behavior of linear systems with random parametric excitations. *J. Math. and Phys.* **41**, 300–318 (1962).
20. F. Kozin and J. L. Bogdanoff, A comment on "The behavior of linear systems with random parametric excitations." *J. Math. and Phys.* **42**, 336–337 (1963).
21. S. T. Ariaratnam and P. W. V. Graeffe, Linear systems with stochastic coefficients. *Internat. J. Control* **1**, 239–250; **2**, 161–169, 205–210 (1965).
22. J. C. Samuels and A. C. Eringen, On stochastic linear systems. *J. Math. and Phys.* **38**, 83–103 (1959).

23. S. N. Chuang and T. T. Soong, On the zero shift problem of a vertical pendulum. *J. Franklin Inst.* **289**, 47–55 (1970).

24. R. A. Van Slooten and T. T. Soong, Buckling of a long, axially compressed, thin cylindrical shell with random initial imperfections. *J. Appl. Mech.* **39**, 1066–1071 (1972).

25. J. B. Keller, Stochastic equations and wave propagation in random media. *Proc. Symp. Appl. Math. 16th*, pp. 145–170. Amer. Math. Soc., Providence, Rhode Island, 1964.

26. F. C. Karal and J. B. Keller, Elastic, electromagnetic, and other waves in a random medium. *J. Math. Phys.* **5**, 537–547 (1964).

27. Y. M. Chen and C. L. Tien, Penetration of temperature waves in a random medium. *J. Math. and Phys.* **46**, 188–194 (1967).

28. J. R. Jokipii, Cosmic-ray propagation I. Charged particles in a random magnetic field. *Astrophys. J.* **146**, 480–487 (1966).

29. C. H. Liu, Propagation of coherent magnetohydrodynamic waves. *Phys. Fluids* **12**, 1642–1647 (1969).

30. K. K. Chen and T. T. Soong, Covariance properties of waves propagating in a random medium. *J. Acoust. Soc. Amer.* **49**, 1639–1642 (1971).

31. J. M. Richardson, The application of truncated hierarchy techniques in the solution of stochastic linear differential equation. *Proc. Symp. Appl. Math.*, *16th*, pp. 290–302. Amer. Math. Soc., Providence, Rhode Island, 1964.

32. C. W. Haines, Hierarchy methods for random vibrations of elastic strings and beams. *J. Engrg. Math.* **1**, 293–305 (1967).

33. J. C. Amazigo, Buckling under axial compression of long cylindrical shells with random axisymmetric imperfections. *Quart. Appl. Math.* **26**, 537–566 (1969).

34. R. C. Bourret, Propagation of randomly perturbed fields. *Canad. J. Phys.* **40**, 782–790 (1962).

35. G. Adomian, Random operator equations in mathematical physics I. *J. Math. Phys.* **11**, 1069–1084 (1970).

36. R. Bellman, T. T. Soong, and R. Vasudevan, On moment properties of a class of stochastic difference equations. *J. Math. Anal. Appl.* **40**, 283–299 (1972).

PROBLEMS

8.1. Consider the first-order scalar equation

$$\dot{X}(t) + A(t)\,X(t) = 0, \qquad t \geq 0; \qquad X(0) = 1$$

where $A(t)$ is a stationary Gaussian process with mean m_A and correlation function $\Gamma_{AA}(\tau)$. Determine the mean and the variance of the solution process. Do they reach stationary values as $t \to \infty$?

8.2. The equation of motion of a linear oscillator is given by

$$\ddot{X}(t) + \Omega^2 X(t) = 0, \qquad t \geq 0, \qquad X(0) = X_0, \qquad \dot{X}(0) = \dot{X}_0$$

In the above, the quantities Ω^2, X_0, and $\dot{X}_0$ are regarded as random variables with joint density function $f_0(x_0, \dot{x}_0, \omega^2)$.

(a) Express the joint density function of $X(t)$, $\dot{X}(t)$, and Ω^2 in terms of $f_0(x_0, \dot{x}_0, \omega^2)$.

(b) Determine the density function of $X(t)$ under the condition that $X_0 \equiv 0$, $\dot{X}_0$ is Gaussian with mean and variance both one, and Ω^2 is independent of $\dot{X}_0$ and is uniformly distributed between ω_1^2 and $\omega_2^2(\omega_1^2, \omega_2^2 > 0)$.

8.3. Consider the first-order differential equation

$$\dot{X}(t) + [a + W_1(t)] X(t) = 0, \qquad t \geq 0; \qquad X(0) = 1$$

where $W_1(t)$ is a Gaussian white noise with zero mean and covariance $2D\,\delta(t - s)$. It is shown in Example 8.4 that the mean solution is given by

$$E\{X(t)\} = e^{-at}$$

Can this result be reproduced from the results of Problem 8.1 after appropriate substitutions? Explain the apparent difference in light of discussions of Ito integrals given in Section 5.2.2.

8.4. Determine the variance of $X(t)$ governed by Eq. (8.42). Discuss its asymptotic behavior.

8.5. Verify Eq. (8.72).

8.6. A linear oscillator is characterized by

$$\ddot{X}(t) + \Omega^2 X(t) = 0, \qquad t \geq 0$$

where Ω, the natural frequency, is a function of the system parameters. Let the randomness in the parameter values be such that

$$\Omega^2 = \omega_0^2[1 + \varepsilon A(t)]$$

where ε is a small parameter and $A(t)$ is wide-sense stationary with mean zero and correlation function

$$\Gamma_{AA}(\tau) = \sigma^2 e^{-\alpha|\tau|}$$

Based upon the perturbation method, determine the effective frequency of the oscillator up to ε^2-terms.

8.7. Determine the effective frequency of the oscillator considered in Problem 8.6 using the truncated hierarchy technique; use the method of correlation discard executed at $n = 3$.

Chapter 9

Stochastic Stability

Stability studies of differential equations are concerned with the qualitative behavior of their solutions. Initiated by Poincaré in about 1880, the stability of systems governed by differential equations has been a subject of numerous studies, leading to results of basic importance from both mathematical and practical points of view; see, for example, Bellman [1] and LaSalle and Lefschetz [2] for detailed discussions on this important topic.

In recent years, the theory of stability of systems governed by differential equations with random coefficients has generated a great deal of interest. This is particularly true on the part of control engineers whose interest in the subject stems from the need for a better understanding of the stability of control systems having random characteristics. The development of stability concepts for stochastic systems has just begun. As these concepts mature and as we continue to construct meaningful stochastic models for physical as well as physiological, economical, biological, and other systems, the theory of stability for stochastic systems will certainly become one basic to all system sciences.

We consider in this chapter the stability of differential equations with random coefficients or, in short, *stochastic stability*. There are a number of possible stability concepts in the study of deterministic systems, and we expect many more in the stochastic case. Since the problem of stability is essentially a problem of convergence, for each deterministic stability definition we find that there are at least four corresponding stochastic stability definitions. These are generated by the four modes of convergence, namely, convergence in distribution, convergence in probability, convergence in mean square, and almost sure convergence. As a result, there has been a great deal of activities in this area. Within the past fifteen years, a large number of stochastic stability concepts have been introduced and studied, and the advances have been formidable.

In what follows, we consider only a selected few stochastic stability concepts consistent with the mean square approach adopted in the development of this book. Hence, our primary concern will be the stability concepts based upon convergence in mean square criterion.

We have pointed out repeatedly in the preceding chapters that the sample function considerations of random differential equations have significant physical implications. Indeed, upon observing a real system in operation, it is the sample solutions that are witnessed and not the averages or the probabilities. In the same vein, recent stability studies of stochastic systems have emphasized sample stabilities and almost sure stabilities based upon almost sure convergence criterion. Our background does not permit us to delve into this important topic. The interested reader is referred to Kushner [3] and a recent comprehensive survey article by Kozin [4].

9.1. Moment Stabilities

It appears that Rosenbloom [5] initiated the stability study of the statistical moments associated with the solution of random differential equations. He studied the asymptotic moment behavior of the solution of a first-order random differential equation through its explicit solution representation. The study of higher-order random differential equations was followed by Samuels and Eringen [6] and Samuels [7], who introduced the concept of *mean square stability*. Let us first present this method of attack. In this context, we are interested in the stability in the mean as well as the stability in mean square.

Let $\mathbf{X}(t)$ be the solution process of a system of random differential equations. We give the following definitions.

Definition. The system is said to be *stable in the mean* if

$$\lim_{t \to \infty} E \mid \mathbf{X}(t) \mid < \mathbf{c} \tag{9.1}$$

where **c** is a finite constant vector. It implies that every input with bounded mean gives rise to an output with bounded mean.

Definition. The system is *asymptotically stable in the mean* if

$$\lim_{t \to \infty} E \mid \mathbf{X}(t) \mid \to \mathbf{0} \tag{9.2}$$

Unless stated otherwise, we shall always refer to the equilibrium solution $\mathbf{X}(t) \equiv \mathbf{0}$ as the solution whose stability is being studied.

Definition. The system is said to be *mean square stable* if

$$\lim_{t \to \infty} E \mid \mathbf{X}(t) \, \mathbf{X}^{\mathrm{T}}(t) \mid < C \tag{9.3}$$

where C is a constant square matrix whose elements are finite. This criterion implies that every mean square bounded input leads to a mean square bounded output.

Definition. The system is *asymptotically mean square stable* if

$$\lim_{t \to \infty} E \mid \mathbf{X}(t) \, \mathbf{X}^{\mathrm{T}}(t) \mid \to 0 \tag{9.4}$$

where 0 is the null matrix.

9.1.1. An Integral Equation Approach

Let us consider now the application of these criteria to the stability studies of some random systems. The method considered by Samuels and Eringen [6] is first presented. In their work, the solutions of random differential equations are successively approximated in terms of integral equations. Under various assumptions imposed upon the coefficient processes, the stability results are obtained with the aid of the deterministic stability theory.

Consider an nth-order random differential equation given by

$$L(t) \, X(t) = \sum_{k=0}^{n} A_k(t) \frac{dX^k(t)}{dt^k} = Y(t), \qquad t \geq 0 \tag{9.5}$$

where the coefficient processes $A_k(t)$ and the input process $Y(t)$ are prescribed second-order stochastic processes. It is assumed that $Y(t)$ is independent of $A_k(t)$. We shall discuss the stability of the solution process $X(t)$ based upon the mean square criterion.

It is convenient to write the coefficient processes $A_k(t)$ in the form

$$A_k(t) = E\{A_k(t)\} + P_k(t)$$
$$= a_k(t) + P_k(t) \tag{9.6}$$

where $P_k(t)$ are zero-mean stochastic processes. Upon substituting Eq. (9.6) into Eq. (9.5), we can write

$$[L_0(a_k(t), t) + L_1(P_k(t), t)] X(t) = Y(t), \qquad t \geq 0 \tag{9.7}$$

where $L_0(a_k(t), t)$ represents the deterministic part of the differential operator $L(t)$ and $L_1(P_k(t), t)$ represents the stochastic part. Clearly,

$$E\{L_1(P_k(t), t)\} = 0 \tag{9.8}$$

Equation (9.7) can be put in the form

$$L_0(t) X(t) = Y(t) - L_1(t) X(t) \tag{9.9}$$

A representation for the solution process $X(t)$ can now be obtained in terms of the integral equation

$$X(t) = L_0^{-1}(t) Y(t) - L_0^{-1}(t) L_1(t) X(t) + \sum_{k=1}^{n} c_k \phi_k(t) \tag{9.10}$$

where $L_0^{-1}(t)$, the inverse of $L_0(t)$, is a deterministic integral operator whose kernel is the weighting function associated with $L_0(t)$. The functions $\phi_k(t)$ form a fundamental set of independent solutions of $L_0(t) \phi(t) = 0$, and c_k are constants to be determined from the initial conditions. We shall assume, for convenience, that these constants are deterministic.

We would like to consider a representation for the second moment of $X(t)$. Let

$$F(t) = L_0^{-1}(t) Y(t) + \sum_{k=1}^{n} c_k \phi_k(t) \tag{9.11}$$

Clearly, the process $F(t)$ is only stochastically dependent upon $Y(t)$. Now, multiplying Eq. (9.10) by itself at two points t_1 and t_2 and taking expectation, we have, upon using Eq. (9.11),

$$\langle X(t_1) X(t_2) \rangle = \langle F(t_1) F(t_2) \rangle - L_0^{-1}(t_1)\langle L_1(t_1) X(t_1) F(t_2) \rangle$$
$$- L_0^{-1}(t_2)\langle L_1(t_2) X(t_2) F(t_1) \rangle$$
$$+ L_0^{-1}(t_1) L_0^{-1}(t_2)\langle L_1(t_1) L_1(t_2) X(t_1) X(t_2) \rangle \tag{9.12}$$

Again, we have used the symbol $\langle \ \rangle$ to indicate mathematical expectation.

At this stage, it is necessary to make some approximations before we can proceed. These approximations are

$$\langle L_1(t_1) \, X(t_1) \, F(t_2) \rangle \cong \langle L_1(t_1) \rangle \langle X(t_1) \, F(t_2) \rangle$$
$$\langle L_1(t_2) \, X(t_2) \, F(t_1) \rangle \cong \langle L_1(t_2) \rangle \langle X(t_2) \, F(t_1) \rangle \qquad (9.13)$$
$$\langle L_1(t_1) \, L_1(t_2) X(t_1) \, X(t_2) \rangle \cong \langle L_1(t_1) \, L_1(t_2) \rangle \langle X(t_1) \, X(t_2) \rangle$$

It is shown by Samuels and Eringen [6] that these relations hold rigorously when the basic differential equation (9.5) contains only one white noise coefficient. They are approximately true when the coefficient processes $P_k(t)$ and the solution process have very widely separated spectra.

Using Eqs. (9.13) and noting from Eq. (9.8) that $\langle L_1(t) \rangle = 0$, Eq. (9.12) simplifies to

$$\langle X(t_1) \, X(t_2) \rangle = \langle F(t_1) \, F(t_2) \rangle + L_0^{-1}(t_1) \, L_0^{-1}(t_2) \langle L_1(t_1) \, L_1(t_2) \rangle \langle X(t_1) \, X(t_2) \rangle \qquad (9.14)$$

In terms of the original quantities in the basic differential equation (9.5), Eq. (9.14) has the explicit form

$$\Gamma_{XX}(t_1, t_2) = \Gamma_{FF}(t_1, t_2) + \int_{-\infty}^{\infty} \int_{-\infty}^{\infty} \langle K(t_1, s_1) \, K(t_2, s_2) \rangle \, \Gamma_{XX}(s_1, s_2) \, ds_1 \, ds_2 \qquad (9.15)$$

where

$$\langle K(t_1, s_1) \, K(t_2, s_2) \rangle$$
$$= \sum_{j,k=0}^{n} (-1)^{j+k} \frac{\partial^{j+k}}{\partial s_1^{\ j} \, \partial s_2^{\ k}} \{ \Gamma_{jk}(s_1, s_2) \, h(t_1, s_1) \, h(t_2, s_2) \} \qquad (9.16)$$

In the above, $h(t, s)$ is the weighting function associated with the deterministic differential operator $L_0(t)$ and

$$\Gamma_{jk}(s_1, s_2) = \langle P_j(s_1) \, P_k(s_2) \rangle \qquad (9.17)$$

Equation (9.15) defines an integral equation satisfied by the correlation function of the solution process $X(t)$. It is a convenient form for mean-square stability studies when the correlation functions $\Gamma_{jk}(s_1, s_2)$ are prescribed. If the cross-power spectral density functions associated with the coefficient processes are given, a more convenient form is the double Fourier transform of Eq. (9.15). We can write it in the form

$$\hat{\Gamma}_{XX}(\omega_1, \omega_2) = \hat{\Gamma}_{FF}(\omega_1, \omega_2) + \sum_{j,k=0}^{\infty} \frac{(-1)^{j+k}}{4\pi^2} \int_{-\infty}^{\infty} \int_{-\infty}^{\infty} \hat{h}(\omega_1) \, \hat{h}(\omega_2) \, v_1^{\ j} v_2^{\ k}$$
$$\times \hat{\Gamma}_{jk}(\omega_1 - v_1, \omega_2 - v_2) \, \hat{\Gamma}_{XX}(v_1, v_2) \, dv_1 \, dv_2 \qquad (9.18)$$

where the double Fourier transform of a function is denoted by a circumflex, for example

$$\hat{\Gamma}_{XX}(\omega_1, \omega_2) = \mathscr{F}\{\Gamma_{XX}(t_1, t_2)\}$$

$$= \int_{-\infty}^{\infty} \int_{-\infty}^{\infty} \exp[i(\omega_1 t_1 + \omega_2 t_2)]\, \Gamma_{XX}(t_1, t_2)\, dt_1\, dt_2 \qquad (9.19)$$

and the inverse transform is

$$\Gamma_{XX}(t_1, t_2) = \mathscr{F}^{-1}\{\hat{\Gamma}_{XX}(\omega_1, \omega_2)\}$$

$$= (1/4\pi^2) \int_{-\infty}^{\infty} \int_{-\infty}^{\infty} \exp[-i(t_1\omega_1 + t_2\omega_2)]\, \hat{\Gamma}_{XX}(\omega_1, \omega_2)\, d\omega_1\, d\omega_2$$

$$(9.20)$$

We note that $\hat{\Gamma}_{jk}(\omega_1, \omega_2)$ is proportional to the cross-power spectral density function of the coefficient processes.

Equation (9.15) and its transform (9.18) provide the basis for investigating the mean square stability of a class of random systems. In order to apply the above results to some practical examples, the analysis will be specialized to the following two cases.

Case 1. $P_k(t) \equiv 0$ for all $k \neq j$ and $P_j(t)$ is a white noise process with mean zero and correlation function $\Gamma_{jj}(\tau) = \pi S_0\, \delta(\tau)$. For this case, we see from Eq. (9.16) that

$$\langle K(t_1, s_1)\, K(t_2, s_2)\rangle = \pi S_0\, \frac{\partial^{2j}}{\partial s_1{}^j\, \partial s_2{}^j}\, \{\delta(s_1 - s_2)\, h(t_1 - s_1)\, h(t_2 - s_2)\} \quad (9.21)$$

Substituting the above into Eq. (9.15) and integrating by parts j times gives

$$\Gamma_{XX}(t_1, t_2) = \Gamma_{FF}(t_1, t_2) + \pi S_0 \int_{-\infty}^{\infty} \int_{-\infty}^{\infty} h(t_1 - s_1)\, h(t_2 - s_2)\, \delta(s_1 - s_2)$$

$$\times \frac{\partial^{2j}}{\partial s_1{}^j\, \partial s_2{}^j}\, [\Gamma_{XX}(s_1, s_2)]\, ds_1\, ds_2 \qquad (9.22)$$

The equation above leads to, upon setting $t_1 = t_2 = t$,

$$\langle X^2(t)\rangle = \langle F^2(t)\rangle + \pi S_0 \int_{-\infty}^{\infty} h^2(t - s)\langle(d^j X(s)/ds^j)^2\rangle\, ds \qquad (9.23)$$

In order to determine the mean square value $\langle X^2(t)\rangle$, we need to evaluate $\langle(d^j X(s)/ds^j)^2\rangle$. This can be found by differentiating Eq. (9.22) j times with respect to t_1 and j times with respect to t_2, setting $t_1 = t_2 = t$, and carrying out integration with respect to s_2. The result is

$$\langle [X^{(j)}(t)]^2\rangle = \langle [F^{(j)}(t)]^2\rangle + \pi S_0 \int_{-\infty}^{\infty} [h^{(j)}(t-s)]^2 \langle [X^{(j)}(s)]^2\rangle \, ds \qquad (9.24)$$

where the superscript (j) represents the jth derivative of a function with respect to its argument. This integral equation has the solution

$$\langle [X^{(j)}(t)]^2\rangle = \int_{-\infty}^{\infty} \langle [F^{(j)}(s)]^2\rangle \, \gamma(t-s) \, ds + \oint \frac{g(s)\exp(-ist)\,ds}{1 - \pi S_0 \mathscr{F}\{[h^{(j)}(t)]^2\}} \tag{9.25}$$

where

$$\gamma(t-s) = \mathscr{F}^{-1}\left\{ \frac{1}{\pi S_0 \mathscr{F}\{[h^{(j)}(t)]^2\}} \right\} \tag{9.26}$$

and $g(s)$ is an analytic function in the domain of analyticity of $\mathscr{F}\{[h^{(j)}(t)]^2\}$, $\oint$ being a closed contour integral in that domain.

The substitution of Eq. (9.25) into Eq. (9.23) gives the solution for $\langle X^2(t)\rangle$ in the form

$$\langle X^2(t)\rangle = \langle F^2(t)\rangle + \pi S_0 \left\{ \int_{-\infty}^{\infty} \frac{\mathscr{F}\{\langle [F^{(j)}(t)]^2\rangle\}\mathscr{F}\{h^2(t)\}\,e^{its}\,ds}{1 - \pi S_0 \mathscr{F}\{[h^{(j)}(t)]^2\}} \right.$$
$$\left. + \oint \frac{g(s)\mathscr{F}\{h^2(s)\}\,e^{its}\,ds}{1 - \pi S_0 \mathscr{F}\{[h^{(j)}(s)]^2\}} \right\} \tag{9.27}$$

While the actual evaluation of $\langle X^2(t)\rangle$ is generally difficult, the mean square stability can be examined in a fairly straightforward fashion based upon Eq. (9.27). It is seen that, assuming that the mean square input is bounded, the mean square output $\langle X^2(t)\rangle$ is a bounded function of t only if the roots of

$$1 - \pi S_0 \mathscr{F}\{[h^{(j)}(t)]^2\} = 0 \tag{9.28}$$

have nonpositive real parts. Hence, we have the following procedure for examining the mean square stability of this class of random systems.

a. Determine the deterministic weight function $h(t)$ associated with the operator $L_0(t)$ and find its jth derivative, $h^{(j)}(t)$.

b. Determine the roots of the algebraic equation (9.28). We note that S_0 is the power spectral density of the coefficient process $P_j(t)$.

c. If any of the roots of Eq. (9.28) have a positive real part, the system is not m.s. stable. Otherwise, the system is m.s. stable.

Example 9.1. Let us consider an RLC circuit whose resistance is perturbed by a white noise process. The circuit to be analyzed is shown in Fig. 9.1

where the randomly varying resistance $R(t)$ is modeled by

$$R(t) = R_0[1 + A(t)] \qquad (9.29)$$

In the above, R_0 is a constant and $A(t)$ is a white noise whose spectral density function is S_0.

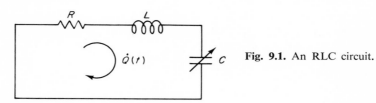

Fig. 9.1. An RLC circuit.

The associated differential equation is of the form

$$L \frac{d^2Q(t)}{dt^2} + R_0[1 + A(t)] \frac{dQ(t)}{dt} + \frac{1}{C} Q(t) = 0, \qquad t \geq 0 \qquad (9.30)$$

Letting $2\beta = R_0/L$ and $\omega_0{}^2 = 1/CL$, Eq. (9.30) can be written as

$$\ddot{Q}(t) + 2\beta \dot{Q}(t) + \omega_0{}^2 Q(t) = -2\beta A(t) \dot{Q}(t) \qquad (9.31)$$

The weighting function $h(t)$ associated with this example is

$$h(t) = i(\lambda_1 - \lambda_2)^{-1}[\exp(-i\lambda_1 t) - \exp(-i\lambda_2 t)], \qquad t \geq 0 \qquad (9.32)$$

where

$$\lambda_{1,2} = -i\beta \pm (\omega_0{}^2 - \beta^2)^{1/2} \qquad (9.33)$$

The m.s. stability of this random circuit can be discussed following the procedure outlined above. The Fourier transform of $[\dot{h}(t)]^2$ is needed and it is given by

$$\mathscr{F}\{[\dot{h}(t)]^2\} = \frac{1}{(\lambda_1 - \lambda_2)^2} \left[\frac{\lambda_1{}^2}{\omega + 2i\lambda_1} - \frac{2\lambda_1\lambda_2}{\omega + i(\lambda_1 + \lambda_2)} + \frac{\lambda_2{}^2}{\omega + 2i\lambda_2} \right] \qquad (9.34)$$

or, upon simplifying,

$$\mathscr{F}\{[\dot{h}(t)]^2\} = \frac{\omega^2 - 2\beta\omega + 2\omega_0{}^2}{\omega^3 + 6\beta\omega^2 + 4(\omega_0{}^2 + 2\beta^2)\omega + 8\beta\omega_0{}^2} \qquad (9.35)$$

In our case,

$$P_1(t) = -2\beta A(t)$$

The spectral density function of $P_1(t)$ is thus $4\beta^2 S_0$. Hence, Eq. (9.28) in this case becomes, upon rearranging,

$$\omega^3 + 2\beta(3 - 2\pi\beta S_0)\,\omega^2 + 4[\omega_0{}^2 + 2\beta^2(1 + 2\pi\beta S_0)]\,\omega$$
$$+ 8\beta\omega_0{}^2(1 + \pi\beta S_0) = 0 \tag{9.36}$$

Based upon the procedure given above, the m.s. stability of this random circuit requires that the roots of the equation above have negative real parts. Applying the Routh–Hurwitz Criterion [8], we find that the system is m.s. stable if and only if the following conditions are satisfied:

$$8\beta\omega_0{}^2(1 + \pi\beta S_0) > 0$$
$$(3 - 2\pi\beta S_0)[\omega_0{}^2 + 2\beta^2(1 + 2\pi\beta S_0)] > \omega_0{}^2(1 + \pi\beta S_0) \tag{9.37}$$

Noting that β is positive, the first condition is always satisfied. The second leads to

$$S_0 < [-(3 - 8x) \pm (158x^2 + 16x + 9)^{1/2}]/16\pi\beta x \tag{9.38}$$

where $x = (\beta/\omega_0)^2$. The m.s. stability region for this example is shown in Fig. 9.2.

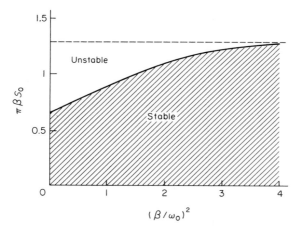

Fig. 9.2. Mean square stability region for random circuit (Example 9.1).

Case 2. $P_k(t) \equiv 0$ for all $k \neq j$ and $P_j(t)$ is a narrow band stochastic process. It is convenient to work with Eq. (9.18) in this case. Since $P_j(t)$ is a narrow band process, let us assume that its power spectral density can be approximated by a delta function. Hence, we can write

$$\hat{\Gamma}_{jj}(\omega_1 - v_1, \omega_2 - v_2) = 4\pi^2\Gamma_{jj}(0)\,\delta(\omega_2 - v_2)\,\delta(\omega_1 - v_1 + \omega_2 - v_2) \tag{9.39}$$

Upon substituting this expression into Eq. (9.18) and carrying out the integration, we have

$$\hat{\Gamma}_{XX}(\omega_1, \omega_2) = \hat{\Gamma}_{FF}(\omega_1, \omega_2) + \Gamma_{jj}(0)\,\omega_1{}^j\omega_2{}^j\,\hat{h}(\omega_1)\,\hat{h}(\omega_2)\,\hat{\Gamma}_{XX}(\omega_1, \omega_2) \quad (9.40)$$

Thus,

$$\hat{\Gamma}_{XX}(\omega_1, \omega_2) = \hat{\Gamma}_{FF}(\omega_1, \omega_2)[1 - \Gamma_{jj}(0)\,\omega_1{}^j\omega_2{}^j\,\hat{h}(\omega_1)\,\hat{h}(\omega_2)]^{-1} \quad (9.41)$$

Its inverse Fourier transform is the particular solution of the integral equation

$$\Gamma_{XX}(t_1, t_2) = \Gamma_{FF}(t_1, t_2) + \Gamma_{jj}(0) \int_{-\infty}^{\infty} \int_{-\infty}^{\infty} h^{(j)}(t_1 - s_1)\,h^{(j)}(t_2 - s_2)$$
$$\times\,\Gamma_{XX}(s_1, s_2)\,ds_1\,ds_2 \quad (9.42)$$

The above integral equation has as its complementary solution

$$[\Gamma_{XX}(t_1, t_2)]_c = \oiint \frac{g(\omega_1, \omega_2)\exp[-i(t_1\omega_1 + t_2\omega_2)]\,d\omega_1\,d\omega_2}{1 - \Gamma_{jj}(0)\,\omega_1{}^j\omega_2{}^j\,\hat{h}(\omega_1)\,\hat{h}(\omega_2)} \quad (9.43)$$

The solution for $\langle X^2(t)\rangle$ is therefore given by, using Eqs. (9.41) and (9.43),

$$\langle X^2(t)\rangle = \frac{1}{4\pi^2} \int_{-\infty}^{\infty} \int_{-\infty}^{\infty} \frac{\hat{\Gamma}_{FF}(\omega_1, \omega_2)\exp[-it(\omega_1 + \omega_2)]\,d\omega_1\,d\omega_2}{1 - \Gamma_{jj}(0)\,\omega_1{}^j\omega_2{}^j\,\hat{h}(\omega_1)\,\hat{h}(\omega_2)}$$
$$+ \oiint \frac{g(\omega_1, \omega_2)\exp[-it(\omega_1 + \omega_2)]\,d\omega_1\,d\omega_2}{1 - \Gamma_{jj}(0)\,\omega_1{}^j\omega_2{}^j\,\hat{h}(\omega_1)\,\hat{h}(\omega_2)} \quad (9.44)$$

It is again difficult to obtain the actual mean square value $\langle X^2(t)\rangle$. However, the form of Eq. (9.44) enables us to develop a simple criterion for m.s. stability. Now, the complementary solution is the important one for stability considerations. Let us write the second term of Eq. (9.44) in the form

$$\sum_{r=1}^{n} \alpha_r \exp[-it(\omega_{1r} + \omega_{2r})]$$

where ω_{1r} and ω_{2r}, $r = 1, 2, \ldots, n$, are the roots of

$$1 - \Gamma_{jj}(0)\,\omega_1{}^j\omega_2{}^j\,\hat{h}(\omega_1)\,\hat{h}(\omega_2) = 0$$
$$\omega_2{}^j\,\hat{h}(\omega_2)[\,j\omega_1{}^{j-1}\hat{h}(\omega_1) + \omega_1{}^j\,d\hat{h}(\omega_1)/d\omega_1] = 0 \quad (9.45)$$

We note that these are algebraic equations which can be solved for ω_1 and ω_2. The m.s. stability then becomes: The system is m.s. unstable if $(\omega_{1r} + \omega_{2r})$ has a positive imaginary part. Otherwise, the system is m.s. stable.

Example 9.2. Consider the RLC circuit again as shown in Fig. 9.1. In this example, we assume that the capacitance is randomly varying and is expressed in the form

$$C(t) = C_0[1 + A(t)]^{-1} \tag{9.46}$$

where C_0 is a constant and $A(t)$ is a narrow band process with correlation function $\Gamma_{AA}(\tau)$. The differential equation in this case is

$$\ddot{Q}(t) + 2\beta \dot{Q}(t) + \omega_0^2 Q(t) = -\omega_0^2 A(t) Q(t) \tag{9.47}$$

where $2\beta = R/L$ and $\omega_0^2 = 1/LC_0$.

The weighting function $h(t)$ is for this example the same as the one given in Example 9.1. Its Fourier transform is

$$\hat{h}(\omega) = -\frac{1}{(\omega - \lambda_1)(\omega - \lambda_2)} \tag{9.48}$$

where λ_1 and λ_2 are defined by Eq. (9.33). For this example, Eqs. (9.45) have the forms

$$1 - \frac{\omega_0^4 \Gamma_{AA}(0)}{(\omega_1 - \lambda_1)(\omega_1 - \lambda_2)(\omega_2 - \lambda_1)(\omega_2 - \lambda_2)} = 0$$

$$\frac{1}{(\omega_2 - \lambda_1)(\omega_2 - \lambda_2)} \left[\frac{2(\omega_1 + i\beta)}{(\omega_1 - \lambda_1)^2 (\omega_1 - \lambda_2)^2} \right] = 0 \tag{9.49}$$

We have, from the second equation,

$$\omega_1 = -i\beta \tag{9.50}$$

The substitution of this relation into the first of Eqs. (9.49) gives

$$\omega_2^2 - (\lambda_1 + \lambda_2) \omega_2 + \lambda_1 \lambda_2 + \omega_0^4 \Gamma_{AA}(0)/(\omega_0^2 - \beta^2) = 0 \tag{9.51}$$

which has the solutions

$$\omega_2 = -i\beta \pm [(\omega_0^2 - \beta^2) - \omega_0^4 \Gamma_{AA}(0)/(\omega_0^2 - \beta^2)]^{1/2} \tag{9.52}$$

Equations (9.50) and (9.52) now give

$$\gamma_{1,2} = -2i\beta \pm [(\omega_0^2 - \beta^2) - \omega_0^4 \Gamma_{AA}(0)/(\omega_0^2 - \beta^2)]^{1/2} \tag{9.53}$$

where $\gamma_r = \omega_{1r} + \omega_{2r}$, $r = 1, 2$. The regions of m.s. stability can now be determined. If either γ_1 or γ_2 has positive imaginary part, the system is

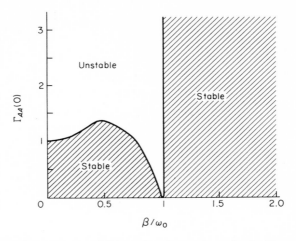

Fig. 9.3. Mean square stability region for random circuit (Example 9.2).

m.s. unstable. Otherwise, the system is m.s. stable. Figure 9.3 shows the stability plot for this example.

We remark that Samuels [7] considered the same example; however, our result is at variance with his. It appears that Samuel's result is invalidated by some algebraic errors in the derivation.

Samuels [9] has also investigated the m.s. stability properties of systems having Gaussian parametric variations. Unfortunately, his results in this investigation are open to question. As pointed out by Bogdanoff and Kozin [10], the moment equations in these cases are not separable; hence, it is not possible to isolate, say, the second-order moments and to discuss its m.s. stability properties. This inseparability of moments was demonstrated in Example 8.7.

9.1.2. Equations of the Ito Type

The m.s. stability of a system having a white noise coefficient was considered in Example 9.1. We observe, however, stability studies of systems of this type can be readily carried out using the moment equations derived for equations of the Ito type. We see from Section 8.4 that, for Ito equations, it is possible to derive a useful set of differential equations satisfied by the moments of the solution processes, and the stability of these moment equations lead directly to moment stabilities of the associated random systems. Hence, moment stability problems can be investigated based upon well developed stability theorems of deterministic systems. Let us consider several specific cases.

Example 9.3. Consider the simple first-order system

$$\dot{X}(t) + [a + W_1(t)] X(t) = 0, \qquad t \geq 0 \qquad (9.54)$$

where $W_1(t)$ is a zero-mean white-noise process with covariance $2D_{11} \, \delta(\tau)$. This system has been considered in Example 8.4. If we let

$$m_j(t) = E\{X^j(t)\} \qquad (9.55)$$

the deterministic equations governing $m_j(t)$ are given in Example 8.4. They are

$$\dot{m}_j(t) = -j[a - (j-1) D_{11}] m_j(t), \qquad j = 1, 2, \ldots \qquad (9.56)$$

Therefore, the moment stability properties for this system can be directly inferred from the equations above. It is stable in the mean if

$$a \geq 0 \qquad (9.57)$$

and stable in mean square if

$$a - D_{11} \geq 0 \qquad (9.58)$$

Conditions for stability in higher-order moments can also be written down easily. Clearly, the system is stable in the jth moment if

$$a - (j-1) D_{11} \geq 0 \qquad (9.59)$$

or

$$1 + a/D_{11} \geq j \qquad (9.60)$$

Example 9.4. Let us now consider a second-order system studied in Example 8.6.

$$\ddot{X}(t) + a_2 \dot{X}(t) + [a_1 + W_1(t)] X(t) = W_2(t) \qquad (9.61)$$

Let

$$\mathbf{X}(t) = \begin{bmatrix} X(t) \\ \dot{X}(t) \end{bmatrix} \qquad (9.62)$$

The moment equations associated with $\mathbf{X}(t)$ have been obtained. If we let

$$m_{jk}(t) = E\{X^j(t) \, \dot{X}^k(t)\} \qquad (9.63)$$

it is seen from Example 8.6 that the equations for the first-order moments are

$$\dot{m}_{10}(t) = m_{01}(t), \qquad \dot{m}_{01}(t) = -a_1 m_{10}(t) - a_2 m_{01}(t) \qquad (9.64)$$

and, for the second-order moments,

$$\dot{m}_{20} = 2m_{11}, \qquad \dot{m}_{11} = -a_1 m_{20} - a_2 m_{11} + m_{02}$$
$$\dot{m}_{02} = 2(D_{11} m_{20} - a_1 m_{11} - a_2 m_{02} + D_{22} - 2D_{12} m_{10}) \tag{9.65}$$

Equations (9.64) lead to the conditions for stability in the mean. Since these equations are linear, the solutions of $m_{10}(t)$ and $m_{01}(t)$ are expressible in the forms

$$m_{10}(t) = \sum_j \alpha_j \exp(\lambda_j t), \qquad m_{01}(t) = \sum_j \beta_j \exp(\lambda_j t) \tag{9.66}$$

where α_j and β_j are constants. In order that the first moments be (asymptotic) stable, we require that the real parts of λ_j be negative. Upon substituting Eqs. (9.66) into Eqs. (9.64), we find that, for nontrivial solutions, λ_j must satisfy the characteristic equation

$$\begin{vmatrix} \lambda & -1 \\ a_1 & a_2 + \lambda \end{vmatrix} = 0 \tag{9.67}$$

or

$$\lambda^2 + a_2 \lambda + a_1 = 0 \tag{9.68}$$

In order that its roots have negative real parts, the Routh–Hurwitz criterion requires that

$$a_1 > 0 \qquad \text{and} \qquad a_2 > 0 \tag{9.69}$$

These are then the requirements for asymptotic mean stability.

Assuming that the conditions (9.69) are satisfied, the asymptotic stability in mean square can be studied by means of Eqs. (9.65) by setting the non-homogeneous terms

$$D_{22} - 2D_{12} m_{10}$$

in the third equation zero. In doing so, we have implicitly assumed that the quantities D_{12} and D_{22} are finite.

The requirements for asymptotic mean square stability can be established in the same manner. The characteristic equation associated with the second-order moments is

$$\begin{vmatrix} \lambda & -2 & 0 \\ a_1 & a_2 + \lambda & -1 \\ -2D_{11} & 2a_1 & 2a_2 + \lambda \end{vmatrix} = 0 \tag{9.70}$$

or

$$\lambda^3 + 3a_2 \lambda^2 + 2(a_2^2 + 2a_1) \lambda + 4(a_1 a_2 - D_{11}) = 0 \tag{9.71}$$

According to the Routh–Hurwitz criterion, the second moments are stable if all the coefficients in the cubic equation above are positive and the inequality

$$2(a_2{}^2 + 2a_1)(3a_2) > 4(a_1a_2 - D_{11}) \tag{9.72}$$

is satisfied. Hence, the conditions for asymptotic mean square stability are

$$a_1 > 0, \qquad a_2 > 0 \tag{9.73}$$

which are conditions for mean stability, and

$$a_1a_2 > D_{11} \tag{9.74}$$

It is clear that inequalities (9.73) and (9.74) imply inequality (9.72).

9.1.3. Some Other Approaches

Some approximate methods were used in Chapter 8 for obtaining moment properties of the solution process. However, the application of these approximations to the study of moment stabilities must be approached with caution. Take, for example, the perturbation technique approach discussed in Section 8.5 and, specifically, the moment equations derived in Section 8.5.2. These moment equations are in general valid only for a finite interval of time. Unless uniform convergence of the perturbation expansion in the time variable can be insured, these equations are not to be relied upon to yield usable moment stability results since moment stabilities deal with *asymptotic* moment behaviors.

This point is illustrated by the example below.

Example 9.5. Consider a scalar first-order equation

$$\dot{X}(t) + (a + \varepsilon^2 A(t))X(t) = 0 \tag{9.75}$$

where ε is a small parameter and $A(t)$ is a stationary Gaussian process with mean zero and covariance

$$\Gamma_{AA}(\tau) = \sigma^2 e^{-|\tau|} \tag{9.76}$$

Let us first discuss its mean stability using the approximate moment equation derived in Section 8.5.2. As seen from Eq. (8.71), the mean $\langle X(t) \rangle$ satisfies

$$[L_0 + \varepsilon\langle L_1 \rangle + \varepsilon^2(\langle L_1 \rangle L_0^{-1}\langle L_1 \rangle - \langle L_1 L_0^{-1}L_1 \rangle + \langle L_2 \rangle) + O(\varepsilon^3)]\langle X(t) \rangle = 0 \tag{9.77}$$

In our case,

$$L_0 = d/dt + a, \qquad L_1 \equiv 0, \qquad L_2 = A(t), \qquad \langle L_2 \rangle = 0 \tag{9.78}$$

Hence, Eq. (9.77) reduces to

$$(d/dt + a + O(\varepsilon^3))\langle X(t) \rangle = 0 \tag{9.79}$$

Clearly, this approximation predicts that, up to the second-order terms in ε, the system is stable in the mean if

$$a \geq 0 \tag{9.80}$$

and the mean stability is independent of the properties of the coefficient process $A(t)$.

We now look at the exact situation. Being a first-order differential equation, an explicit solution for $X(t)$ is available and it takes the form

$$X(t) = \exp\left[-at - \varepsilon^2 \int_0^t A(\tau)\, d\tau\right] \tag{9.81}$$

where the initial condition $X(0) \equiv 1$ is used for convenience. Since

$$E\left\{\int_0^t A(\tau)\, d\tau\right\} = 0 \qquad \text{and} \qquad E\left\{\left(\int_0^t A(\tau)\, d\tau\right)^2\right\} = 2\sigma^2[t + (e^{-t} - 1)] \tag{9.82}$$

we have

$$\langle X(t) \rangle = \exp[-at + \varepsilon^4\sigma^2 t + \varepsilon^4\sigma^2(e^{-t} - 1)] \tag{9.83}$$

Therefore, this equation indicates that for stability in the mean we require

$$a - \varepsilon^4\sigma^2 \geq 0 \tag{9.84}$$

which is not predicted by the approximate equation (9.79).

A more dramatic result can be achieved by letting $A(t)$ in Eq. (9.75) be a Wiener process, that is, $A(t)$ is Gaussian with mean zero and covariance

$$\Gamma_{AA}(t_1, t_2) = \sigma^2 \min(t_1, t_2) \tag{9.85}$$

For this case, the exact solution yields

$$\langle X(t) \rangle = \exp\left[-at + \frac{\varepsilon^4\sigma^2}{6} t^3\right] \tag{9.86}$$

which implies that the mean is always unstable.

The results given above do not, of course, completely invalidate the perturbation procedure in stability studies. In fact, we observe that the results given by Eqs. (9.80) and (9.84) are consistent since the difference is in the fourth-order terms in ε. On the other hand, the result using perturbation approach can be misleading in the example above if we compare it with Eq. (9.86).

Some moment stability results for random systems can also be obtained using moment equation derived through truncated hierarchy techniques discussed in Section 8.6. However, because of a lack of sound mathematical basis in the development, the merit of hierarchy methods in stability studies is not clear at this time.

Finally, we remark in passing that approximation schemes have been used extensively in the study of moment stabilities of nonlinear systems, mainly with Gaussian white noise coefficients. The application of the method of averaging of Bogoliubov (see Bogoliubov and Mitropolskii [11]) and the linearization techniques due to Krylov and Bogoliubov [12] to these problems have been explored by Kuznetsov *et al.* [13].

9.2. Lyapunov Stability

It appears that the use of the concept of Lyapunov stability in stochastic stability studies was initiated independently by Bertram and Sarachik [14] and Kats and Krasovskii [15]. Motivated by the powerful direct method of Lyapunov, they have considered the stabilities of stochastic systems in the Lyapunov sense by carrying out a transition from the deterministic Lyapunov stability concepts to their stochastic counterparts.

We follow the work of Bertram and Sarachik in our development. In order to show the analogy between deterministic and stochastic stability definitions, let us first review the concepts of Lyapunov stability for deterministic systems.

Consider a system of deterministic differential equations represented by

$$\dot{\mathbf{x}}(t) = \mathbf{f}(\mathbf{x}, t) \tag{9.87}$$

where $\mathbf{x}(t)$ is an n-dimensional vector. The function $\mathbf{f}$ is a continuous vector function satisfying a Lipschitz condition and

$$\mathbf{f}(\mathbf{0}, t) = \mathbf{0} \tag{9.88}$$

for all t. Hence, $\mathbf{x}(t) \equiv \mathbf{0}$ is an equilibrium or null solution whose stability properties are under discussion.

In the second method of Lyapunov, a central role is played by a scalar function $v(\mathbf{x}, t)$ having the following properties:

a. $v(\mathbf{x}, t)$ is continuous in both $\mathbf{x}$ and t and has partial derivatives in these variables in a certain region R about the equilibrium point.

b. $v(\mathbf{x}, t)$ is nonnegative in the region R and vanishes only at the origin. The origin is thus an isolated minimum of $v(\mathbf{x}, t)$.

c. The total time derivative of $v(\mathbf{x}, t)$,

$$\dot{v} = \partial v/\partial t + \sum_{i=1}^{n} f_i \, \partial v/\partial x_i \tag{9.89}$$

is nonpositive in R. The quantities f_i and x_i denote, respectively, the components of $\mathbf{f}$ and $\mathbf{x}$.

The function $v(\mathbf{x}, t)$ defined above is called a *Lyapunov function*. The basic idea of the Lyapunov's direct method is to investigate the stability of the equilibrium solution by examining the properties of a Lyapunov function without the knowledge of the solutions. We first give some definitions.

Definition. Let the initial condition be $\mathbf{x}(t_0) = \mathbf{x}_0$. The equilibrium solution, $\mathbf{x}(t) = \mathbf{0}$, of Eq. (9.87) is *stable* if, given $\varepsilon > 0$, there exists a $\delta(\varepsilon, t_0) > 0$ such that for any initial condition whose norm satisfies

$$\| \mathbf{x}_0 \| < \delta \tag{9.90}$$

the norm of the solution satisfies the relation

$$\sup_{t \geq t_0} \| \mathbf{x}(t) \| < \varepsilon \tag{9.91}$$

In what follows and unless stated otherwise, the norm $\| \mathbf{x} \|$ will stand for

$$\| \mathbf{x} \| = (\mathbf{x} \cdot \mathbf{x})^{1/2}, \tag{9.92}$$

the simple distance norm. This definition states essentially that a system is stable if it remains close to the equilibrium state under a slight perturbation from the equilibrium state.

Definition. The equilibrium solution is *asymptotically stable* if, in addition to being stable, for every t_0 there exists a $\delta'(t_0) > 0$ such that $\| \mathbf{x}_0 \| \leq \delta'$ implies

$$\lim_{t \to \infty} \| \mathbf{x}(t) \| = 0 \tag{9.93}$$

Definition. If, in addition to being stable, the equilibrium solution has the property (9.93) for any finite initial state $\mathbf{x}_0$, then it is said to be *globally stable*.

We state below without proof the basic stability theorems due to Lyapunov. (For proof, see, for example, La Salle and Lefschetz [2]).

Theorem 9.2.1 (Stability Theorem). If a Lyapunov function $v(\mathbf{x}, t)$ exists in some region R of the origin, then the origin or the equilibrium solution is stable.

Intuitively, this theorem resembles somewhat the energy concept in stability studies. If the Lyapunov functions are viewed as extensions of the energy idea, Theorem 9.2.1 states that the equilibrium state of a system is stable if its energy close to the equilibrium state is always decreasing.

Theorem 9.2.2 (Asymptotic Stability Theorem). The stability is asymptotic if a Lyapunov function $v(\mathbf{x}, t)$ exists in some region R of the origin and $\dot{v}(\mathbf{x}, t)$ is negative definite, that is, $\dot{v}(\mathbf{x}, t)$ is nonpositive everywhere in R and vanishes only at the origin.

Depending upon the choice of the mode of stochastic convergence, the transition of the stability concepts outlined above to stochastic stability concepts can be accomplished in a variety of ways. Stability definitions related to the convergence in mean square or in the norm defined by Eq. (9.92) were considered by Bertram and Sarachik [14].

Let us now consider a system of random differential equations described by

$$\dot{\mathbf{X}}(t) = \mathbf{f}(\mathbf{X}, \mathbf{A}(t), t) \tag{9.94}$$

where $\mathbf{A}(t)$ represents random parametric variations. We shall assume that $\mathbf{f}(\mathbf{0}, \mathbf{A}(t), t) \equiv \mathbf{0}$ so that $\mathbf{X}(t) \equiv \mathbf{0}$ is again our equilibrium solution.

Definition. The equilibrium solution $\mathbf{X}(t) = \mathbf{0}$ of Eq. (9.94) is *stable in the norm* if, given $\varepsilon > 0$, there exists a $\delta(\varepsilon) > 0$ such that for any (deterministic) initial condition whose norm satisfies

$$\| \mathbf{x}_0 \| < \delta \tag{9.95}$$

the norm of the solution process satisfies

$$E\{\| \mathbf{X}(t) \|\} < \varepsilon \tag{9.96}$$

for all $t \geq t_0$. It is clear that this definition is a simple stochastic translation of the stability definition for deterministic systems.

The definition for asymptotic stability in the norm and global stability in the norm follow analogously.

Definition. The equilibrium solution is *asymptotically stable in the norm* if, in addition to being stable in the norm, for every t_0 there exists a $\delta'(t_0) > 0$ such that $\| \mathbf{x}_0 \| < \delta'$ implies

$$\lim_{t \to \infty} E\{\| \mathbf{X}(t) \|\} = 0 \tag{9.97}$$

Definition. If, in addition to being stable in the norm, the equilibrium solution has the property (9.97) for any finite initial state $\mathbf{x}_0$, then it is *globally stable in the norm*.

Let us assume that $\mathbf{A}(t)$ in Eq. (9.94) possesses well behaved sample functions. The following theorems are obtained.

Theorem 9.2.3 (Stability in the Norm). If there exists a Lyapunov function $V(\mathbf{X}, t)$ defined over the state space and it satisfies the conditions

1. $V(\mathbf{X}, t)$ is continuous in both $\mathbf{X}$ and t and its first partial derivatives in these variables exist.
2. $V(\mathbf{0}, t) = 0$ and $V(\mathbf{X}, t) \geq a \| \mathbf{X} \|$ for some $a > 0$. $\qquad$ (9.98)
3. $E\{\dot{V}(\mathbf{X}, t)\} \leq 0$.

then the equilibrium solution of Eq. (9.94) is stable in the norm.

Proof. Let $v_0 = V(\mathbf{x}_0, t)$. We may write

$$V(\mathbf{X}, t) = v_0 + \int_{t_0}^{t} \dot{V}(\mathbf{X}, \tau) \, d\tau \tag{9.99}$$

where v_0 is deterministic since $\mathbf{X}(t_0) = \mathbf{x}_0$ is assumed to be deterministic. Taking expectation of Eq. (9.99) yields

$$E\{V(\mathbf{X}, t)\} = v_0 + \int_{t_0}^{t} E\{\dot{V}(\mathbf{X}, \tau)\} \, d\tau \tag{9.100}$$

It follows from Eq. (9.98) that

$$aE\{\| \mathbf{X} \|\} \leq v_0 + \int_{t_0}^{t} E\{\dot{V}(\mathbf{X}, \tau)\} \, d\tau \tag{9.101}$$

and, from the condition $E\{\dot{V}(\mathbf{X}, t)\} \leq 0$,

$$aE\{\| \mathbf{X} \|\} \leq v_0 \qquad (9.102)$$

In view of the continuity and the existence of the partial derivatives for $V(\mathbf{X}, t)$, this function satisfies a local Lipschiz condition, that is

$$V(\mathbf{X}, t) \leq k \| \mathbf{X} \| \qquad (9.103)$$

for some positive and finite k and $\mathbf{X}$ such that $0 < \| \mathbf{X} \| < \gamma$. Then, for any $\varepsilon > 0$, if we choose $\delta(\varepsilon)$ in such a way that

$$\delta(\varepsilon) = \inf(a\varepsilon/k, \gamma) \qquad (9.104)$$

It follows that for $\| \mathbf{x}_0 \| < \delta$ we have

$$a\varepsilon \geq k\delta > k \| \mathbf{x}_0 \| \geq v_0 \geq aE\{\| \mathbf{X} \|\} \qquad (9.105)$$

Hence,

$$E\{\| \mathbf{X} \|\} < \varepsilon \qquad (9.106)$$

and the proof is complete.

Bertram and Sarachik also give sufficient conditions for asymptotic stability in the norm and global stability in the norm. Without repeating the proofs, we state the following results.

Theorem 9.2.4 (Asymptotic Stability in the Norm). If there exists a Lyapunov function $V(\mathbf{X}, t)$ which satisfies, in addition to Conditions 1, 2, and 3 of Theorem 9.2.3, the condition

$$E\{\dot{V}(\mathbf{X}, t)\} < -g(\| \mathbf{X} \|) \qquad (9.107)$$

where $g(0) = 0$ and $g(\| \mathbf{X} \|)$ is an increasing function, then the equilibrium solution is asymptotically stable in the norm.

Theorem 9.2.5 (Global Stability in the Norm). If there exists a Lyapunov function $V(\mathbf{X}, t)$ which satisfies all the conditions in Theorem 9.2.4 and the additional condition

$$V(\mathbf{X}, t) < h(\| \mathbf{X} \|) \qquad (9.108)$$

where $h(0) = 0$ and $h(\| \mathbf{X} \|)$ is an increasing function, then the equilibrium solution is globally stable in the norm.

Although the theorems stated above hold for general random systems characterized by Eq. (9.94), useful results for explicit systems are still difficult to obtain without imposing restrictions on the general system. Bertram and Sarachik applied their results to a system with piecewise constant random parameters as described below.

Example 9.6. Consider a linear system characterized by the n-dimensional vector equation

$$\dot{\mathbf{X}}(t) = A_k \mathbf{X}(t), \qquad t_k \leq t < t_{k+1}, \qquad k = 0, 1, 2, \ldots \quad (9.109)$$

where A_k, $k = 0, 1, 2, \ldots$, is an $n \times n$ matrix with random but constant elements. Thus, the solution of Eq. (9.109) has the form

$$\mathbf{X}(t) = \Phi_k(t - t_k) \mathbf{X}(t_k), \qquad t_k \leq t < t_{k+1}, \quad (9.110)$$

where the transition matrix can be written as

$$\Phi_k(t - t_k) = \exp[A_k(t - t_k)] = \sum_{j=0}^{\infty} (A_k)^j (t - t_k)^j / j! \quad (9.111)$$

In terms of the initial condition $\mathbf{X}(t_0) = \mathbf{x}_0$, Eq. (9.110) can also be written in the form

$$\mathbf{X}(t) = \Phi_k(t - t_k) \Phi_{k-1}(t_k - t_{k-1}) \cdots \Phi_0(t_1 - t_0) \mathbf{X}(t_0), \qquad t_k \leq t < t_{k+1} \quad (9.112)$$

Let us now define a Lyapunov function by

$$V(\mathbf{X}, t) = \mathbf{X}^{\mathrm{T}}(t) Q \mathbf{X}(t) \quad (9.113)$$

where Q is a constant positive-definite matrix. We have

$$\dot{V}(\mathbf{X}, t) = \dot{\mathbf{X}}^{\mathrm{T}}(t) Q \mathbf{X}(t) + \mathbf{X}^{\mathrm{T}}(t) Q \dot{\mathbf{X}}(t) \quad (9.114)$$

and, using Eq. (9.109),

$$\dot{V}(\mathbf{X}, t) = \mathbf{X}^{\mathrm{T}}(t)[A_k^{\mathrm{T}}(t) Q + Q A_k(t)] \mathbf{X}(t), \qquad t_k \leq t < t_{k+1} \quad (9.115)$$

Hence,

$$E\{\dot{V}(\mathbf{X}, t)\} = E\{\mathbf{X}^{\mathrm{T}}(t)[A_k^{\mathrm{T}}(t) Q + Q A_k(t)] \mathbf{X}(t)\}$$
$$= E\{\mathbf{X}^{\mathrm{T}}(t_k) \Phi_k^{\mathrm{T}}(t - t_k)(A_k^{\mathrm{T}}(t) Q + Q A_k(t)) \Phi_k(t - t_k) \mathbf{X}(t_k)\}$$
$$t_k \leq t < t_{k+1} \quad (9.116)$$

Let the initial state $\mathbf{x}_0$ be finite, then a sufficient condition for (global) stability in the norm is just that

$$E\{\Phi_k{}^{\mathrm{T}}(t - t_k)(A_k{}^{\mathrm{T}}(t) Q + Q A_k(t)) \Phi_k(t - t_k)\} \tag{9.117}$$

be negative definite for each k.

As an illustration, consider a system which is governed in each interval $t_k \leq t < t_{k+1}$ by the differential equations

$$\dot{X}_1(t) = -X_1(t) + X_2(t), \qquad \dot{X}_2(t) = -X_2(t) \tag{9.118}$$

with probability p and

$$\dot{X}_1(t) = X_2(t), \qquad \dot{X}_2(t) = 0 \tag{9.119}$$

with probability $1 - p$. We assume that the selection of Eq. (9.118) and Eq. (9.119) is independent from one time interval to another. It is clear that Eq. (9.119) represents an unstable system.

In the form of Eq. (9.109), the situation above corresponds to the case where

$$A_k = \begin{bmatrix} -1 & 1 \\ 0 & -1 \end{bmatrix} \tag{9.120}$$

with probability p and

$$A_k = \begin{bmatrix} 0 & 1 \\ 0 & 0 \end{bmatrix} \tag{9.121}$$

with probability $(1 - p)$.

A sufficient condition for global stability in the norm can be found by means of Eq. (9.117). In this case, the transition matrix is

$$\Phi_k(t) = \begin{bmatrix} e^{-t} & te^{-t} \\ 0 & e^{-t} \end{bmatrix} \tag{9.122}$$

with probability p and

$$\Phi_k(t) = \begin{bmatrix} 1 & t \\ 0 & 1 \end{bmatrix} \tag{9.123}$$

with probability $(1 - p)$. The substitution of Eqs. (9.120), (9.121), (9.122), and (9.123) into Eq. (9.117), with $Q = I$, the identity matrix, shows that Eq. (9.117) is negative definite if

$$p > 1/(1 + e^{-2\tau}), \qquad \tau = t - t_k \tag{9.124}$$

which guarantees global stability in the norm. This example illustrates that, even though the system is unstable during those time intervals in which it is governed by Eqs. (9.119), it is still possible to have global stability in the norm.

We remark that a sharper result than that given by Eq. (9.124) may be obtained with a more suitable choice of Q. One can show, in fact, that the system considered here is global stable in the norm whenever τ is finite and $p > 0$. The proof of this result is left as an exercise.

References

1. R. Bellman, *Stability Theory of Differential Equations*. Dover, New York, 1969.
2. J. La Salle and S. Lefschetz, *Stability by Lyapunov's Direct Method with Applications*. Academic Press, New York, 1961.
3. H. J. Kushner, *Stochastic Stability and Control*. Academic Press, New York, 1967.
4. F. Kozin, A survey of stability of stochastic systems. *Automatica Internat. J. Automat. Control Automation* **5**, 95–112 (1969).
5. A. Rosenbloom, Analysis of linear systems with randomly time-varying parameters. *Proc. Symp. Information Networks, Polytech. Inst. of Brooklyn, 1954*, pp. 145–153.
6. J. C. Samuels and A. C. Eringen, On stochastic linear systems. *J. Math. and Phys.* **38**, 93–103 (1969).
7. J. C. Samuels, On the mean square stability of random linear systems. *IRE Trans. Circuit Theory* **CT-6** (Special Suppl.), 248–259 (1959).
8. D. K. Cheng, *Analysis of Linear Systems*, p. 280. Addison-Wesley, Reading, Massachusetts, 1961.
9. J. C. Samuels, Theory of stochastic linear systems with Gaussian parameter variations. *J. Acoust. Soc. Amer.* **33**, 1782–1786 (1961).
10. J. L. Bogdanoff and F. Kozin, Moments of the output of linear random systems. *J. Acoust. Soc. Amer.* **34**, 1063–1068 (1962).
11. N. N. Bogoliubov and Y. A. Mitropolskii, *Asymptotic Methods in the Theory of Non-linear Oscillations*. Gordon & Breach, New York, 1961.
12. N. Krylov and N. N. Bogoliubov, *Introduction to Non-linear Mechanics: Asymptotic and Approximate Methods* (Ann. Math. Studies No. 11). Princeton Univ. Press, Princeton, New Jersey, 1947.
13. P. I. Kuznetsov, R. Stratonovich, and V. Tikhonov, *Non-linear Transformations of Stochastic Processes*. Pergamon, Oxford, 1965.
14. J. E. Bertram and P. E. Sarachik, Stability of circuits with randomly time-varying parameters. *IRE Trans. on Circuit Theory* **CT-6** (Special Suppl.), 260–270 (1959).
15. I. I. Kats and N. N. Krasovskii, On the stability of systems with random parameters. *Prikl. Mat. Meh.* **24**, 809–815 (1960).
16. A. Z. Akcasu, Mean square instability in boiling reactors. *Nucl. Sci. Eng.* **10**, 337–345 (1961).

PROBLEMS

9.1. Problem 8.1 is essentially a problem on moment stability. Discuss the solution of Problem 8.1 in this light. Is mean stability a necessary condition for m.s. stability?

9.2. Verify the result of Example 9.1 using the Ito equation approach discussed in Section 9.1.2.

9.3. Starting from Eq. (9.53), give details of derivation leading to the stability plot shown in Fig. 9.3.

9.4. Using the integral equation approach discussed in Sec. 9.1.1, consider an RLC circuit represented by

$$L\ddot{Q}(t) + R_0[1 + A(t)] \dot{Q}(t) + (1/C) Q(t) = 0, \qquad t \geq 0$$

where $A(t)$ is a narrow band stochastic process. Show that the system is always m.s. unstable.

9.5. Consider a second-order system characterized by

$$\ddot{X}(t) + 2A(t) \dot{X}(t) + X(t) = W(t), \qquad t \geq 0$$

where $W(t)$ is a Gaussian white noise with mean zero and correlation function $2D\,\delta(\tau)$. The process $A(t)$ is Gaussian with mean a and covariance $\sigma^2 \exp(-|\tau|)$. We assume that $W(t)$ and $A(t)$ are uncorrelated.

If $A(t)$ is considered small and slowly time varying, show that $X(t)$ can be represented by (see Akcasu [16])

$$X(t) = U(t) \exp\left[-\int_0^t A(s)\,ds\right]$$

where, as an approximation, $U(t)$ satisfies

$$\dot{U}(t) + U(t) = W(t) \exp\left[\int_0^t A(s)\,ds\right]$$

Determine the conditions for m.s. stability of the solution process $X(t)$.

9.6. Consider a random system represented by

$$\ddot{X}(t) - [a + W(t)] \dot{X}(t) + X(t) = 0, \qquad t \geq 0$$

where $a > 0$ and $W(t)$ is a Gaussian white noise. It is noted that, without the white noise term, the deterministic system is unstable.

Discuss the m.s. stability of the system. Show that, in the mean square sense, the deterministic system cannot be stabilized with the addition of $W(t)$.

9.7. Let

$$Q = \begin{bmatrix} 1 & b \\ b & a^2 \end{bmatrix}$$

with $a^2 - b^2 > 0$. Show that, with suitable choices of a and b, the system considered in Example 9.6 is globally stable in the norm whenever τ is finite and $p > 0$.

A Sample Treatment of Random Differential Equations

The purpose of this appendix is to present a different approach to the study of random differential equations, namely, a sample-function approach, and to show how it is related to the mean square treatment. The motivation for doing this is clear. In the analysis of real systems, stochastic processes describing physical phenomena represent families of individual realizations. Each realization of a stochastic system leads to a deterministic sample differential equation with a unique solution trajectory if the problem is well posed. Hence, to add physical meaning to the mean square solution of a random differential equation, it is constructive to show that these mean square results are consistent with the sample theoretic approach where the solution is produced as a collection of trajectories. This collection of trajectories can be investigated with the tools of probability theory if it possesses certain measurability properties. In other words, the collection of all solution trajectories should be a stochastic process.

The sample treatment of random differential equations in this appendix depends heavily on the theory of differential equations in real analysis. Hence, all stochastic processes contained in the random differential equations are assumed to possess continuous trajectories. This means, quite naturally, that physical random systems described in terms of random differential equations should have continuous realizations.

A brief review of some basic measure theoretic considerations of stochastic processes is given in Section A.1. This is followed in Section A.2 by the development of a sample calculus confined to stochastic processes of which almost all trajectories are continuous functions of time. It is seen that under these conditions the stochastic processes involved in the random differential equations can be treated as functions of two variables, t and ω, where ω is the "hidden parameter."

A stochastic sample solution is defined in Section A.3. Its existence and uniqueness are results based upon the theory of real functions. We also show that the stochastic sample solution has meaning in the probabilistic sense.

The bulk of the material presented in this appendix is contained in the work of Ruymgaart and Soong [1, 2]. An application of the sample treatment presented here to the study of a class of Ito equations is given by Ruymgaart and Soong [2].

A.1. Stochastic Processes with Continuous Trajectories

Throughout our development the measure theoretic approach of probability theory will be followed. In this section, a brief review of this approach is given. The proofs of all theorems stated in this section are omitted and the reader is referred to the standard texts [3–5] for details.

A stochastic process $X(t)$, $t \in T$, is considered given when the class of (consistent and symmetric) joint distribution functions of all finite sets of random variables of $X(t)$ is given. The interval $T = [0, a]$, $a > 0$, is an interval of the real line.

There always exist suitable probability spaces $\{\Omega, \mathscr{B}, P\}$ such that $X(t)$ can be represented by a function $F(\omega, t)$, $(\omega, t) \in \Omega \times T$, having the following properties:

1. at fixed $t \in T$, $F(\omega, t)$ is a $\mathscr{B}$-measurable function of ω, and
2. at all finite sets $\{t_1, \ldots, t_n\} \subset T$, the distribution function induced by $\{F(\omega, t_1), \ldots, F(\omega, t_n)\}$ is identical to the prescribed distribution function of $\{X(t_1), \ldots, X(t_n)\}$.

Functions with Property (1) are called representations. A function $F(\omega, t)$ with Properties (1) and (2) is called a representation of $X(t)$ on $\{\Omega, \mathscr{B}, P\}$. The underlying probability space $\{\Omega, \mathscr{B}, P\}$ is called the representation space.

Conversely, it is seen that any representation specifies the distribution of a stochastic process.

The $\mathscr{B}$-measurable ω-functions $F(\omega, t)$ at fixed $t \in T$ are called the sections at t of $F(\omega, t)$. Occasionally they are denoted by $F_t(\omega)$. The sections at ω are the functions of $t \in T$ represented by $F(\omega, t)$ at fixed ω. They are occasionally denoted by $F_\omega(t)$.

Given the distribution of a stochastic process, there is a great deal of ambiguity in its representations in that there exist infinitely many suitable representation spaces and the representation of the stochastic process on a given suitable representation space is not necessarily unique. Two representations on a same probability space $\{\Omega, \mathscr{B}, P\}$ are called equivalent if their sections at t differ only at sets $N_t \in \mathscr{B}$, $P(N_t) = 0$, where N_t may depend on t. It is seen that equivalent representations represent one and the same stochastic process. All representations on a fixed probability space with t in the same time interval form a set of disjoint classes consisting of equivalent representations. Different classes may represent the same stochastic process.

Let us illustrate these situations by two examples given below. In these examples we let Ω in $\{\Omega, \mathscr{B}, P\}$ be the segment $[0, 1]$ of the real line, $\mathscr{B}$ the class of Borel sets of $[0, 1]$, and P is the Lebesgue measure.

Example A.1. The s.p. $X(t)$, $t \in T$, defined by $P\{X(t) = 0\} = 1$ at each $t \in T$ can be represented by

   (a) $F(\omega, t) = 0,$ $(\omega, t) \in [0, 1] \times T$

or equivalently by

   (b) $F(\omega, t) = 0,$ if $(\omega, t) \in [0, 1] \times T,$ $\omega \neq t + r$
          $F(\omega, t) = 1,$ if $(\omega, t) \in [0, 1] \times T,$ $\omega = t + r$

where r represents any rational number.

Example A.2. The functions

$$F(\omega, t) = \begin{cases} C, & \text{if } t > \omega \\ 0, & \text{if } 0 \leq t \leq \omega \end{cases}$$

and

$$G(\omega, t) = \begin{cases} C, & \text{if } t \geq 1 - \omega \\ 0, & \text{if } 0 \leq t < 1 - \omega, \end{cases} \quad C \neq 0,$$

are representations. They represent the same stochastic process although they are not equivalent.

Definition. A *probabilistic property* of a stochastic process is a property common to all of its representations.

Theorem A.1.1. A property of a random system is probabilistic if it can be described in terms of the distribution function of a Borel function defined on this random system.

For example, continuity in probability, continuity in mean square, and differentiability in mean square are probabilistic properties of a stochastic process. If $\{X_n; n = 1, 2, 3, \ldots\}$ is a random sequence,

$$\inf_n X_n, \qquad \sup_n X_n, \qquad \liminf_{n \to \infty} X_n, \qquad \limsup_{n \to \infty} X_n$$

are Borel functions defined on this sequence. Their distribution functions are independent of the representation in which they are derived. On the other hand, in the case of a s.p. $X(t)$, $t \in T$, the functions

$$\inf_{t \in T} X(t) \qquad \text{and} \qquad \sup_{t \in T} X(t)$$

for example, fail to be Borel functions on $X(t)$, because the number of random variables involved is no longer countable. In general, they are not measurable in arbitrary representations; and if they happen to be measurable in some representations, their distribution functions are not necessarily identical. It follows that the notion of "almost sure (a.s.) continuity" cannot be defined as a probabilistic property.

We also see that the sample functions (or trajectories) of a stochastic process on T, which we take to be the sections at ω of its representations, cannot be defined within this framework of probability theory. This is clearly shown in Example A.1 where we have two equivalent representations of a stochastic process such that in the first all sections at ω are continuous functions of t on T, whereas in the second these sections are nowhere continuous.

In order to improve the mathematical model of stochastic processes and to extend the domain of random analysis, Doob [3] introduced the idea of separability. He proved that, to each representation, there is an equivalent separable representation. Loosely speaking, separable representations are the smoothest ones in their equivalent classes. If $F(\omega, t)$, $(\omega, t) \in \Omega \times T$, is a separable representation and if I is an interval in T, the ω-functions

$$\inf_{t \in I} F(\omega, t) \qquad \text{and} \qquad \sup_{t \in I} F(\omega, t)$$

for example, represent Borel functions defined on a well chosen countable number of random variables of the represented process. They are measurable in all separable representations and have unique distribution functions, independent of the separable representations in which they are derived.

With the concept of separability, we distinguish two types of properties associated with a stochastic processes. They are the strong properties which are common to all of its representations and the weak properties which are common to all of its separable representations.

The strong properties of a stochastic process are its usual probabilistic properties. They constitute a subclass of the class of weak properties of this stochastic process. In the sequel we will write 'property in the weak sense,' 'weak distribution function,' 'weak function,' and so forth, without further explanation. Weak random variables associated with a s.p. $X(t)$, $t \in T$, are, for instance,

$$\inf_{t \in I} X(t) \qquad \text{and} \qquad \sup_{t \in I} X(t)$$

where I is an interval of T. Continuity a.s. at $t \in T$ or on T, if present, is a weak property of $X(t)$.

It is noteworthy that all properties of a separable representation need not belong to the class of the weak properties of a given process. Example A.2 illustrates this very well. It can be shown that both representations in this example are separable. In the first, all sections at ω are continuous from below. In the second, they are all continuous from above. Still F and G represent the same stochastic process.

Hence, even with the aid of separable representations it is in general not possible to give a unique picture of the sections at ω of the representations of a given stochastic process. In other words, even in terms of weak properties it is in general not possible to define the trajectories of a stochastic process.

As we have explained above, in this appendix we will be only concerned with processes whose trajectories are continuous functions of time. Here we need separability in its most elementary form.

Definition. A representation $F(\omega, t)$ on $\{\Omega, \mathscr{B}, P\}$, $(\omega, t) \in \Omega \times T$, is called *sample continuous* if its sections $F_\omega(t)$ are continuous functions of t on T for all $\omega \in \Omega - N$, $N \in \mathscr{B}$ and $P(N) = 0$. N is called the exceptional set.

We remark that any sample continuous representation is separable; to any sample continuous representation there are equivalent sample continuous representations with empty exceptional sets; and any sample continuous representation is a.s. continuous at each fixed $t \in T$. However, as we have seen from Example A.2, the converse is not true.

If a stochastic process possesses sample continuous representations, there is no longer any lack of uniqueness in weak probabilistic sense concerning the sections at ω of its representations. This is shown in the following theorem.

Theorem A.1.2

1. If one representation of a stochastic process is sample continuous, then all of its separable representations, represented on whatever suitable representation space, are sample continuous.

2. If a stochastic process possesses a sample continuous representation, the sections at ω of this representation can be seen as the elementary events of a probability space. The distribution function of the sections at ω is independent of the sample continuous representation in which they are defined. It only depends on the distribution of the given stochastic process.

3. On a suitable representation space $\{\Omega, \mathcal{B}, P\}$, two equivalent sample continuous representations of a s.p. $X(t)$, $t \in T$, are identical outside a set $N \times T$, $N \in \mathcal{B}$ and $P(N) = 0$.

The following definition is now admissible.

Definition

1. If the separable representations of a stochastic process are sample continuous, it is called *sample continuous*.

2. If a stochastic process is sample continuous, its *trajectories* are the sections at ω of one of its sample continuous representations. The trajectories are also called *realizations* or *sample functions*.

Theorem A.1.3. Let $\mathcal{A}$ be the Borel field of T. If $F(\omega, t)$, $(\omega, t) \in \Omega \times T$, is a sample continuous representation on $\{\Omega, \mathcal{B}, P\}$, it is an $\mathcal{B} \times \mathcal{A}$-measurable function.

A.2. A Sample Calculus

A sample calculus with respect to sample continuous stochastic processes is now developed. The results given below are consistent with the weak uniqueness of trajectories of sample continuous processes.

Consider first an integral definition. Let $X(t)$, $t \in T$, be a sample continuous process of which $F(\omega, t)$, $(\omega, t) \in \Omega \times T$, is a sample continuous representation on a suitable representation space $\{\Omega, \mathcal{B}, P\}$. To facilitate the exposition, it can be assumed without loss of generality that the exceptional set of $F(\omega, t)$ is empty. At each $\omega \in \Omega$ the Riemann integral $\int_0^t F_\omega(s)\, ds$ exists for all $t \in T$. The collection of these integrals at all $\omega \in \Omega$ is denoted by

$$G(\omega, t) = \int_0^t F(\omega, s)\, ds, \qquad (\omega, t) \in \Omega \times T \qquad (A.1)$$

On account of the sample continuity of $F(\omega, t)$, the sections $G_\omega(t)$ are continuous functions of $t \in T$. The sections at t of $G(\omega, t)$ can be established as limits of sequences of ($\mathcal{B}$-measurable) Riemann sums

$$\sum_k F_{s_k}(\omega)(t_k - t_{k-1}) \qquad (A.2)$$

since a uniform sequence of partitions of T and uniform values s_k can be used in obtaining $\int_0^t F_\omega(s)\, ds$ at all $\omega \in \Omega$ with t fixed. The sections $G_t(\omega)$ are thus $\mathcal{B}$-measurable and they are independent of the sequences of partitions and the values s_k involved in the limiting procedures. It follows that the integral (A.1) is a sample continuous representation.

In arbitrary representations the Riemann sums corresponding to Eq. (A.2) converge a.s. This is seen by replacing the arbitrary representation by an equivalent sample continuous representation. The number of sections at t of the representation of $X(t)$ involved in each sequence of Riemann sums is countable. Therefore, the Riemann sums of any sequence in the arbitrary representation are identical to the corresponding Riemann sums in the sample continuous representation outside a countable union N of sets of probability zero. It follows that $P(N) = 0$ and that each sequence of Riemann sums in the arbitrary representation converges outside a set of probability zero. At each fixed $t \in T$, the resulting a.s. limits are Borel functions strongly defined on $X(s)$, $s \in T$.

Definition. If $X(t)$, $t \in T$, is a sample continuous stochastic process, its *stochastic Riemann sample integral* is the s.p. represented by

$$\int_0^t F(\omega, s)\, ds, \qquad (\omega, t) \in \Omega \times T$$

where $F(\omega, s)$, $(\omega, s) \in (\Omega - N) \times T$ and $P(N) = 0$, is any sample continuous representation of $X(t)$ on any suitable representation space $\{\Omega, \mathcal{B}, P\}$.

The integral is denoted by

$$\int_0^t X(s)\,ds, \qquad t \in T$$

We have proved the following theorem.

Theorem A.2.1. If the s.p. $X(t)$, $t \in T$, possesses the weak property of sample continuity, it possesses the strong property that, at fixed $t \in T$, $Y(t) = \int_0^t X(s)\,ds$ is a Borel function uniquely defined on $X(s)$, $s \in [0, t]$. The process $Y(t)$, $t \in T$, possesses the weak property of sample continuity.

Theorem A.2.2. In any sample continuous representation $\{F(\omega, t),$ $G(\omega, t)\}$, $(\omega, t) \in \Omega \times T$, on $\{\Omega, \mathscr{B}, P\}$ of $\{X(t), Y(t) = \int_0^t X(s)\,ds\}$, $t \in T$, where $X(t)$ is a sample continuous stochastic process and $Y(t)$ its stochastic Riemann sample integral, the relation $\int_0^t F(\omega, s)\,ds = G(\omega, t)$ holds for all $(\omega, t) \in (\Omega - N) \times T$, $P(N) = 0$.

Proof. It follows from Theorem A.2.1 that $\int_0^t F(\omega, s)\,ds$ and $G(\omega, t)$ are equivalent on $\Omega \times T$. Moreover, since both representations are sample continuous, the assertion follows from Theorem A.1.2.

Stochastic Riemann–Stieltjes sample integrals can be treated in an analogous way.

Definition. A representation $F(\omega, t)$, $(\omega, t) \in \Omega \times T$, on $\{\Omega, \mathscr{B}, P\}$ is *continuously sample differentiable* if its sections $F_\omega(t)$ are continuously differentiable on T at all $\omega \in \Omega - N$ with $P(N) = 0$.

It is seen that continuously sample differentiable representations are separable. To any continuously sample differentiable representation, there are equivalent continuously sample differentiable representations with empty exceptional sets.

Let the s.p. $X(t)$, $t \in T$, possess a continuously sample differentiable representation $F(\omega, t)$, $(\omega, t) \in \Omega \times T$, on $\{\Omega, \mathscr{B}, P\}$. It can be assumed without loss of generality that its exceptional set is empty. Hence, $dF_\omega(t)/dt$ exists at each ω and is a continuous function of t on T. The collection of these derivatives at all $\omega \in \Omega$ is denoted by

$$K(\omega, t) = dF(\omega, t)/dt, \qquad (\omega, t) \in \Omega \times T$$

The sections $K_t(\omega)$ are $\mathscr{B}$-measurable, since they can be established as limits of ($\mathscr{B}$-measurable) difference quotients

$$[F(\omega, s_n) - F(\omega, t)]/(s_n - t) \tag{A.3}$$

where $\{s_n\} \subset T$ is any sequence approaching t as $n \to \infty$. Moreover, as the sections $K_\omega(t)$ are continuous on T, $K(\omega, t)$ is a sample continuous representation. It represents a s.p. $U(t)$, $t \in T$.

In any arbitrary representation equivalent to $F(\omega, t)$, the sequences of difference quotients corresponding to (A.3) converge a.s. to $K(\omega, t)$. The number of sections at t involved in each limiting procedure is countable. Hence, the corresponding difference quotients are identical outside a countable union N on sets of probability zero. Hence, outside N, $P(N) = 0$, they converge to $K(\omega, t)$. As an a.s. limit in arbitrary representation, $U(t)$ at fixed $t \in T$ is a Borel function strongly defined on $X(t)$.

We have shown the admissibility of the following definition.

Definition. If a s.p. $X(t)$, $t \in T$, possesses a continuously sample differentiable representation $F(\omega, t)$, $(\omega, t) \in \Omega \times T$, the stochastic process represented by the sample continuous representation

$$K(\omega, t) = \frac{d}{dt} F(\omega, t), \qquad (\omega, t) \in \Omega \times T$$

is called the *stochastic sample derivative* of $X(t)$. It is denoted by dX/dt, $t \in T$.

We have thus proved the following theorem.

Theorem A.2.3. If a s.p. $X(t)$, $t \in T$, possesses a continuously sample differentiable representation, then at each fixed $t \in T$ its stochastic sample derivative dX/dt is strongly defined on $X(t)$ as a Borel function. The derivative is a sample continuous stochastic process.

Theorem A.2.4. Continuous sample differentiability of the representations is a weak property of the represented stochastic process, that is, if one of its representations is continuously sample differentiable, all of its separable representations possess this property.

Proof. Let the s.p. $X(t)$, $t \in T$, possess a continuously sample differentiable representation $F(\omega, t)$, $(\omega, t) \in \Omega \times T$, on $\{\Omega, \mathscr{B}, P\}$. It is assumed that the exceptional set of $F(\omega, t)$ is empty. Then $K(\omega, t) = dF(\omega, t)/dt$ is a sample continuous representation of $U(t) = dX/dt$, $t \in T$. The Riemann integral of $K(\omega, t)$ exists and is given by

$$\int_0^t K(\omega, s) \, ds = F(\omega, t) - F(\omega, 0) \tag{A.4}$$

it follows that

$$X(t) = X(0) + \int_0^t U(s) \, ds \tag{A.5}$$

where $\int_0^t U(s) \, ds$ is a stochastic Riemann sample integral. Let $F^*(\omega^*, t)$, $(\omega^*, t) \in \Omega^* \times T$, be a separable and thus sample continuous representation of $X(t)$ on $\{\Omega^*, \mathscr{B}^*, P^*\}$. At each $t \in T$

$$\text{a.s. } \lim_{n \to \infty} \frac{F^*(\omega^*, s_n) - F^*(\omega^*, t)}{s_n - t} = K^*(\omega^*, t) \tag{A.6}$$

exists and represents $U(t) = dX/dt$, where $\{s_n\}$ is any sequence in T approaching t as $n \to \infty$. As $U(t)$ is sample continuous, there is a sample continuous representation $K^{**}(\omega^*, t)$ equivalent to $K^*(\omega^*, t)$. From Eq. (A.5) we see that $X(t)$ is represented by

$$F^*(\omega^*, 0) + \int_0^t K^{**}(\omega^*, s) \, ds$$

It is continuously sample differentiable and equivalent to $F^*(\omega^*, t)$. By virtue of Theorem A.1.2,

$$F^*(\omega^*, t) = F^*(\omega^*, 0) + \int_0^t K^{**}(\omega^*, s) \, ds \tag{A.7}$$

outside a set $N^* \times T$, $P^*(N^*) = 0$. Therefore, $F^*(\omega^*, t)$ is continuously sample differentiable outside $N^* \times T$.

Corollary. If $X(t)$ is continuously sample differentiable, then in each separable representation $\{F(\omega, t), K(\omega, t)\}$ of $\{X(t), U(t)\}$ on whatever suitable probability space $\{\Omega, \mathscr{B}, P\}$, the relations

$$dF(\omega, t)/dt = K(\omega, t) \quad \text{and} \quad F(\omega, t) = F(\omega, 0) + \int_0^t K(\omega, s) \, ds$$

hold outside a set $N \times T$, $P(N) = 0$.

It should be noted that the existence of a.s. limits such as (A.3) at all $t \in T$ does not imply sample differentiability. For example, we see from Example A.2 that

$$\lim_{s \to t} \frac{F(\omega, s) - F(\omega, t)}{s - t} = 0 \text{ a.s.}$$

at all fixed $t > 0$, independent of the value of the constant C. Hence, if the differential quotient dX/dt had been defined in this way, the differential equation

$$P\{dX/dt = 0\} = 1$$

with initial condition $P\{X(0) = 0\} = 1$ would not have possessed a unique solution.

It is also noteworthy that, if the integrals (differential quotients) in mean square of a sample continuous (sample differentiable) second-order stochastic process exist, they are a.s. equal to the sample integral (sample derivative) at fixed t. This is due to the fact that in both cases the same sequences of type (A.2) [type (A.3)] can be used. As they converge a.s. as well as in mean square and therefore in probability, their limits are a.s. identical. However, given a second-order stochastic process, the existence of its stochastic sample derivative does not imply the existence of its derivative in mean square, or vice versa.

Finally, since the stochastic Riemann sample integral and the stochastic continuous sample derivative are sample continuous, it follows from Theorem A.1.3 that their separable representations on any suitable $\{\Omega, \mathscr{B}, P\}$ are $\mathscr{B} \times \mathscr{A}$-measurable, where $\mathscr{A}$ is the Borel field of the t-segment T.

A.3. Stochastic Sample Solutions

Consider a system of random variables $A_1, \ldots, A_n$ and of sample continuous s.p. $Y_1(t), \ldots, Y_m(t)$, $t \in T$, specified by their joint distribution functions at all $\{t_1, \ldots, t_k\} \subset T$. Let $g_i(a_1, \ldots, a_n; y_1, \ldots, y_m; t)$, $i = 1, \ldots, n$, be a system of continuous mappings of $R^{n+m} \times T$ into R^1.

If $X_1(t), \ldots, X_n(t)$, $t \in T$, is a system of stochastic processes, it follows from the continuity of g_i that at each $t \in T$

$$g_i(X_1(t), \ldots, X_n(t); Y_1(t), \ldots, Y_m(t); t), \qquad i = 1, \ldots, n \qquad \text{(A.8)}$$

is a system of Borel functions defined on the random variables $X_1(t), \ldots, X_n(t)$; $Y_1(t), \ldots, Y_m(t)$. If $X_1(t), \ldots, X_n(t)$ are also sample continuous, the functions g_i given above define a system of sample continuous stochastic processes.

Let us examine the following system of random differential equations:

$$dX_i(t)/dt = g_i(X_1(t), \ldots, X_n(t); Y_1(t), \ldots, Y_m(t); t), \qquad t \in T \qquad \text{(A.9a)}$$

with initial conditions

$$X_i(0) = A_i, \qquad i = 1, \ldots, n \tag{A.9b}$$

In the above, the meaning of the symbol d/dt depends upon the type of solution desired. We use it here to mean stochastic continuous sample derivative.

Definition. A stochastic sample solution of Eq. (A.9) is a system of s.p.'s $X_i(t)$, $i = 1, \ldots, n$, satisfying the following conditions:

1. Almost all trajectories of the processes $X_i(t)$ are defined on an interval $[0, S] \subset T$, where $S > 0$ and is independent of the trajectories.

2. At each fixed $t \in [0, S]$ the processes $X_i(t)$ are Borel functions strongly defined on the random system $\{A_1, \ldots, A_n; Y_1(s), \ldots, Y_m(s), s \in [0, t]\}$.

3. The stochastic continuous sample derivatives of the processes $X_i(t)$ exist on $[0, S]$ and satisfy Eq. (A.9).

It follows that the stochastic sample solution is independent of the representations of the random system.

In order to show that the definition above is admissible, let $\{H_1(\omega), \ldots, H_n(\omega); G_1(\omega, t), \ldots, G_m(\omega, t), t \in T\}$ be a separable representation on a suitable probability space $\{\Omega, \mathscr{B}, P\}$ of the above random system $\{A_1, \ldots, A_n; Y_1(t), \ldots, Y_m(t), t \in T\}$. Again, we may assume that the exceptional sets of the resulting sample continuous representations $G_j(\omega, t), j = 1, \ldots, m$, are empty. Let us consider the system of differential equations and initial conditions

$$dF_i/dt = g_i(F_1, \ldots, F_n; G_1(\omega, t), \ldots, G_m(\omega, t); t), \qquad t \in T$$
$$F_i = H_i(\omega) \qquad \text{at} \quad t = 0, \qquad i = 1, \ldots, n \tag{A.10}$$

At each fixed ω, Eq. (A.10) represents a system of ordinary differential equations in real analysis. Since the functions g_i depend continuously on their variables and since at fixed ω the functions $G_j(\omega, t)$ depend continuously on t, the right hand sides of Eq. (A.10) are continuous functions of $F_1, \ldots, F_n$ and t with ω fixed. Moreover, if at each fixed ω the g_i's satisfy a Lipschitz condition, it follows from the theorem of Cauchy that there exists a unique sample solution $F_i(\omega, t)$, $i = 1, \ldots, n$ of Eq. (A.10) at each fixed ω. The t-domain of such a solution at ω is an interval $[0, S(\omega)]$ $\subset T$ which depends on ω.

If Condition (1) of the definition is satisfied, $F_i(\omega, t)$, $i = 1, \ldots, n$, is a system of functions with the property that their sections at all $\omega \in \Omega$ satisfy Eq. (A.10) with $t \in [0, S]$. Again, according to the theorem of Cauchy, the functions $F_i(\omega, t)$ can be established as limits of the sequences $\{F_i^{(k)}(\omega, t)\}$ defined as follows:

$$F_i^{(1)}(\omega, t) = H_i(\omega)$$

$$F_i^{(2)}(\omega, t) = H_i(\omega) + \int_0^t g_i(F_1^{(1)}(\omega, s), \ldots, F_n^{(1)}(\omega, s);$$
$$\vdots \qquad\qquad G_1(\omega, s), \ldots, G_m(\omega, s); s)\, ds$$

$$\text{(A.11)}$$

$$F_i^{(k+1)}(\omega, t) = H_i(\omega) + \int_0^t g_i(F_1^{(k)}(\omega, s), \ldots, F_n^{(k)}(\omega, s);$$
$$\vdots \qquad\qquad G_1(\omega, s), \ldots, G_m(\omega, s); s)\, ds$$

$$t \in [0, S], \qquad i = 1, \ldots, n$$

Because of the continuity of all functions involved, the integrals in Eq. (A.11) exist as Riemann integrals. Hence, we see that all members of the sequences defined by Eq. (A.11) are Borel functions strongly defined on $\{H_1(\omega), \ldots, H_n(\omega); G_1(\omega, s), \ldots, G_m(\omega, s), s \in [0, t]\}$ and so are the limits $F_i(\omega, t)$, $i = 1, \ldots, n$. As seen from Section A.2, $F_i(\omega, t)$ represent stochastic processes whose stochastic sample derivatives exist and satisfy Condition (3) of the definition.

We have shown that it is constructive to consider stochastic sample solutions associated with random differential equations. The existence and uniqueness are results of the theory of real functions. Furthermore, we have exhibited several measurability properties associated with these solutions. It is instructive to state the following theorem which singles out these measurability properties.

Theorem A.3.1. Consider the system given by Eq. (A.10) where the functions g_i are continuous mapping of $R^{n+m} \times T$ into R^1 and the representations $G_i(\omega, t)$ are sample continuous. Assume that at (almost) each fixed ω there is a unique solution $F_i(\omega, t)$, $i = 1, 2, \ldots, n$ and that $t \in [0, S]$ with $S > 0$ and S independent of ω. Then at each fixed $t \in [0, S]$ the functions $F_i(\omega, t)$ are $\mathscr{B}$-measurable functions of ω.

Proof. Let $[0, S]$ be divided into 2^k parts of equal length h_k. To each k, $k = 1, 2, 3, \ldots$, we define the functions $F_{ik}(\omega, t)$, $i = 1, \ldots, n$, as

$$F_{ik}(\omega, t) = \begin{cases} H_i(\omega) + g_i(H_i(\omega), \ldots, H_n(\omega); G_1(\omega, 0), \ldots, G_m(\omega, 0); 0) \cdot t \\ \quad \text{if} \quad 0 \leq t \leq h_k; \\[4pt] F_{ik}(\omega, h_k) + g_i(F_{1k}(\omega, h_k), \ldots, F_{nk}(\omega, h_k); \\ \qquad\qquad G_1(\omega, h_k), \ldots, G_m(\omega, h_k); h_k)(t - h_k) \\ \quad \text{if} \quad h_k \leq t \leq 2h_k; \\ \quad \vdots \\[4pt] F_{ik}(\omega, S - h_k) + g_i(F_{1k}(\omega, S - h_k), \ldots, F_{nk}(\omega, S - h_k); \\ \qquad\qquad G_1(\omega, S - h_k), \ldots, G_m(\omega, S - h_k); \\ \qquad\qquad S - h_k)[t - (S - h_k)] \\ \quad \text{if} \quad S - h_k \leq t \leq S \end{cases}$$

Again we may suppose that all exceptional sets are empty. Following the methods of real analysis, it can be shown that for each i, $\{F_{ik}(\omega, t)\}$, $k = 1, 2, \ldots$, is a sequence of uniformly bounded, equicontinuous functions on $[0, S]$ if ω is fixed, and that this sequence contains convergent subsequences each of which uniformly converges to a solution of Eq. (A.10) at each fixed ω. As it is assumed that there is just one solution, the sequences $\{F_{ik}(\omega, t)\}$, $k = 1, 2, \ldots$, themselves converge to a solution trajectory at (almost) each fixed $\omega \in \Omega$. Since the same partitions of $[0, S]$ are used at each ω and since the functions g_i, being continuous functions of their arguments, are Borel functions strongly defined on the given random system, so are the functions $F_{ik}(\omega, t)$ and so are their limits as $k \to \infty$.

Since the representations $F_i(\omega, t)$ are sample continuous, it follows from Theorem A.1.3 that they are $\mathscr{B} \times \mathscr{A}$-measurable, where $\mathscr{A}$ is the Borel field of the t-segment $[0, S]$.

In closing, we remark that the techniques of the calculus in mean square are applicable if the stochastic sample solution is a second-order process. If all functions contained in Eq. (A.8) are of second order and mean square continuous and if the functions g_i satisfy a suitable Lipschitz condition in mean square, the system (A.9) possesses a unique solution in the mean square sense provided that the derivatives are interpreted as mean square derivatives and that the initial values have finite second moments. This solution can also be established as a limit in mean square of the sequences (A.11) as the integrals in Eq. (A.11) can now be interpreted as mean square integrals. Hence, if both the sample conditions and the mean square conditions are satisfied, the sequences (A.11) converge a.s. as well as in mean square. It follows then that the system (A.9) possesses a unique stochastic sample solution and a unique mean square solution, and that both solutions are one and the same stochastic process uniquely defined on the given random system.

References

1. P. A. Ruymgaart and T. T. Soong, A sample theory of ordinary stochastic differential equations. *Proc. Hawaii Internat. Conf. System Sci., 2nd, Honolulu, Hawaii, 1969,* pp. 603–606. Western Periodicals Co., Honolulu, 1969.
2. P. A. Ruymgaart and T. T. Soong, A sample treatment of Langevin-type stochastic differential equations. *J. Math. Anal. Appl.* **34,** 325–338 (1971).
3. J. L. Doob, *Stochastic Processes.* Wiley, New York, 1953.
4. M. Loéve, *Probability Theory.* Van Nostrand-Reinhold, Princeton, New Jersey, 1960.
5. J. Neveu, *Bases Mathématiques du Calcul des Probabilités.* Masson, Paris, 1964.

Appendix B

Some Useful Properties of the Solution Process

As we have stated in Chapter 5, "solving" a random differential equation usually means the determination of a limited amount of the information about the solution process. In many applied problems, it is usually the properties of certain random variables associated with the solution process that are of practical importance. In this appendix, we introduce several of these random variables and consider briefly the problem of relating the properties of these random variables to those of the solution process.

B.1. Zero and Threshold Crossings

The expected number of zeroes of a s.p. $X(t)$ within a given interval $[t_1, t_2]$ or, more generally, the expected number of crossings of $X(t)$ at some arbitrary level x_0 in $[t_1, t_2]$ is of considerable interest. This information is useful, for example, for examining certain modes of failure in engineering systems and structural components. These are called, respectively, zero-crossing and threshold-crossing problems, and they were originally studied by Rice [1].

Let $X(t)$, $t \in T$, be a m.s. differentiable stochastic process. The formulation of the threshold-crossing problem for $X(t)$ is facilitated by defining

296

a new process (see Middleton [2, p. 426])

$$Y(t) = u[X(t) - x_0], \qquad t \in T \tag{B.1}$$

where $u[\ ]$ is a unit step function. It is seen that the formal derivative of $Y(t)$ can be written in the form

$$\dot{Y}(t) = \dot{X}(t)\,\delta[X(t) - x_0], \qquad t \in T \tag{B.2}$$

where $\delta[\ ]$ is the Dirac delta function. We note that, as a function of t, the delta function has the weight $1/\dot{X}(t)$. A typical sample function of $X(t)$ along with its corresponding sample functions of $Y(t)$ and $\dot{Y}(t)$ are shown in Fig. B.1. We see that a sample function of $\dot{Y}(t)$ consists of unit impulses.

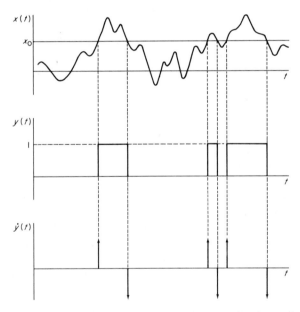

Fig. B.1. A typical sample function of $X(t)$ and its associated sample functions $y(t)$ and $\dot{y}(t)$.

These impulses are directed upward or downward depending upon whether the crossings of its associated $x(t)$ at x_0 occur with positive slopes or negative slopes.

If we denote by $N(x_0; t_1, t_2)$ the random variable whose value is the number of crossings of $X(t)$ at x_0 within the interval $[t_1, t_2]$, we may thus express it by

$$N(x_0; t_1, t_2) = \int_{t_1}^{t_2} |\dot{X}(t)| \, \delta[X(t) - x_0] \, dt \tag{B.3}$$

The number of zeros of $X(t)$ within the interval $[t_1, t_2]$ is simply $N(0; t_1, t_2)$.

Hence, the expected number of crossings of $X(t)$ at x_0 within the interval $[t_1, t_2]$ is given by

$$E\{N(x_0; t_1, t_2)\} = \int_{t_1}^{t_2} \int_{-\infty}^{\infty} \int_{-\infty}^{\infty} |\dot{x}| \, \delta(x - x_0) f(x, t; \dot{x}, t) \, dx \, d\dot{x} \, dt$$

$$= \int_{t_1}^{t_2} \int_{-\infty}^{\infty} |\dot{x}| f(x_0, t; \dot{x}, t) \, d\dot{x} \, dt \tag{B.4}$$

We thus see that the information on expected number of zero-crossings and threshold-crossings within an interval is contained in the joint density function of $X(t)$ and $X(t)$ at t over the interval.

Expressions for the second and higher moments of $N(x_0; t_1, t_2)$ can be written down following the same procedure. For example, the second moment is

$$E\{N^2(x_0; t_1, t_2)\}$$

$$= \int_{t_1}^{t_2} \int_{t_1}^{t_2} \int_{-\infty}^{\infty} \int_{-\infty}^{\infty} |\dot{x}_1| |\dot{x}_2| f(x_0, \tau_1; \dot{x}_1; \tau_1; x_0, \tau_2; \dot{x}_2, \tau_2)$$

$$\times d\dot{x}_1 \, d\dot{x}_2 \, d\tau_1 \, d\tau_2 \tag{B.5}$$

The computational labor for higher moments rapidly becomes formidable. Some discussions on the computation of the variance of N can be found in Steinberg *et al.* [3] and Miller and Freund [4]. For the case where $X(t)$ is a Gaussian process, Helstrom [5] also derives the distribution associated with the number of threshold crossings.

In certain cases one might be interested in the number of threshold crossings with positive slopes only. If $N_+(x_0; t_1, t_2)$ represents the number of crossings of $X(t)$ at x_0 with positive slopes within an interval $[t_1, t_2]$, we easily see from Eqs. (B.4) and (B.5) that

$$E\{N_+(x_0; t_1, t_2)\} = \int_{t_1}^{t_2} \int_{0}^{\infty} \dot{x} f(x_0, t; \dot{x}, t) \, d\dot{x} \, dt \tag{B.6}$$

$$E\{N_+^2(x_0; t_1, t_2)\} = \int_{t_1}^{t_2} \int_{t_1}^{t_2} \int_{0}^{\infty} \int_{0}^{\infty} \dot{x}_1 \dot{x}_2 f(x_0, \tau_1; \dot{x}_1, \tau_1; x_0, \tau_2; \dot{x}_2, \tau_2)$$

$$\times d\dot{x}_1 \, d\dot{x}_2 \, d\tau_1 \, d\tau_2 \tag{B.7}$$

For the problem of threshold crossings with negative slopes we easily get

$$E\{N_-(x_0; t_1, t_2)\} = -\int_{t_1}^{t_2} \int_{-\infty}^{0} \dot{x} f(x_0, t; \dot{x}, t) \, d\dot{x} \, dt \tag{B.8}$$

$$E\{N_-{}^2(x_0; t_1, t_2)\} = \int_{t_1}^{t_2} \int_{t_1}^{t_2} \int_{-\infty}^{0} \int_{-\infty}^{0} \dot{x}_1 \dot{x}_2 f(x_0, \tau_1; \dot{x}_1, \tau_1; x_0, \tau_2; \dot{x}_2, \tau_2)$$
$$\times d\dot{x}_1 \, d\dot{x}_2 \, d\tau_1 \, d\tau_2 \tag{B.9}$$

As seen from Eq. (B.4), the integral

$$r(x_0, t) = \int_{-\infty}^{\infty} |\dot{x}| f(x_0, t; \dot{x}, t) \, d\dot{x} \tag{B.10}$$

may be regarded as the expected rate of crossings at x_0. This rate becomes independent of t if, for example, the s.p. $X(t)$ is stationary, and $r(x_0, t) = r(x_0)$ gives the expected number of crossings at x_0 per unit time in this case.

Example B.1. As an illustration, consider the case where $X(t)$ is a stationary, once m.s. differentiable, and Gaussian process with mean zero and correlation function $\Gamma_{XX}(\tau)$. The joint density function of $X(t)$ and $\dot{X}(t)$ at t then takes the form (see Problem 4.13)

$$f(x, t; \dot{x}, t) = f(x, \dot{x}) = \frac{1}{2\pi\sigma_X\sigma_{\dot{X}}} \exp\left[-\tfrac{1}{2}\left(\frac{x^2}{\sigma_X{}^2} + \frac{\dot{x}^2}{\sigma_{\dot{X}}{}^2}\right)\right] \tag{B.11}$$

where $\sigma_X{}^2 = \Gamma_{XX}(0)$ and

$$\sigma_{\dot{X}}{}^2 = \Gamma_{\dot{X}\dot{X}}(0) = -d^2\Gamma_{XX}(\tau)/d\tau^2 \,|_{\tau=0} \tag{B.12}$$

The expected rate of crossings at x_0 is constant and, upon substituting Eq. (B.11) into Eq. (B.10), we have

$$r(x_0) = (\sigma_{\dot{X}}/\pi\sigma_X) \exp(-x_0{}^2/2\sigma_X{}^2) \tag{B.13}$$

The expected rate of zero crossings has the simple form

$$r(0) = \sigma_{\dot{X}}/\pi\sigma_X, \qquad r_+(0) = r_-(0) = r(0)/2 \tag{B.14}$$

In order to develop an appreciation for this result, consider the case where $X(t)$ is narrow-band, that is, its spectral density function $S_{XX}(\omega)$ has significant values only in a frequency band whose width is small compared with the value of the mid-band frequency. As an example, a power

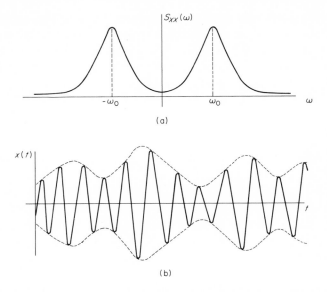

Fig. B.2. A narrow-band stationary process $X(t)$. (a) Spectral density function $S_{XX}(\omega)$. (b) A sample function of $X(t)$.

spectral density $S_{XX}(\omega)$ along with a possible sample function $x(t)$ of $X(t)$ is shown in Fig. B.2. The s.p. $X(t)$ oscillates with frequencies grouped around ω_0, with slowly varying random amplitude and with random phase.

For this case, the zero crossing information reveals some of the oscillatory properties of $X(t)$. Specifically, we expect that the expected rate of zero crossings *with positive slopes* gives the number of cycles per second in an average sense. Thus, it is meaningful to regard

$$\omega_0{}^* = 2\pi r_+(0) = \sigma_{\dot{X}}/\sigma_X \tag{B.15}$$

as the "equivalent frequency" of the narrow-band stationary Gaussian process.

In terms of the spectral density function, Eq. (B.15) takes the form

$$\omega_0^{*2} = \int_0^\infty \omega^2\, S_{XX}(\omega)\, d\omega \bigg/ \int_0^\infty S_{XX}(\omega)\, d\omega \tag{B.16}$$

It is seen that the squared equivalent frequency of $X(t)$ is simply a weighted sum of its squared frequency content, the weighting function being $S_{XX}(\omega)$. This simple and important result was first established by Rice.

Let us make some calculations. For ease in computation we shall approximate the spectral density function $S_{XX}(\omega)$ shown in Fig. B.2 by a "box" spectrum as shown in Fig. B.3. The height S_0 is taken to be equal to $S_{XX}(\omega_0)$ and the width $\Delta\omega$ is such that the box spectrum contains the same amount of power as that contained in $S_{XX}(\omega)$.

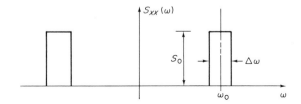

Fig. B.3. An approximation of $S_{XX}(\omega)$.

For this case

$$\int_0^\infty \omega^2 \, S_{XX}(\omega) \, d\omega = S_0 \, \Delta\omega(\omega_0^2 + \Delta\omega^2/12) \tag{B.17}$$

$$\int_0^\infty S_{XX}(\omega) \, d\omega = S_0 \, \Delta\omega \tag{B.18}$$

and

$$\omega_0^{*2} = \omega_0^2 + \Delta\omega^2/12 \tag{B.19}$$

The equivalent frequency is seen to be larger than the mid-band frequency for this case.

As $\Delta\omega \to 0$ and $S_0 \to \infty$ in such a way that $S_0 \, \Delta\omega \to$ const, Eq. (B.19) gives

$$\omega_0^* = \omega_0 \tag{B.20}$$

The equivalent frequency is then the actual frequency with which the limiting stochastic process $X(t)$ oscillates.

B.2. Distribution of Extrema

There is a direct analogy between the zero-crossing problem and the problem of determining the expected number of extrema (maxima and minima) of a s.p. $X(t)$ within a given interval. A maximum in a sample function $x(t)$ of $X(t)$ occurs when its first derivative $\dot{x}(t)$ is zero and the second derivative $\ddot{x}(t)$ is negative; a minimum occurs when $\dot{x}(t) = 0$ and

$\ddot{x}(t)$ is positive. The number of extrema of the sample function within a given interval is then equal to the number of zero crossings of $\dot{x}(t)$ in that interval.

Let $X(t)$, $t \in T$, be at least twice m.s. differentiable. As in the zero-crossing problem, we consider a new stochastic process defined by

$$Y(t) = u[\dot{X}(t)], \qquad t \in T \tag{B.21}$$

Formally differentiating $Y(t)$ we obtain

$$\dot{Y}(t) = \ddot{X}(t)\,\delta[\dot{X}(t)], \qquad t \in T \tag{B.22}$$

Figure B.4 gives a typical sample function $x(t)$ of $X(t)$ and the corresponding sample functions of $Y(t)$ and $\dot{Y}(t)$. In this case, the unit impulses of $\dot{y}(t)$ occur at the extrema of $x(t)$. They are directed upward or downward depending upon whether the extrema of $x(t)$ are minima or maxima.

Hence, the number of extrema of $X(t)$ within the interval $[t_1, t_2]$ is given by

$$\int_{t_1}^{t_2} |\ddot{X}(t)|\,\delta[\dot{X}(t)]\,dt \tag{B.23}$$

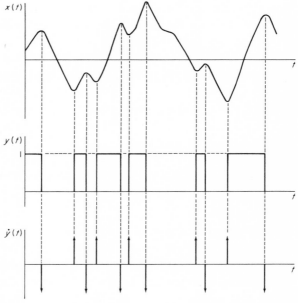

Fig. B.4. A typical sample function $x(t)$ of $X(t)$ and its associated sample functions $y(t)$ and $\dot{y}(t)$.

Equation (B.23) can be easily modified to give other useful information concerning the extrema. In certain applications one requires the information about the number of extrema of $X(t)$ in the interval $[t_1, t_2]$ *only* above a certain level x_0. Then, if we let the random variable $M(x_0; t_1, t_2)$ take this value, we have

$$M(x_0; t_1, t_2) = \int_{t_1}^{t_2} | \ddot{X}(t) | \, \delta[\dot{X}(t)] \, u[X(t) - x_0] \, dt \qquad (B.24)$$

Consistent with this notation, we may denote the integral in Eq. (B.23) by $M(-\infty; t_1, t_2)$.

Formulas for determining the statistical moments of $M(x_0; t_1, t_2)$ can be set up in a straightforward fashion. The mean, for example, is given by

$$E\{M(x_0; t_1, t_2)\}$$

$$= \int_{t_1}^{t_2} \int_{-\infty}^{\infty} \int_{-\infty}^{\infty} \int_{-\infty}^{\infty} | \ddot{x} | \, \delta(\dot{x}) \, u(x - x_0) \, f(x, t; \dot{x}, t; \ddot{x}, t) \, dx \, d\dot{x} \, d\ddot{x} \, dt$$

$$= \int_{t_1}^{t_2} dt \int_{-\infty}^{\infty} d\ddot{x} \int_{x_0}^{\infty} | \ddot{x} | \, f(x, t; 0, t; \ddot{x}, t) \, dx \qquad (B.25)$$

We thus see that the information about the expected number of extrema in a given interval is in general contained in the joint density function of $X(t)$, $\dot{X}(t)$, and $\ddot{X}(t)$ at t over the interval. However, the computation for even the mean is in general quite involved.

We remark that if $M_+(x_0; t_1, t_2)$ denotes the number of *maxima* of $X(t)$ above x_0 in the interval $[t_1, t_2]$, the formula for its expectation is

$$E\{M_+(x_0; t_1, t_2)\} = - \int_{t_1}^{t_2} dt \int_{-\infty}^{0} d\ddot{x} \int_{x_0}^{\infty} \ddot{x} f(x, t; 0, t; \ddot{x}, t) \, dx \qquad (B.26)$$

As seen from the above, the integral

$$q_+(x_0, t) = - \int_{-\infty}^{0} d\ddot{x} \int_{x_0}^{\infty} \ddot{x} f(x, t; 0, t; \ddot{x}, t) \, dx \qquad (B.27)$$

gives the number of maxima of $X(t)$ above x_0 per unit time. It is independent of t if $X(t)$ is stationary.

A heuristic definition of the distribution function for the maxima of $X(t)$ at time t has found wide applications in applied problems. The quantity

$$\frac{q_+(-\infty, t) - q_+(x_0, t)}{q_+(-\infty, t)}$$

gives the ratio of the expected number of maxima per unit time *below* x_0

to the expected total number of maxima per unit time. Let us define the random variable $Z(t)$ as one which takes the values of the maxima of $X(t)$ at t. A widely used definition for the distribution function of $Z(t)$ is

$$G(z, t) = \text{prob}\{Z(t) \leq z\}$$

$$= \frac{q_+(-\infty, t) - q_+(z, t)}{q_+(-\infty, t)} = 1 - \frac{q_+(z, t)}{q_+(-\infty, t)} \quad \text{(B.28)}$$

The corresponding density function is given by

$$g(z, t) = \frac{\partial}{\partial z} G(z, t)$$

$$= -\frac{1}{q_+(-\infty, t)} \int_{-\infty}^{0} \ddot{x} f(z, t; 0, t; \ddot{x}, t)\, d\ddot{x} \quad \text{(B.29)}$$

Strictly speaking, the expressions given by Eqs. (B.28) and (B.29) are incorrect since they make use of the ratio of the expectations and not the expectation of the ratio. The exact forms of the distribution and density functions for the maxima, however, are difficult to obtain.

Example B.2. Let us again consider a stationary, twice m.s. differentiable, and Gaussian process $X(t)$ with mean zero. The joint density function of $X(t)$, $\dot{X}(t)$, and $\ddot{X}(t)$ takes the form (see Problem 4.13)

$$f(\mathbf{x}, t) = (2\pi)^{-3/2} |\varLambda|^{-1/2} \exp[-\tfrac{1}{2}\mathbf{x}^T \varLambda^{-1}\mathbf{x}] \quad \text{(B.30)}$$

where $\mathbf{x}^T = [x\ \dot{x}\ \ddot{x}]$ and

$$\varLambda = \begin{bmatrix} \sigma_X{}^2 & 0 & -\sigma_{\dot{X}}{}^2 \\ 0 & \sigma_{\dot{X}}{}^2 & 0 \\ -\sigma_{\dot{X}}{}^2 & 0 & \sigma_{\ddot{X}}{}^2 \end{bmatrix} \quad \text{(B.31)}$$

Various statistical properties associated with the extrema of $X(t)$ can be obtained by substituting Eq. (B.30) into appropriate formulas derived above. Upon substituting Eq. (B.30) into Eq. (B.27) we obtain

$$q_+(x_0, t) = q_+(x_0) = (2\pi)^{-3/2}(\sigma_X\sigma_{\dot{X}})^{-2} \int_{x_0}^{\infty} \left\{ |\varLambda|^{-1/2} \exp\left[-\frac{1}{2|\varLambda|}\sigma_{\dot{X}}{}^2\sigma_{\ddot{X}}{}^2 x^2 \right] \right.$$

$$+ (\pi/2)^{1/2}(\sigma_{\dot{X}}{}^3 x/\sigma_X) \exp(-x^2/2\sigma_X{}^2)$$

$$\times \left. [1 + \text{erf}(\sigma_{\dot{X}}{}^3 x/\sqrt{2}\,\sigma_X |\varLambda|^{1/2})] \right\} dx \quad \text{(B.32)}$$

where erf( ) is the error function, defined by

$$erf(x) = (2/\sqrt{\pi}) \int_0^x \exp(-u^2)\, du \qquad (B.33)$$

The evaluation of the integral in Eq. (B.32) is in general difficult. We do, however, have a simple expression for $q_+(-\infty)$, the expected total number of maxima per unit time. It has the form

$$q_+(-\infty) = (1/2\pi)(\sigma_{\ddot{X}}/\sigma_{\dot{X}}) \qquad (B.34)$$

We remark that this result can also be obtained easily from Eq. (B.14) by noting the equivalence between the zero crossings of $\dot{X}(t)$ with negative slopes and the maxima of $X(t)$. Equations (B.32) and (B.34) specify the distribution function and density function of the maxima defined by Eqs. (B.28) and (B.29). The density function $g(z, t) = g(z)$ can be written as

$$g(z) = \frac{(1 - \nu^2)^{1/2}}{(2\pi)^{1/2}\sigma_X} \exp\left[-\frac{z^2}{2\sigma_X^2(1 - \nu^2)}\right]$$

$$+ \frac{\nu z}{2\sigma_X^2} \exp\left(-\frac{z^2}{2\sigma_X^2}\right)\left[1 + erf\left(\frac{\nu z}{\sigma_X[2(1 - \nu^2)]^{1/2}}\right)\right] \qquad (B.35)$$

where

$$\nu = \sigma_{\dot{X}}^2/\sigma_X\sigma_{\ddot{X}} = r_+(0)/q_+(-\infty) \qquad (B.36)$$

is the ratio of the expected number of zero crossings with positive slopes to the expected number of maxima. The values of ν thus lie within the interval $[0, 1]$. It is of interest to consider the forms of $g(z)$ for some limiting cases.

The case $\nu \to 1$ can occur when the s.p. $X(t)$ is narrow-band. In the limiting case we expect that one zero crossing with positive slope leads to one and only one maximum. As seen from Eq. (B.35), the limiting density function $g(z)$ for $\nu = 1$ is

$$g(z) = (z/2\sigma_X^2) \exp(-z^2/2\sigma_X^2), \qquad z \geq 0 \qquad (B.37)$$

The random variable Z thus has a Rayleigh distribution.

The case $\nu \to 0$ represents the situation where many peaks are present between two successive zero crossings with positive slopes. This can occur, for example, when oscillations of high frequencies and small amplitudes are superimposed upon a dominant oscillation with low frequency. A typical sample function of $X(t)$ in this situation is shown in Fig. B.5.

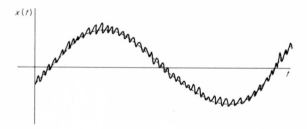

Fig. B.5. A typical sample function of $X(t)$ when $\nu \to 0$.

In this limiting case, the s.p. $X(t)$ at almost all t is a local maximum, and the density function of the maxima thus tends to the density function of $X(t)$ itself. We observe from Eq. (B.35) that, as $\nu = 0$, $g(z)$ takes the Gaussian form

$$g(z) = (1/(2\pi)^{1/2}\sigma_X) \exp(-z^2/2\sigma_X^2) \tag{B.38}$$

as we expected.

The distribution of the random variable Z is neither Gaussian nor Rayleigh for intermediate values of ν. The density function $g(z)$ is plotted in Fig. B.6 for several values of ν between zero and one.

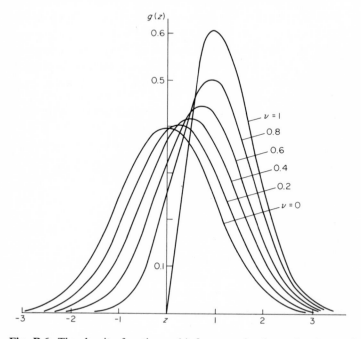

Fig. B.6. The density functions $g(z)$ for several values of ν ($\sigma_X = 1$).

B.3. Distribution of Intervals between Zeros

Another statistical problem closely related to the problems of extrema and threshold crossings is the determination of statistical properties of zero-crossing intervals. This problem was studied by Rice in connection with random noise in electrical circuits, followed by extensive investigations performed by workers in communication theory, biophysics, control, and other areas. The development here follows in the main the work of Mc-Fadden [6].

Given a s.p. $X(t)$, $t \in T$, we are interested in the statistical properties of a random variable T_n representing the length of the sum of $n + 1$ successive zero-crossing intervals associated with $X(t)$. The main problem we set out to solve is the determination of its probability distribution or density function.

Let us assume at the outset that the s.p. $X(t)$ is stationary. The determination of the mean values of T_n is relatively easy. To find the distribution of T_n, however, is considerably more difficult.

We will need the following Lemma.

Lemma. Consider a second-order, at least once m.s. differentiable, and stationary s.p. $X(t)$, $t \in T$. Let $p(n, \tau)$ be the probability that a given interval $[t, t + \tau]$ contains *exactly* n zero crossings of $X(t)$, and let $f_n(\tau)$ be the probability density function of the sum of $n + 1$ successive zero-crossing intervals. If the first two derivatives of $p(n, \tau)$ with respect to τ exist, then

$$p''(0, \tau) = r(0) f_0(\tau), \qquad p''(1, \tau) = r(0)[f_1(\tau) - 2f_0(\tau)]$$
$$p''(n, \tau) = r(0)[f_n(\tau) - 2f_{n-1}(\tau) + f_{n-2}(\tau)], \qquad n \geq 2 \tag{B.39}$$

where $r(0)$ is the expected number of zero crossings per unit time and the primes denote derivatives with respect to τ.

Proof. Let $A(n, \tau)$ be the event that there are exactly n zero crossings in the interval $[t, t + \tau]$, and let $A(n, \tau; m, d\tau)$ be the event that there are exactly n zero crossings in $[t, t + \tau]$ and exactly m zero crossings in $[t + \tau, t + \tau + d\tau]$. We have, by the conservation of probability and for sufficiently small $d\tau$,

$$P\{A(n, \tau)\} = P\{A(n, \tau; 0, d\tau)\} + P\{A(n, \tau; 1, d\tau)\} \tag{B.40}$$

$$P\{A(n, \tau + d\tau)\} = P\{A(n, \tau; 0, d\tau)\} + P\{A(n - 1, \tau; 1, d\tau)\} \tag{B.41}$$

Upon subtracting Eq. (B.40) from Eq. (B.41) we have

$$P\{A(n, \tau+d\tau)\} - P\{A(n, \tau)\} = P\{A(n-1, \tau; 1, d\tau)\} - P\{A(n, \tau; 1, d\tau)\}$$
(B.42)

The left-hand side of the equation above can be written in the form, upon neglecting higher-order terms in the Taylor expansion,

$$P\{A(n, \tau + d\tau)\} - P\{A(n, \tau)\} = p(n, \tau + d\tau) - p(n, \tau)$$
$$= p'(n, \tau)\, d\tau$$
(B.43)

For the right-hand side of Eq. (B.42) let us construct two more events. If we let A_1 be the event that there are no more than n zero crossings in $[t, t + \tau]$ and there is one in $[t + \tau, t + \tau + d\tau]$, and let A_2 be the event that there are no more than $(n - 1)$ zero crossings in $[t, t + \tau]$ and there is one in $[t + \tau, t + \tau + d\tau]$, we can write

$$P\{A(n, \tau; 1, d\tau)\} = P\{A_1\} - P\{A_2\}$$
(B.44)

Now, the quantity

$$r(0)\, d\tau$$

is the probability of finding one zero crossing in $[t + \tau, t + \tau + d\tau]$, and the integral

$$\int_\tau^\infty f_n(t)\, dt$$

can be interpreted as the conditional probability that, looking backward in time, the $(n + 1)$st zero crossing prior to the one in $[t + \tau, t + \tau + d\tau]$ occurred before time t, given that there is a zero crossing in $[t + \tau, t + \tau + d\tau]$. The product

$$r(0)\, d\tau \int_\tau^\infty f_n(t)\, dt$$

is thus the joint probability of realizing the event A_1. Hence,

$$P\{A_1\} = r(0)\, d\tau \int_\tau^\infty f_n(t)\, dt$$
(B.45)

Similarly

$$P\{A_2\} = r(0)\, d\tau \int_\tau^\infty f_{n-1}(t)\, dt$$
(B.46)

and, from Eq. (B.44),

$$P\{A(n, \tau; 1, d\tau)\} = r(0) \, d\tau \int_{\tau}^{\infty} [f_n(t) - f_{n-1}(t)] \, dt \qquad (B.47)$$

Following similar procedures we also have

$$P\{A(n - 1, \tau; 1, d\tau)\} = r(0) \, d\tau \int_{\tau}^{\infty} [f_{n-1}(t) - f_{n-2}(t)] \, dt \qquad (B.48)$$

The substitution of Eqs. (B.43), (B.47), and (B.48) into Eq. (B.42) hence gives

$$p'(n, \tau) = -r(0) \int_{\tau}^{\infty} [f_n(t) - 2f_{n-1}(t) + f_{n-2}(t)] \, dt \qquad (B.49)$$

The third of Eqs. (B.39) is now obtained by differentiating the above once with respect to τ. The first two of Eqs. (B.39) also follow directly if we impose the condition that

$$f_n(\tau) \equiv 0 \qquad (B.50)$$

if $n < 0$. The Lemma is thus proved.

In view of Eqs. (B.39), the density functions $f_n(\tau)$, $n = 0, 1, 2, \ldots$, of T_n can be determined provided that the probabilities $p(n, \tau)$, $n = 0, 1, 2, \ldots$, can be related to the statistical properties of $X(t)$. In order to do this, let us construct a new stochastic process $Y(t)$ based upon the definition

$$Y(t) = \begin{cases} +1, & X(t) > 0 \\ -1, & X(t) < 0 \end{cases} \qquad (B.51)$$

In terms of the unit step function, we may express $Y(t)$ by

$$Y(t) = 2u[X(t)] - 1 \qquad (B.52)$$

Consider the product $Y(t) \, Y(t + \tau)$. It is equal to one if there are even numbers of zero crossings in the interval $[t, t + \tau]$ and is equal to minus one if the number is odd. Hence, in terms of $p(n, \tau)$, we can write

$$\Gamma_{YY}(\tau) = E\{Y(t) \, Y(t + \tau)\} = \sum_{n=0}^{\infty} (-1)^n \, p(n, \tau) \qquad (B.53)$$

This equation relates $p(n, \tau)$ to the correlation function of $Y(t)$ which is in turn related to the statistical properties of $X(t)$ through Eq. (B.52) in the form

$$\Gamma_{YY}(\tau) = E\{(2u[X(t)] - 1)(2u[X(t + \tau)] - 1)\}$$

$$= 1 + 4\left[\int_0^\infty \int_0^\infty f(x_1, t; x_2, t + \tau)\, dx_1\, dx_2 - \int_0^\infty f(x, t)\, dx\right] \quad \text{(B.54)}$$

Unfortunately, Eq. (B.53) is only one of the desired relations; many others are needed to find the probabilities $p(n, \tau)$ explicitly. Hence, the density functions $f_n(\tau)$ cannot be derived explicitly in general. Equation (B.53), however, can help us to derive $f_n(\tau)$ approximately under certain conditions. We shall consider two cases here and we shall be concerned only with $f_0(\tau)$, the density function of the interval between *two* successive zero crossings.

Case 1. An approximate solution for $f_0(\tau)$ is possible when τ is small. We assume that the first two derivatives of $\Gamma_{YY}(\tau)$ exist for $\tau > 0$.

Differentiating Eq. (B.53) twice with respect to τ and substituting Eqs. (B.39) into the resulting equation yields

$$\Gamma''_{YY}(\tau)/4r(0) = \sum_{n=0}^\infty (-1)^n f_n(\tau) \quad \text{(B.55)}$$

If τ is sufficiently small, the first term on the right-hand side is dominant. We thus have, on neglecting $f_n(\tau)$ for $n \geq 1$ in the equation above,

$$f_0(\tau) \cong \frac{\Gamma''_{YY}(\tau)}{4r(0)} \quad \text{(B.56)}$$

This approximation becomes better as $\tau \to 0$.

Example B.3. Let $X(t)$ be a zero-mean stationary Gaussian process with correlation function $\Gamma_{XX}(\tau)$. The second density function of $X(t)$ is given by

$$f(x_1, t; x_2, t + \tau)$$

$$= \frac{1}{2\pi\sigma_X^2[1 - \varrho^2(\tau)]^{1/2}} \exp\left[-\frac{(x_1^2 + x_2^2 - 2\varrho(\tau)x_1 x_2)}{2\sigma_X^2[1 - \varrho^2(\tau)]}\right] \quad \text{(B.57)}$$

where

$$\sigma_X^2 = \Gamma_{XX}(0), \qquad \varrho(\tau) = \Gamma_{XX}(\tau)/\Gamma_{XX}(0) = \Gamma_{XX}(\tau)/\sigma_X^2 \quad \text{(B.58)}$$

The correlation function $\Gamma_{YY}(\tau)$ is obtained by substituting Eq. (B.57) into Eq. (B.54). It has the form

$$\Gamma_{YY}(\tau) = (2/\pi) \sin^{-1} \varrho(\tau) \qquad (B.59)$$

Hence, for the Gaussian case and for small τ, Eqs. (B.56) and (B.59) lead to

$$f_0(\tau) \cong \frac{1}{2\pi r(0)} \frac{\varrho(\tau)\,\varrho'^2(\tau) + [1 - \varrho^2(\tau)]\,\varrho''(\tau)}{[1 - \varrho^2(\tau)]^{3/2}} \qquad (B.60)$$

where, from Eq. (B.14),

$$r(0) = \sigma_{\dot{X}}/\pi\sigma_X = (1/\pi)[-\varrho''(0)]^{1/2} \qquad (B.61)$$

Case 2. A solution for $f_0(\tau)$ can be found under the assumption that successive zero-crossing intervals are statistically independent.

Let us assume that the functions $\Gamma''_{YY}(\tau)$ and $f_n(\tau)$ are Laplace transformable and let

$$u(s) = \mathcal{L}\left[\frac{\Gamma''_{YY}(\tau)}{4r(0)}\right] = \frac{1}{4r(0)} \int_0^\infty e^{-s\tau}\,\Gamma''_{YY}(\tau)\,d\tau \qquad (B.62)$$

$$v_n(s) = \mathcal{L}[f_n(\tau)] = \int_0^\infty e^{-s\tau} f_n(\tau)\,d\tau, \qquad n = 0, 1, 2, \ldots \quad (B.63)$$

The Laplace transform of Eq. (B.55) takes the form

$$u(s) = \sum_{n=0}^\infty (-1)^n\, v_n(s) \qquad (B.64)$$

If we assume that successive zero-crossing intervals are statistically independent, the density function $f_n(\tau)$ of the sum of $n + 1$ successive zero-crossing intervals is given by the $(n + 1)$-fold convolution of $f_0(\tau)$. In terms of the Laplace transforms we have

$$v_n(s) = v_0^{n+1}(s) \qquad (B.65)$$

Equation (B.64) then becomes

$$u(s) = \sum_{n=0}^\infty - [-v_0(s)]^{n+1}$$
$$= v_0(s)/[1 + v_0(s)] \qquad (B.66)$$

Thus,

$$v_0(s) = u(s)/[1 - u(s)] \qquad (B.67)$$

and, finally, we have the solution

$$f_0(\tau) = \mathscr{L}^{-1}[v_0(s)] = \mathscr{L}^{-1}\{u(s)/[1 - u(s)]\} \tag{B.68}$$

Recalling that T_0 is the successive zero-crossing interval, its statistical moments can also be found directly from the formulation above under general conditions. We first note from Eqs. (B.62) and (B.63) that

$$v_0(0) = 1, \qquad v_0'(0) = -E\{T_0\}, \qquad v_0''(0) = E\{T_0^2\}, \ldots \tag{B.69}$$

$$u(0) = -\frac{\Gamma_{YY}'(0^+)}{4r(0)}, \quad u'(0) = \frac{1}{4r(0)}, \quad u''(0) = \frac{1}{2r(0)} \int_0^\infty \Gamma_{YY}(\tau)\, d\tau, \ldots \tag{B.70}$$

where $\Gamma_{YY}'(0^+)$ is the right-hand derivative of $\Gamma_{YY}(\tau)$ at $\tau = 0$. Equations (B.70) are obtained by assuming that $\Gamma_{YY}(\tau)$ is sufficiently smooth so that the quantities $\Gamma_{YY}(\tau)$, $\Gamma_{YY}'(\tau)$, $\tau\Gamma_{YY}(\tau)$, and $\tau^2\Gamma_{YY}'(\tau)$ vanish as $\tau \to \infty$.
Upon expanding $u(s)$ and $v(s)$ into Maclaurin's series, we have

$$v_0(s) = v_0(0) + v_0'(0)s + v_0''(0)\, s^2/2! + \cdots$$
$$= 1 - E\{T_0\}s + E\{T_0^2\}\, s^2/2! + \cdots \tag{B.71}$$

$$u(s) = u(0) + u'(0)s + u''(0)\, s^2/2! + \cdots$$
$$= \frac{1}{4r(0)}\left[-\Gamma_{YY}'(0^+) - s + s^2 \int_0^\infty \Gamma_{YY}(\tau)\, d\tau + \cdots\right] \tag{B.72}$$

Substituting Eqs. (B.71) and (B.72) into Eq. (B.67) and equating terms of equal powers of s yields

$$u(0) = \tfrac{1}{2} \tag{B.73}$$

$$E\{T_0\} = -2u'(0)/[1 - u(0)] \tag{B.74}$$

$$E\{T_0^2\} = 2[1 - u(0)]^{-2}\{u''(0)[1 - u(0)] + 2u'^2(0)\} \tag{B.75}$$

Equation (B.73) gives

$$r(0) = -\tfrac{1}{2}\Gamma_{YY}'(0^+) \tag{B.76}$$

an interesting result in itself. With the aid of this result, Eq. (B.74) becomes

$$E\{T_0\} = 1/r(0) \tag{B.77}$$

a result in agreement with our intuition, and Eq. (B.75) takes the form

$$E\{T_0^2\} = [1/r^2(0)]\left[1 + 2r(0) \int_0^\infty \Gamma_{YY}(\tau)\,d\tau\right] \qquad \text{(B.78)}$$

The variance of T_0 is seen to be

$$\text{Var}\{T_0\} = [2/r(0)] \int_0^\infty \Gamma_{YY}(\tau)\,d\tau \qquad \text{(B.79)}$$

We see that the determination of the moments of T_0 is straightforward in this case. In practice, however, it is not simple to determine the density function $f_0(\tau)$ as the inverse Laplace transform, indicated by Eq. (B.68), presents great computational difficulty.

In addition to the two cases that we have taken up here, slightly more general cases are also considered by McFadden. A very intricate analysis of mean square distances between zeros for a class of stationary stochastic processes is presented by Kac [7], a detailed account of which is beyond the scope of this book.

B.4. The First Passage Time Probability

As our last topic in this appendix, we consider briefly the important problem of finding the first passage time probability associated with a stochastic process. This probability has important applications to a number of problems in physics, astronomy, engineering, and other areas. For example, Wang and Uhlenbeck [8] considered the first passage time probability associated with the velocity of a free particle in Brownian motion; Chandrasekhar [9] estimated the escape rate of stars from clusters by determining the first passage time probabilities of their initial velocities. In engineering applications, the first passage time problem is intimately related to the failure problem of engineering systems and structures when the failure is due to first large excursions exceeding a given level.

The problem of first passage time is unsolved under general conditions. In what follows we shall restrict ourselves to the case where the stochastic process under consideration is a stationary Markov process. The treatment outlined below follows that of Siegert [10] and Darling and Siegert [11]. Let $X(t)$, $t \geq 0$, be a s.p. and $b < X(0) \equiv x_0 < a$ where a and b are arbitrary constants which will be called absorbing barriers. The first passage time is a r.v., $T_{ab}(x_0)$, defined as the time at which $X(t)$ crosses a or b for the first time. A typical sample function of $X(t)$ is shown in Fig. B.7. Let us

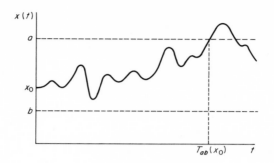

Fig. B.7. A sample function of $X(t)$.

assume that $X(t)$ is a stationary, Markov process and is continuous with probability 1, and let $F_{ab}(t \mid x_0) = P\{T_{ab}(x_0) \leq t\}$ denote the distribution function of $T_{ab}(x_0)$ and $f_{ab}(t \mid x_0)$ the corresponding density function. The transition distribution and density functions are denoted as $F(y, t \mid x_0) = P\{X(t) \leq y \mid X(0) = x_0\}$ and $f(y, t \mid x_0)$, respectively.

A special case of the above formulation occurs when there is only one absorbing barrier, that is, either $a = +\infty$ or $b = -\infty$. If $T_c(x_0)$ is the r.v. denoting the time at which the first passage at c occurs, it is easily seen that

$$T_c(x_0) = \begin{cases} T_{\infty,c}(x_0) & \text{if } x_0 > c \\ T_{c,-\infty}(x_0) & \text{if } x_0 < c \end{cases} \tag{B.80}$$

The functions $F_c(t \mid x_0)$ and $f_c(t \mid x_0)$ will denote respectively the distribution and density function of $T_c(x_0)$.

To develop a computational method for determining the density function for the first passage times, $f_{ab}(t \mid x_0)$, the following lemma will be needed. In the lemma, the Laplace transform of a function is denoted by a circumflex, for example,

$$\hat{f}(y, s \mid x_0) = \mathscr{L}[f(y, t \mid x_0)] = \int_0^\infty e^{-st} f(y, t \mid x_0)\, dt \tag{B.81}$$

Lemma.

$$\hat{f}(y, s \mid x_0) = \begin{cases} u(x_0)\, u_1(y), & y > x_0 \\ v(x_0)\, v_1(y), & y < x_0 \end{cases} \tag{B.82}$$

and

$$\hat{f}_c(s \mid x_0) = \begin{cases} u(x_0)/u(c), & x_0 < c \\ v(x_0)/v(c), & x_0 > c \end{cases} \tag{B.83}$$

Proof. Since $X(t)$ is stationary and Markovian, the transition probabilities satisfy the Smoluchowski–Chapman–Kolmogorov equation (Eq. (3.105)). This fact, together with the continuity assumption on $X(t)$, can be used to show that, if $y > c > x_0$, then

$$f(y, t \mid x_0) = \int_0^t f_c(\tau \mid x_0) f(y, t - \tau \mid c) \, d\tau \tag{B.84}$$

This integral is the convolution of $f_c(t \mid x_0)$ with $f(y, t \mid c)$ and therefore

$$\hat{f}(y, s \mid x_0) = \hat{f}_c(s \mid x_0) \hat{f}(y, s \mid c), \qquad y > c > x_0 \tag{B.85}$$

Thus, $\hat{f}(y, s \mid x_0)$ is a function of x_0 times a function of y, say $u(x_0) u_1(y)$, and hence for $y > c > x_0$,

$$\hat{f}_c(s \mid x_0) = u(x_0) u(y)/u(c) u(y) = u(x_0)/u(c)$$

Similarly, for $y < c < x_0$, $\hat{f}(y, s \mid x_0) = v(x_0) v_1(y)$ and $\hat{f}_c(s \mid x_0) = v(x_0)/v(c)$ and the lemma is proved. It should be noted that $u(x_0)$ and $v(x_0)$ are uniquely determined to within a constant.

Theorem B.4.1.

$$f_{ab}(s \mid x_0) = \frac{v(x_0)[u(a) - u(b)] - u(x_0)[v(a) - v(b)]}{u(a) v(b) - u(b) v(a)} \tag{B.86}$$

where $u(\ )$ and $v(\ )$ are the functions developed in the preceding lemma.

Proof. Let $F_{ab}^+(t \mid x_0)$ be the distribution function of $T_{ab}(x_0)$ given that absorption takes place at barrier a and $F_{ab}^-(t \mid x_0)$ be the distribution function of $T_{ab}(x_0)$ given that absorption takes place at barrier b. The density functions associated with the distributions above are $f_{ab}^+(t \mid x_0)$ and $f_{ab}^-(t \mid x_0)$, respectively. Heuristically, it is clear that there are two ways in which absorption can take place at barrier a, namely, $X(t)$ crosses a at time t and does not cross b prior to t or $X(t)$ crosses b at time t and then crosses a at a time $t - \tau$ after the crossing at b. Mathematically, this heuristic argument can be put in the rigorous form

$$f_a(t \mid x_0) = f_{ab}^+(t \mid x_0) + \int_0^t f_{ab}^-(\tau \mid x_0) f_a(t - \tau \mid b) \, d\tau \tag{B.87}$$

Similarly

$$f_b(t \mid x_0) = f_{ab}^-(t \mid x_0) + \int_0^t f_{ab}^+(\tau \mid x_0) f_b(t - \tau \mid a) \, d\tau \tag{B.88}$$

Taking Laplace transform of each equation gives

$$\hat{f}_a(s \mid x_0) = \hat{f}_{ab}^+(s \mid x_0) + \hat{f}_{ab}^-(s \mid x_0)\hat{f}_a(s \mid b)$$
$$\hat{f}_b(s \mid x_0) = \hat{f}_{ab}^-(s \mid x_0) + \hat{f}_{ab}^+(s \mid x_0)\hat{f}_b(s \mid a)$$

(B.89)

Solving the above equations simultaneously for $\hat{f}_{ab}^+(s \mid x_0)$ and $\hat{f}_{ab}^-(s \mid x_0)$ and using the previous results that

$$\hat{f}_b(s \mid x_0) = v(x_0)/v(b), \qquad x > b$$
$$\hat{f}_b(s \mid a) = v(a)/v(b), \qquad a > b$$
$$\hat{f}_a(s \mid x_0) = u(x_0)/u(a), \qquad x < a$$
$$\hat{f}_a(s \mid b) = u(b)/u(a), \qquad a < b$$

(B.90)

gives

$$\hat{f}_{ab}^+(s \mid x_0) = \frac{v(b)\, u(x_0) - v(x_0)\, u(b)}{u(a)\, v(b) - u(b)\, v(a)}$$

(B.91)

and

$$\hat{f}_{ab}^-(s \mid x_0) = \frac{u(a)\, v(x_0) - v(a)\, u(x_0)}{u(a)\, v(b) - u(b)\, v(a)}$$

(B.92)

Using the fact that $\hat{f}_{ab}(s \mid x_0) = \hat{f}_{ab}^+(s \mid x_0) + \hat{f}_{ab}^-(s \mid x_0)$ completes the proof of the theorem.

Therefore, if $u(x)$ and $v(x)$ are known, Theorem B.4.1 shows that the density function associated with the first passage time is just the inverse Laplace transform of $\hat{f}_{ab}(s \mid x_0)$. The functions $u(x)$ and $v(x)$ can be obtained by solving an associated Kolmogorov equation as is illustrated by the next theorem.

Theorem B.4.2. The functions $u(x)$ and $v(x)$ can be obtained as any two linearly independent solutions of

$$\tfrac{1}{2}\alpha_2(x)\, d^2w/dx^2 + \alpha_1(x)\, dw/dx - sw = 0$$

(B.93)

where $\alpha_1(x)$ and $\alpha_2(x)$ are the first and second derivate moments associated with $X(t)$, respectively.

Proof. Let $f(y, t \mid x, t_0)$ be the transition probability density of the stationary Markov process $X(t)$. The Kolmogorov equation satisfied by $f(y, t \mid x, t_0)$ is (Eq. (7.145))

$$\partial f(y, t \mid x, t_0)/\partial t_0 = -\alpha_1(x, t_0)(\partial f/\partial x) - \tfrac{1}{2}\alpha_2(x, t_0)(\partial^2 f/\partial x^2)$$

(B.94)

Since $X(t)$ is stationary, the transition probability density is a function of $t - t_0$ only. Hence

$$\partial f / \partial t_0 = -\partial f / \partial t \tag{B.95}$$

and Eq. (B.94) can be written as, with $t_0 = 0$,

$$\frac{\partial f(y, t \mid x)}{\partial t} = \alpha_1(x) \, (\partial f / \partial x) + \tfrac{1}{2}\alpha_2(x) \, (\partial^2 f / \partial x^2) \tag{B.96}$$

with initial and boundary conditions

$$f(y, 0 \mid x) = \delta(x - y)$$
$$f(y, t \mid \pm\infty) = 0 \tag{B.97}$$

Taking Laplace transforms of both sides of Eq. (B.96) gives (with $x \neq y$)

$$s\hat{f}(y, s \mid x) = \alpha_1(x)(d\hat{f}/dx) + \tfrac{1}{2}\alpha_2(x)(d^2\hat{f}/dx^2) \tag{B.98}$$

thus $\hat{f}(y, s \mid x)$ satisfies Eq. (B.93). However, as we see from Eqs. (B.82),

$$\hat{f}(y, s \mid x) = \begin{cases} v(x)u(y), & y > x \\ v(y)u(x), & y < x \end{cases} \tag{B.99}$$

Substituting this into Eq. (B.93) proves the assertion.

Thus, one procedure to find the first passage time density function is to solve Eq. (B.93) for the functions $u(x)$ and $v(x)$ and then use Eq. (B.86) to find $\hat{f}_{ab}(s \mid x_0)$. The required density function is then obtained as the inverse Laplace transform of $\hat{f}_{ab}(s \mid x_0)$, given by

$$f_{ab}(t \mid x_0) = (1/2\pi i) \int_{\gamma - i\infty}^{\gamma + i\infty} e^{st} \hat{f}_{ab}(s \mid x_0) \, ds \tag{B.100}$$

Example B.4. Let $X(t)$, $t \geq 0$, be the Weiner process with zero mean and covariance $E\{X(s) X(t)\} = \min(s, t)$. For this case, $\alpha_1 = 0$ and $\alpha_2 = 1$, and Eq. (B.93) becomes in this case

$$\frac{d^2 w}{dx^2} - 2sw = 0 \tag{B.101}$$

Two independent solutions of this equation are

$$u(x) = \exp[-(2s)^{1/2}x], \qquad v(x) = \exp[(2s)^{1/2}x] = u(-x) \tag{B.102}$$

Consider the case of symmetrical barriers, that is, barriers at a and $-a$, and let the initial starting point be $X(0) = x_0$. Substituting $u(x)$ and $v(x)$ given above into Eq. (B.86) and simplifying gives

$$\hat{f}_{a,-a}(s \mid x_0) = \cosh[(2s)^{1/2}x_0]/\cosh[(2s)^{1/2}a] \tag{B.103}$$

Taking the inverse Laplace transform gives

$$f_{a,-a}(t \mid x_0) = \frac{\pi}{a^2} \sum_{j=0}^{\infty} (-1)^j (j + \tfrac{1}{2}) \cos\left\{(j + \tfrac{1}{2}) \frac{\pi x_0}{a}\right\} \exp\left[-\frac{(j+\frac{1}{2})^2\pi^2 t}{2a^2}\right] \tag{B.104}$$

Integrating the above expression with respect to t gives the distribution function

$$F_{a,-a}(t \mid x_0) = 1 - \frac{2}{\pi} \sum_{j=0}^{\infty} \frac{(-1)^j}{(j + \tfrac{1}{2})} \cos\left\{(j + \tfrac{1}{2}) \frac{\pi x_0}{a}\right\} \exp\left[-\frac{(j+\frac{1}{2})^2\pi^2 t}{2a^2}\right] \tag{B.105}$$

This is a complicated expression but is easily amendable to machine computation. This result can be extended to general barriers by noting that any interval can be shifted to make it symmetrical about the origin as is shown in Fig. B.8. The new starting point will be

$$x' = [x_0 - (a + b)/2]$$

Thus,

$$F_{ab}(t \mid x_0) = F_{(a-b)/2, -(a-b)/2}(t \mid x') \tag{B.106}$$

and this completely solves the first passage time problem for the Weiner process.

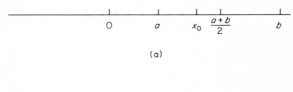

(a)

(b)

Fig. B.8. The shift of general barriers to symmetrical barriers. (*a*) Original barriers. (*b*) Shifted barriers.

References

1. S. O. Rice, Mathematical analysis of random noise. *Bell System Tech. J.* **23**, 282–332 (1944); **24**, 46–156 (1945). Reprinted in *Selected Papers on Noise and Stochastic Processes* (N. Wax, ed.). Dover, New York, 1954.

2. D. Middleton, *An Introduction to Statistical Communication Theory*. McGraw-Hill, New York, 1960.

3. H. Steinberg, P. M. Schultheiss, C. A. Wogrin, and F. Zweig, Short-time frequency measurement of narrow-band random signals by means of zero counting processes. *J. Appl. Phys.* **26**, 195–201 (1955).

4. I. Miller and J. E. Freund, Some results on the analysis of random signals by means of a cut-counting process. *J. Appl. Phys.* **27**, 1290–1293 (1956).

5. C. W. Helstrom, The distribution of the number of crossings of a Gaussian stochastic process. *IRE Trans. Information Theory* **IT-3**, 232–237 (1957).

6. J. A. McFadden, The axis-crossing intervals of random functions. *IRE Trans. Information Theory* **IT-2**, 146–150 (1956); **IT-4**, 14–24 (1958).

7. M. Kac, Probability theory as a mathematical discipline and as a tool in engineering and science. *Proc. Symp. Eng. Appl. of Random Function Theory and Probability, 1st* (J. L. Bogdanoff and F. Kozin, ed.), pp. 31–67. Wiley, New York, 1963.

8. M. C. Wang and G. E. Uhlenbeck, On the theory of the Brownian motion II. *Rev. Modern Phys.* **17**, 323–342 (1945). Reprinted in *Selected Papers on Noise and Stochastic Processes* (N. Wax, ed.), pp. 113–132. Dover, New York, 1954.

9. S. Chandrasekhar, Stochastic problems in physics and astronomy. *Rev. Modern Phys.* **19**, 1–89 (1943). Reprinted in *Selected Papers on Noise and Stochastic Processes* (N. Wax, ed.), pp. 3–92. Dover, New York, 1954.

10. A. J. F. Siegert, On the first passage time probability problems. *Phys. Rev.* **81**, 617–623 (1951).

11. D. A. Darling and A. J. F. Siegert, The first passage problem for a continuous Markov process. *Ann. Math. Statist.* **24**, 624–632 (1953).

Author Index

Numbers in parentheses are reference numbers and indicate that an author's work is referred to although his name is not cited in the text. Numbers in italics show the page on which the complete reference is listed.

A

Adomian, G., 250, 251, *253*
Akcasu, A. Z., *278*, 279
Amazigo, J. C., 248(33), *253*
Aoki, M., 120, *135*
Astrom, K. J., 120, *135*
Ariaratnam, S. T., 229(21), *252*

B

Bartlett, M. S., 73, *112*, 170, *212*
Bell, D. A., 120(5), *135*
Bellman, R., 204, 205, *213*, 251, *253*, 255, *278*
Bergman, P. G., 217(1), *252*
Bernard, M. C., 164(3), 165(3), *212*
Bertram, J. E., 271, 273, *278*
Bharucha-Reid, A. T., 58, *68*
Bochner, S., 48, *67*
Bogdanoff, J. L., 164, 202, *212*, *213*, 222 (7, 8, 10, 11), 229(18, 20), *252*, 266, *278*
Bogoliubov, N., 208, *213*, 271, *278*
Booten, R. C., Jr., 210(20), *213*
Bourret, R. C., 249, *253*
Boyce, W. E., 218, *252*

Braham, H. S., 138, *150*
Bucy, R. S., 120, *135*

C

Caughey, T. K., 162, 210(21), *212*, *213*, 229(19), *252*
Chandrasekhar, S., 313, *319*
Chen, K. K., 238(30), 239(30), 242(30), 243(30), *253*
Chen, Y. M., 236, *253*
Chenea, P. F., 222(7), 222(8), *252*
Chuang, S. N., 228, 236(23), *252*, *253*
Chung, K. L., 6(1), *28*
Coddington, E. A., 118, *134*, 139(5), *150*, 154(1), *212*
Collins, J. D., 222(13), *252*
Cozzarelli, F. A., 222(12, 14), *252*
Cramér, K., 6(2), *28*, 42, 48, *67*
Crandall, S. K., 206, *213*
Cumming, I. G., 202, *213*

D

Darling, D. A., 313, *319*
Dienes, J. K., 229(19), *252*

321

Subject Index